Headache

Series editors
Paolo Martelletti
Roma, Italy

Rigmor Jensen
Glostrup, Denmark

The huge importance of headache in public health arises from its causal association with personal and societal burdens of pain, disability, damaged quality of life, and financial costs. Headache disorders are in fact common and ubiquitous. They have a neurological basis, but rarely they are due to serious underlying illness. The primary headache disorders – migraine, tension-type headache, and cluster headache — are easily seen by family physicians or GPs; however, a relatively small number of secondary headache disorders could also be encountered in primary care. It is important that they are recognized and treated in the most appropriate way because of their potentially dangerous underlying causes; moreover, mismanagement and overuse of medications to treat acute headache are major risk factors for disease aggravation. Purpose of this Series, endorsed by the European Headache Federation – EHF, is to provide a detailed description of all aspects of headache disorders that are common and relevant both in primary care and in hospital setting.

More information about this series at http://www.springer.com/series/11801

Messoud Ashina • Pierangelo Geppetti
Editors

Pathophysiology of Headaches

From Molecule to Man

Springer

Editors
Messoud Ashina, MD, PhD, DMSc
Professor of Neurology
Department of Neurology
Danish Headache Center
Rigshospitalet and Glostrup Hospital
Glostrup
Denmark

Faculty of Health and Medical Sciences
University of Copenhagen
Copenhagen
Denmark

Pierangelo Geppetti, MD
Professor of Clinical Pharmacology
Department of Health Sciences
University of Florence
Florence
Italy

Head, Headache Center
University Hospital – Careggi
Florence
Italy

ISSN 2197-652X ISSN 2197-6538 (electronic)
Headache
ISBN 978-3-319-15620-0 ISBN 978-3-319-15621-7 (eBook)
DOI 10.1007/978-3-319-15621-7

Library of Congress Control Number: 2015936082

Springer Cham Heidelberg New York Dordrecht London

Printed on acid-free paper

Springer International Publishing AG Switzerland is part of Springer Science+Business Media (www.springer.com)

Foreword

The Headache Series is enhanced by this second volume, edited by Messoud Ashina and Pierangelo Geppetti, both well-known scientists in the headache pathophysiology area, to whom a delicate task has been assigned: to render the most updated picture of pathophysiology of headache disorders involving all the many prestigious actors who dominate the scenes of this shimmering theatre. Their different disciplinary backgrounds, primarily neurology but also pharmacology, guaranteed a combined-vision, factual planning of this important chapter in Headache Medicine.

The operation has been brilliantly carried out and the completeness of the volume is now in front of the readers' eyes and is available indiscriminately to everyone with a clinical-scientific-educational interest towards Headache Medicine.

The European Headache Federation thanks the Editors for their essential contribution to this complex and fine volume and every author of each chapter for playing an important role in spreading headache culture through the medical world.

The education/dissemination path within the EHF project of the Headache Series has concluded its second important step.

Paolo Martelletti
Sapienza University, Rome, Italy

Rigmor Høiland Jensen
University of Copenhagen, Copenhagen, Denmark

Preface

The purpose of the second *Headache Series* book endorsed by the European Headache Federation is to bring together in one book much of the diverse body of work on the pathophysiology of headaches.

We hoped to provide the reader with recent advances in the pathophysiology of migraine, cluster headache, and tension-type headache. The topics we have chosen for the book should be of interest to headache basic and clinical science researchers as well as researchers in the pharmaceutical industry. We also believe that this book will also be attractive to senior undergraduate and graduate students in the health sciences, whose interests include headache.

The book is organized into an introductory chapter on the anatomy of headache, which is essential to understand the pathophysiology of headaches, followed by topical chapters focusing on animal and human models, genetics, imaging, neurophysiology, and biochemistry. We also asked experts in the field to provide the current status and future perspectives on the pathophysiology of headaches.

We hope that this book will not only provide interesting reading but also serve as a useful reference in the field of headache research.

Copenhagen, Denmark — Messoud Ashina
Florence, Italy — Pierangelo Geppetti

Contents

Chapter 1
Anatomy of Headache

Hayrunnisa Bolay, Karl Messlinger, Mária Dux, and Didem Akcali

1.1 Introduction

Clinical and experimental observations provide evidence for an essential contribution of peripheral, intracranial, as well as extracranial nociceptive processes in the generation of headaches [1]. A large body of evidence supports the hypothesis that most types of headaches, including migraine, are of trigeminovascular origin, caused or influenced by nociceptive afferents innervating the cranial meninges, particularly the dura mater encephali and large intracerebral blood vessels [2]. The primary role of the meningeal sensory innervation in generating headaches fits well to the intraoperative studies of Ray and Wolff and other investigators [3, 4], who demonstrated that headache-like pain, but not other sensations, can be evoked by electrical, mechanical, thermal, or chemical stimulation of dural blood vessels and sinuses or large intracerebral arteries. Importantly, the painful sensations were referred to the trigeminal dermatomes where typically headaches are localized [5]. These early studies formed the basis of many anatomical and physiological examinations in animals regarding the pathophysiology of headaches. Recordings of

H. Bolay, MD, PhD (✉) • D. Akcali, MD, PhD
Department of Neurology and Algology, Neuropsychiatry Centre, Gazi University, Beşevler, Ankara 06510, Turkey
e-mail: hbolay@gazi.edu.tr; didemakcali@yahoo.com

K. Messlinger (✉)
Institute of Physiology and Pathophysiology, University of Erlangen-Nürnberg, Universitätsstr. 17, Erlangen D-91054, Germany
e-mail: Karl.messlinger@fau.de

M. Dux
Department of Physiology, University of Szeged, Dóm tér 10., Szeged H-6720, Hungary
e-mail: dux.maria@med.u-szeged.hu

M. Ashina, P. Geppetti (eds.), *Pathophysiology of Headaches: From Molecule to Man*, Headache, DOI 10.1007/978-3-319-15621-7_1

action potentials from trigeminal nerves [6] and the trigeminal ganglion [7] as well as higher neurons in the spinal trigeminal nucleus [8, 9] and in the thalamus [10, 11] provided further evidence for an important role of the trigeminovascular system in meningeal nociception.

1.2 Peripheral Structures Involved in Headache States

1.2.1 Trigeminal and Other Cranial Nerves Associated with Headache

1.2.1.1 Organization of the Trigeminal Ganglion and Meningeal Representation

Due to the limited experimental access to intracerebral arteries, most of the morphological and nearly all functional studies have focused on the innervation of the cranial dura mater and the dural venous sinuses. Using neuronal tracing, afferents around the middle meningeal artery have been found predominantly originating in the ophthalmic division (V1) of the ipsilateral trigeminal ganglion but to a minor extent also in the maxillary (V2) and mandibular (V3) divisions [12, 13]. The basal dura mater in the middle cranial fossa was represented mainly in V3. New retrograde tracings in the rat confirmed that meningeal nerves innervating the territorium of the middle cranial fossa, which is mainly supplied by the middle meningeal artery, origin predominantly in V3 and to a lesser extent in V2 [14]. The finding that all three divisions of the trigeminal nerve, though not equally, contribute to the innervation of the meninges is in accordance with old anatomic observations in primates [15]. Moreover, retrograde labeling of nerve fibers around basal intracranial arteries and the superior sagittal sinus, from which in humans headache can be provoked [3], appeared not only in the rat trigeminal ganglia but also in the first and second spinal ganglia [16] projecting to the cervical dorsal horn.

1.2.1.2 Afferent Innervation of the Meninges and Intracerebral Arteries

The innervation of the human cranial dura mater, which has firstly been described centuries ago by the anatomists Arnold [17] and Luschka [18], is regarded as pivotal for the generation or aggravation of headaches. Neuroanatomical studies demonstrated the close relationship between meningeal blood vessels and nerve fibers of different origin (Fig. 1.1a). Besides the trigeminal fibers originating in the ipsilateral trigeminal ganglion [19, 20], a network of sympathetic fibers mainly from the superior cervical ganglion [21, 22] and a comparatively sparse innervation by parasympathetic fibers originating in the sphenopalatine and otic ganglia has been described [23, 24]. The innervation of intracerebral (pial) blood vessels is similarly organized [25] but with a higher proportion of parasympathetic fibers coming mainly from the internal carotid and sphenopalatine ganglia [26].

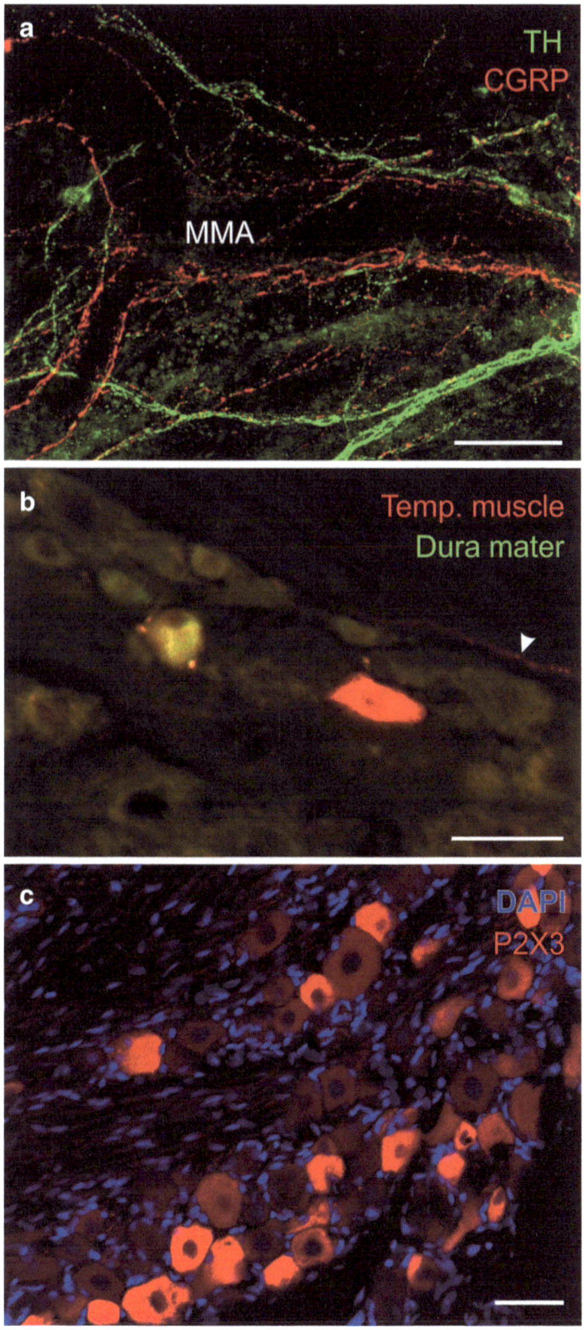

Fig. 1.1 Histochemical demonstration of afferent and efferent innervation of the rat dura mater and markers of trigeminal ganglion neurons. (**a**) Confocal image of putative sympathetic and afferent nerve fibers in the rat dura mater labeled by tyrosine hydroxylase (*TH, green*) and calcitonin gene-related peptide (*CGRP, red*) immunofluorescence. Both TH- and CGRP-immunoreactive fibers form a dense network around the middle meningeal artery (MMA). Scale bar 200 μm. (**b**) Confocal image of trigeminal ganglion neurons labeled by retrograde tracing with Texas red from the temporal muscle (see red nerve fiber, *arrowhead*) and with Rhodamine green from the parietal dura mater. The yellow neuron with *mixed red* and *green* innervates both temporalis muscle and dura mater by afferent collaterals. Scale bar 50 μm. (**c**) Confocal image of a trigeminal ganglion section immunohistochemically stained for purinergic (*$P2X_3$*) receptor channels and nuclei (*DAPI*). The majority of neuronal cell bodies, mostly small ones, are $P2X_3$ immunopositive. Scale bar 50 μm (Courtesy of S. Vilotti, SISSA Trieste)

Several immunohistochemical studies described neuropeptide-immunoreactive nerve fibers in the dura mater [27–29] and around cerebral (pial) blood vessels in different species including humans [30, 31]. Meningeal nerve fibers immunoreactive for substance P (SP), neurokinin A (NKA), and calcitonin gene-related peptide

(CGRP) are thought to belong to the afferent (trigeminal and spinal sensory) system, while nerve fibers immunopositive for neuropeptide Y (NPY) are most likely of sympathetic and those immunoreactive for vasoactive intestinal polypeptide (VIP) of parasympathetic origin [21, 19]. The peptidergic nerve fibers form a dense network around blood vessels but can also be found in nonvascular regions [29, 20]. On the light microscopic level, sympathetic nerve fibers can be labeled by tyrosine hydroxylase (TH) immunoreactivity and thereby discriminated from peptidergic afferents (Fig. 1.1a). It is important to note, however, that a major proportion of trigeminal afferents does not express neuropeptides [32]. For differentiation of sensory fibers, other markers have been used such as neurofilament 200, which is present in myelinated fibers, and isolectin B4, which characterizes mostly unmyelinated, non-peptidergic fibers [33].

Electron microscopic examinations revealed myelinated (Aδ possibly Aβ) and unmyelinated (C) nerve fibers in the cranial dura mater [29, 34, 35]. An attempt was made to classify the C fibers according to their three-dimensional structure and their content of different kinds of vesicles into afferent and autonomic fibers [36]. The majority of meningeal C and Aδ fibers terminate as free nerve endings, but encapsulated Ruffini-like receptors and lamellated nerve terminals have additionally been described in higher vertebrates including man, particularly at sites where cerebral veins enter the sagittal sinus [34]. Myelinated and unmyelinated axons terminate also within the arachnoid granulations at different tissue structures suggesting that they have different mechano- and chemoreceptive functions [37].

1.2.1.3 Extracranial Collaterals of Meningeal Afferent Innervation

Long ago anatomical studies by Luschka [18] on the primate and human meningeal innervation reported on nerves that penetrate the skull, believed to innervate extracranial tissues. Recently a role for pericranial afferents in headache generation is again a matter of discussion [1]. Histological examinations in the mouse have revealed peripherin- and CGRP-immunopositive nerve fibers traversing the bones of the calvaria between the galea aponeurotica and the meninges [38]. The historical intraoperative data from Wolff's group, who observed that noxious stimulation not only of dural but also extracranial structures like pericranial muscles and arteries can cause headache, support this concept [3]. Likewise, further experimental and clinical observations indicated that noxious activation of afferents in pericranial tissues, particularly in the temporal and occipital–cervical regions, can contribute to headache generation [39, 40] and peripheral sensitization in migraine pain [41].

A couple of new studies have been made using in vitro and in vivo neuronal tracing and electron microscopy in rodents and human skulls to investigate extracranial projections from meningeal nerves and their origin in the trigeminal ganglion [14]. In particular, anterograde and retrograde neuronal in vitro tracing with DiI revealed nerve fiber bundles leaving the skull through emissary canals and fissures to innervate the pericranial temporal, parietal, and occipital periosteum as well as deep layers of the temporal and upper neck muscles. A variety of functional measurements in rats confirmed the afferent nature of extracranial afferent collaterals and the impact of their activation on the intracranial secretion of neuropeptides and their vasodilatory function [42]. Following in vivo tracing with different dextran amines applied to the dura mater and the pericranial muscles, some neurons were detected in the trigeminal ganglion containing tracer from both structures (Fig. 1.1b). These data affirmed functional afferent connections between intra- and pericranial tissues and provide a new view on the influence of extracranial meningeal afferent projections on meningeal nociception and headache generation.

1.2.2 Molecular Signature of Trigeminal Afferents and Neurogenic Inflammation of the Dura Mater

Meningeal afferents convey the nociceptive information to the central nervous system, but through the antidromic release of vasoactive peptides from their perivascular peripheral terminals, they also promote a sterile "neurogenic inflammation" in the meningeal tissue characterized by vasodilatation and increased permeability of blood vessels [43, 44]. Neurogenic inflammation is considered to contribute to the peripheral mechanisms in the pathophysiology of headaches.

1.2.2.1 Receptors, Transduction, and Conduction Channels

Although headache-like pain is the only sensation induced by activation of intracranial afferents, regardless of the mode of their stimulation [5, 4], the nerve fibers innervating meningeal tissues consist of a heterogenous population based on their morphological and immunohistochemical properties (Fig. 1.2).

TRP Channels

Chemosensitive meningeal afferents likely contribute to sensitization of the nociceptive pathway [45]. Successful prevention of cluster headache and migraine attacks with the topical, desensitizing application of capsaicin to the patients' nasal mucosa has focused attention to the significant role of capsaicin-/chemosensitive population of trigeminal afferents [46, 47]. Chemosensitive meningeal afferents express different members of the transient receptor potential (TRP) channel family. Sixteen percent of the neurons in the trigeminal ganglion of humans and 21–31 % of dural afferent neurons in rodents have been found to express the transient receptor potential vanilloid 1 (TRPV1) channel [48, 32]. The TRPV1 receptor is a nonspecific cation channel, which can be activated by noxious heat, acidic pH ($pH < 5.3$), and different compounds like some endogenous membrane lipid metabolites (anandamide, *N*-arachidonoyl dopamine) and exogenous capsaicin or resiniferatoxin.

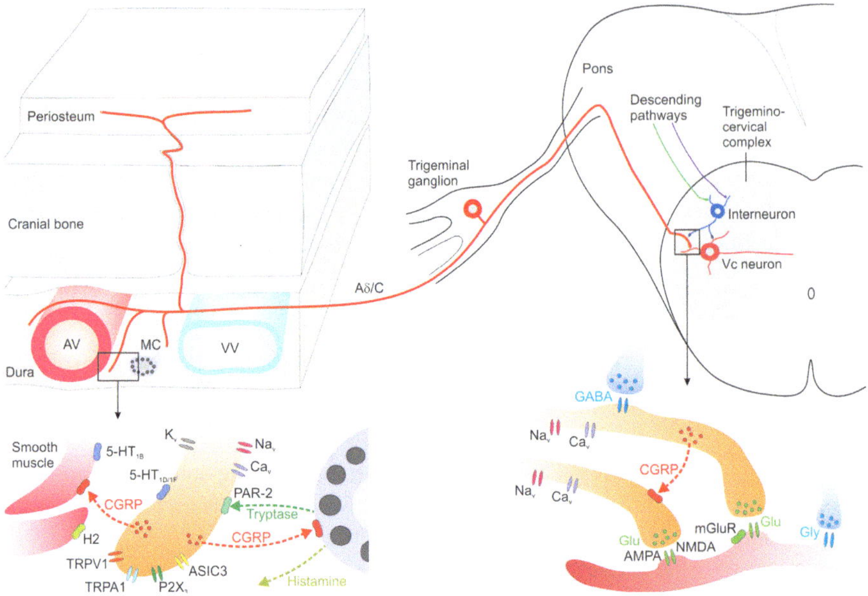

Fig. 1.2 Schematic representation of the trigeminovascular system of the cranial dura mater with arterial vessel (*AV*), venous vessel (*VV*), central canal (*circle*) and mast cell (*MC*) and the afferent projection to the trigeminocervical complex with a second-order neuron in the subnucleus caudalis (*Vc*) and an inhibitory interneuron. One and the same single (Aδ or C) fiber may innervate the dura mater and, with collaterals projecting through the skull, the periosteum. Afferent fibers may contain and release CGRP in the periphery and the CNS. The *left inset* shows some important transduction channels (TRPV1, TRPA1, $P2X_3$, ASIC3), receptors ($5\text{-}HT_1$, PAR-2), and voltage-gated conduction channels (Ca_v, Na_v, K_v) and the proposed signaling between the afferent ending, AV and MC. The *right inset* shows the proposed nociceptive transmission in superficial laminae of the trigeminocervical complex to Vc neurons expressing glutamate (*Glu*) receptor channels (NMDA, AMPA) and metabotropic glutamate receptors (*mGluR*). CGRP signaling is probably between terminals of primary afferents facilitating neurotransmitter release. Inhibitory neurons act pre- and postsynaptically on GABA and glycine (*Gly*) receptors, targeted by descending serotonergic and noradrenergic inhibitory pathways and segmental connections

The transient receptor potential ankyrin 1 (TRPA1) ion channel, another member of the TRP receptor superfamily, has recently emerged as another important receptor activated by noxious chemical agents [49]. TRPA1 receptors can be activated by noxious cold, different environmental irritants like acrolein, and also by pungent ingredients of plant origin, like cinnamaldehyde and umbellulone that is the major volatile constituent of the "headache tree" *Umbellularia californica* [50, 51]. Endogenous activators of the receptor recently defined are some prostaglandin metabolites, hydrogen peroxide and nitroxyl (HNO), the one-electron-reduced sibling of nitric oxide (NO) [52]. Both endogenous and exogenous activators of the TRPA1 modify cysteine residues of the receptor (e.g., by forming disulfide bonds). Similar to TRPV1, activation of trigeminal afferents through TRPA1 receptors induces nociceptive responses and release of CGRP (Fig. 1.2). Histological and functional observations have revealed colocalization of TRPV1 and TRPA1 in trigeminal ganglion neurons [50, 53].

In humans, inhaled irritants may stimulate TRPA1 receptors of extracranial trigeminal afferents that innervate the nasal mucosa and may project collaterals to meningeal blood vessels. Nociceptive stimulation of extracranial tissues may activate intracranial collaterals by an axon reflex mechanism, releases vasoactive neuropeptides in meningeal tissue, and increases intracranial blood flow [42].

ASIC Channels

Some dural afferents express acid-sensing ion channels (ASICs), predominantly the ASIC3 subtype responding to low meningeal pH [54]. ASICs belong to the ENaC/DEG (epithelial amiloride-sensitive Na^+ channel and degenerin) family of ion channels [55]. Relative small changes in the meningeal proton concentration activate the ASIC3 channel initiating an afferent signal in the trigeminal nociceptive pathway. The reason for an acidification of the local meningeal pH can be ischemia of the dura mater possibly developing as a consequence of cortical spreading depression that has been linked to the aura phase of migraine attacks [56, 57]. Degranulation of dural mast cells as a result of neurogenic inflammation of the dura mater can be an additional source of acidic metabolites leading to the activation of sensory nerve endings.

Purinergic Receptors

Purinergic ($P2X_2$ and $P2X_3$) receptor channels that may contribute to the transduction of nociceptive signals have been localized to trigeminal ganglion neurons innervating the dura mater (Fig. 1.1c). Purine receptor immunoreactivity is present predominantly in medium- and small-sized neurons that are mainly non-peptidergic and unmyelinated [33]. An important pathophysiological function of purinergic receptors may be peripheral sensitization of the trigeminal nociceptive pathway through the communication between different clusters of neurons within the trigeminal ganglion. In vitro studies have provided evidence that CGRP release from neurons stimulates the ERK1/2 MAP kinase signaling pathway in surrounding satellite glial cells and increases P2Y1,2 receptor-mediated intracellular calcium responses, which leads to the release of inflammatory cytokines from the activated satellite cells. Increased levels of cytokines have been shown to result in local inflammatory reactions and modulation of the neuronal function [58, 59]. By this way, CGRP may function as a paracrine factor to stimulate adjacent glial cells within a cluster and to cause excitation of more distant neurons and glial cells located in other clusters, thereby propagating an inflammatory signal across the entire ganglion [33, 60].

5-HT Receptors

One of the most effective classes of drugs for the treatment of migraine pain is the triptans, serotonin 1B/1D/1F ($5\text{-HT}_{1B/1D/1F}$) receptor agonists. While the 5-HT_{1B} receptors appear to be located primarily on vascular smooth muscle mediating vasoconstriction, the 5-$HT_{1D/1F}$ receptors are located on the peripheral and central terminals of meningeal afferents [61, 62] (Fig. 1.2). Activation of these G protein-coupled receptors inhibits the release of transmitters from the trigeminal afferents leading to the attenuation of the central transmission of nociceptive signals.

The presence of 5-HT_7 receptors on trigeminal nerve endings and middle meningeal arteries has been demonstrated recently. Vasodilatation induced by the activation of trigeminal 5-HT_7 receptors seems to be the result of CGRP release from the nerve terminals [63].

Calcium Channels

Activation of voltage-gated P/Q-type and N-type calcium channels has a key role in the regulation of synaptic function. Clinical and experimental data provide evidence that changes in the channel structure influence the release of neurotransmitters in the nociceptive transmission (Fig. 1.2). A rare hereditary form of migraine with aura and hemiparesis is the familial hemiplegic migraine type 1 (FHM-1). The FHM-1 gene encodes the pore-forming $Ca_V2.1$ subunit of P/Q-type Ca^{2+} channels. The mutation of the channel structure results in a gain of P/Q-type channel activity in trigeminal neurons and a selective increase in low-voltage-activated T-type currents in the small ($IB4^-$) neuron population. This condition may lead to hyperexcitability of small, probably peptidergic, trigeminal neurons [64, 65].

Clinical and experimental observations provide evidence for N-type voltage-gated calcium channels ($Ca_V2.2$) as therapeutic targets for chronic pain conditions. N-type calcium channels are present in the presynaptic terminals of primary afferent sensory neurons, especially in Aδ and C fibers [66]. Blocking the channel function has a significant antinociceptive effect. In an experimental migraine model, inhibition of the channel function reduced neurotransmitter release from the primary sensory neurons and decreased the excitability of second-order neurons in the trigeminal brainstem [67].

Sodium Channels

The voltage-gated sodium channels with the pore-forming alpha-subunits $Na_V1.7$ and $Na_V1.8$ have emerged as molecules involved in peripheral pain processing and in the development of an increased pain sensitivity associated with inflammation [68]. In experimental models of meningeal nociception, amitriptyline, a tricyclic antidepressant, which is used to prevent migraine attacks, blocked the $Na_V1.8$ currents in trigeminal ganglion neurons and alleviated nociceptive behavior induced by electrical stimulation of the superior sagittal sinus. These results strongly support the contribution of $Na_V1.8$ channels to the pathophysiology of migraine and provide a novel guideline to migraine prophylaxis [69].

Potassium Channels

A subtype of voltage-gated K^+ channels, K_V7 expressed in nociceptors, is recognized to be one of the most important regulators of resting membrane potential and action potential firing threshold. Expression of this K^+ channel contributes strongly to the excitability of the nociceptors. The analgesic drug flupirtine opens K_V7 channels and by this way exerts an analgesic effect in migraine, chronic musculoskeletal pain, and neuralgia [70, 71]. Some of the nonsteroidal anti-inflammatory drugs such as diclofenac used in migraine therapy have also strong K_V7 channel opener activity, which may at least partly be responsible for their analgesic effect [72]. Recent observations indicate an NO-mediated K_V7 channel inhibition in trigeminal ganglion neurons that correlate with increased excitability and release of CGRP in nociceptors. It was suggested to contribute to excitatory effects of NO in headaches [73].

1.2.2.2 Mast Cells and Immunocytes in the Dura Mater

Activated meningeal nociceptors releasing the neuropeptides CGRP and SP, which induce direct vascular effects, can also activate and degranulate dural mast cells [74]. The local release of inflammatory mediators such as histamine from activated mast cells is believed to further stimulate meningeal nociceptors possibly promoting headache (Fig. 1.2). Clinical observations have shown that infusion of histamine induces headaches preferentially in migraineurs [75, 76].

The cranial dura mater is rich in connective tissue-type mast cells in both humans [77] and rodents [78] that are in close apposition to meningeal nociceptive nerve fibers and blood vessels [78]. This anatomical situation allows a multidirectional communication between blood vessels, nociceptors, and mast cells leading to the sensitization or activation of the trigeminal nociceptive pathway and changes in meningeal blood flow [79]. Direct vascular effects of histamine released by activated mast cells are mediated by multiple receptors localized on different histological components of the arterial vessel wall (Fig. 1.2). Relaxation of dural arteries is mediated by H2 receptors of vascular smooth muscle cells and by endothelial H1 receptors. In addition, H1 receptors on smooth muscle cells may mediate vasoconstriction [80].

Additional molecules known to be released from activated mast cells such as prostaglandins, leukotrienes, cytokines, and tryptase may also take part in the sensitization or activation of meningeal nociceptors [81]. The serine protease tryptase released from mast cells upon stimulation cleaves and activates the proteinase-activated receptor 2 (PAR-2) of meningeal nociceptors amplifying the initial vasodilation caused by sensory neuropeptides and possibly also the central transmission of nociceptive signals [82] (Fig. 1.2).

Resident macrophages of the dura mater expressing the inducible nitric oxide synthase (iNOS) are considered to play a significant role in delayed headache induced by infusion of the so-called NO-donor nitroglycerin. Following administration of nitroglycerin, a strong activation of iNOS was observed in the macrophages along the branches of the middle meningeal artery together with an upregulation of pro-inflammatory cytokines. Cytokines and NO synthesized by macrophages can sensitize small unmyelinated trigeminal afferents and generate headache. Activation of trigeminal afferents, in turn, promotes neuropeptide release and local blood flow changes in the meninges [83]. Nuclear factor kappa B (NF-κB) seems to mediate the transcriptional signal to iNOS and inflammatory cytokines. Since NF-κB can be activated by diverse pathological and inflammatory stimuli such as oxidative stress and bacterial and viral metabolites, its activation may provide the substrate within meningeal macrophages that contributes to headaches in response to different exogenous agents in susceptible individuals [84].

1.3 Central Structures Involved in Headache States

1.3.1 Subcortical Structures Implicated in Headaches

1.3.1.1 Morphofunctional Organization of the Trigeminocervical Complex and Meningeal Representation

The central processes of trigeminal ganglion neurons forming the trigeminal nerve enter the brainstem at the pontine level and terminate in the trigeminocervical complex, which consists of the pontine principal sensory nucleus (Vp) and the spinal trigeminal nucleus (Vsp) (Fig. 1.2). Basically, the thick myelinated mechanoreceptive trigeminal afferents terminate in the Vp, whereas both large-diameter and small-diameter fibers descend in the spinal trigeminal tract (SVT) projecting to the Vsp, which is subdivided into three subnuclei [85]: a rostral subnucleus oralis (Vo), a middle subnucleus interpolaris (Vi), and a caudal subnucleus caudalis (Vc). The Vc is often referred to as the medullary dorsal horn (MDH) because of the smooth transition to the anatomically and functionally similar spinal dorsal horn. Olszewski [85] identified three histologically different regions in the MDH: an outer marginal region, the substantia gelatinosa, and a deep magnocellular region. Later Gobel et al. [86] proposed a laminar subdivision of the MDH similar to Rexed's nomenclature of the spinal dorsal horn [87] in which lamina I corresponds to the marginal layer, lamina II to the substantia gelatinosa, and laminae III and IV to the magnocellular region. The most ventral lamina V merges with the medullary reticular formation [88] without clear boundary. Groups of neurons intermingled in the spinal trigeminal tract down to the transition of Vi and Vc are referred to as the interstitial islands of Cajal or as the paratrigeminal or interstitial nucleus [89]. The neurons of these islands are regarded as nociceptive, similar to the neurons in laminae I and II of the Vc [10, 90].

Anatomical and electrophysiological studies [91, 92] revealed that the TBNC is topographically organized in ventrodorsal and in rostrocaudal direction. Mandibular afferents terminate preferentially in the dorsal region of each trigeminal subnucleus (dorsomedial in the MDH), ophthalmic afferents terminate ventrally (ventrolateral in MDH), and maxillary terminals are interposed. The rostrocaudal organization of the trigeminocervical complex is less clear, but within the Vc the rostrocaudal axis of the face is represented from rostral to caudal [93]. Early anatomical [94] and neurophysiological [95] studies suggest that each subnucleus receives information from all parts of the head. Jacquin et al. found mandibular nerves in the rat projecting to all trigeminal subnuclei, although the anterior oral afferents tended to terminate most heavily in the rostral trigeminocervical complex, whereas the posterior perioral–auricular afferents terminated preferentially in the caudal aspect of the complex [96]. Tracing from the superficial temporal artery revealed afferent terminals mainly in the rostral cervical spinal dorsal horn and sparsely in the Vi and Vc of the spinal trigeminal nucleus [97]. Because of a lack of tracing studies, it is not clear if a similar somatotopic distribution in ventrodorsal and rostrocaudal directions exists for intracranial trigeminal structures.

Also based on clinical observations and animal studies, it has been recognized that the Vc is primarily responsible for processing nociceptive and thermoreceptive information from the face and head, whereas the Vp is involved in processing tactile information (Fig. 1.2). Isolated lesions of the Vc caused ipsilaterally complete or partial loss of pain and temperature sensation, whereas tactile sensations remained nearly intact [98]. This clinical experience led Sjoqvist [99] to develop the method of trigeminal tractotomy for the relief of facial pain, in which the spinal trigeminal tract at the level of the obex was transected. The clinical data were supplemented with a large body of neurophysiologic evidence based on trigeminal tractotomy or experimental lesions of different subnuclei demonstrating that the Vc in the perception of pain in trigeminal tissues is essential, whereas for the processing of nociceptive information from intraoral and orofacial tissues, more rostral regions of the trigeminocervical complex are important as well [100, 101].

Transganglionic cholera toxin and HRP tracing of afferents innervating the rat superior sagittal sinus labeled central terminals in the ipsilateral Vc and Vi but also in the ventrolateral area of the C1–C3 spinal dorsal horn on both sides [16]. Labeling was seen in laminae I and II with HRP but in laminae III and IV with cholera toxin [16]. These findings are largely confirmed by electrophysiological recordings from second-order neurons with afferent input from the dura mater in rat [9, 102].

Apart from the abovementioned study [97], the projection of nociceptive afferents to specific laminae of the trigeminocervical complex has again mainly been studied for facial inputs using axonal tracing in cat and rat. Hayashi and Jacquin et al. found high-threshold mechanoreceptive (nociceptive) Aδ afferents forming extensive terminal arbors in the superficial Vi and, most pronounced, in lamina I and, to a lesser extent, outer lamina II of Vc [96, 103]. In the rat, a second termination area was localized in laminae III to V of Vc [96]. Corneal afferents, which are thought to be mainly nociceptive, terminate mainly in the outer laminae of Vc [104]. Corresponding to the distribution of nociceptive afferent terminals in the Vsp, SP- and CGRP-immunoreactive nerve fibers have been demonstrated in different species preferentially around the substantia gelatinosa of Vc and the transition zone between Vi and Vc (Vi/Vc) [105, 106]. Colocalization of immunoreactivity for TRPV1 receptors with SP and CGRP was found in axon collaterals in the dorsal parts of Vp, Vo, and Vi and in terminals and fibers throughout lamina I and the outer zone of lamina II (IIo) of the Vc [107]. Trigeminal rhizotomy in the cat caused disappearance of most of the CGRP-immunoreactive fibers throughout the trigeminocervical complex, whereas a considerable number of SP-immunoreactive fibers remained intact [108, 109] suggesting that these are of central origin.

Electron microscopic immunohistochemistry in the cat Vsp revealed CGRP immunoreactivity within the substantia gelatinosa in axon terminals which were presynaptic to dendritic profiles and postsynaptic to other fibers [110]. Likewise, the presence of CGRP receptors in rat and human Vc was found by immunohistochemistry restricted to trigeminal afferent endings in superficial layers (laminae I and II) of the Vc [111, 112]. This implies that CGRP as a central neuromodulator may exert its effects presynaptically to spinothalamic and other second-order neurons (Fig. 1.2).

1.3.1.2 Projections to Thalamic and Other Subcortical Nuclei

Nociceptive information from cranial and upper cervical structures is transmitted via the trigeminal and spinal afferent system to the higher diencephalic and cortical pain-associated areas. According to the well-delineated conventional pathway, the cell bodies of second-order neurons in the trigeminocervical complex mainly project to third-order neurons in the thalamus and then carry information to somatosensory cortical areas. The majority of the caudal part of the trigeminal nucleus (Vc) and the cervical dorsal horn neurons send axons to two contralateral thalamic nuclei that relay ascending somatosensory information to the primary somatic sensory cortex: the ventroposteromedial thalamic nucleus (VPM) and the posterior thalamic nuclear complex (Po) [113–115]. Trigeminal Vc neurons also project to the posterior part of the ventral medial nucleus (VMpo), the ventral caudal part of the medial dorsal nucleus (MDvc), and the nucleus submedius (Sm) of the thalamus [116] (Fig. 1.3). Projections to the thalamus have distinct molecular characteristics as glutamatergic trigeminothalamic projection neurons in the Vc; Vi and Vo mainly express the glutamate transporter VGLUT2. The sensitization of second-order and/or third-order neurons within the system is an important factor in the development of allodynia during the migraine attack. A recent study revealed that the thalamic reticular nucleus (TRN) was activated following cortical spreading depression (CSD) ipsilaterally in freely moving conscious rodents [117, 118]. Unilateral TRN activation during CSD implicates its potential role in lateralized headache perception and attention, regarding the fact that the parabrachial nuclei (PBN) and amygdala have direct projections on the TRN involved in drawing attention to emotional stimuli [119].

Tracing studies revealed that the Vc also projects to other brainstem and diencephalic structures such as the brainstem reticular formation, nucleus of the solitary tract, superior salivatory nucleus, A5 cell group region, lateral periaqueductal gray matter, inferior colliculus, parabrachial nuclei, hypothalamus, and cerebellum [114, 120]. Trigeminal neurons send axons to the ipsilateral cerebellum, and trigeminocerebellar projection neurons predominantly express the glutamate transporter

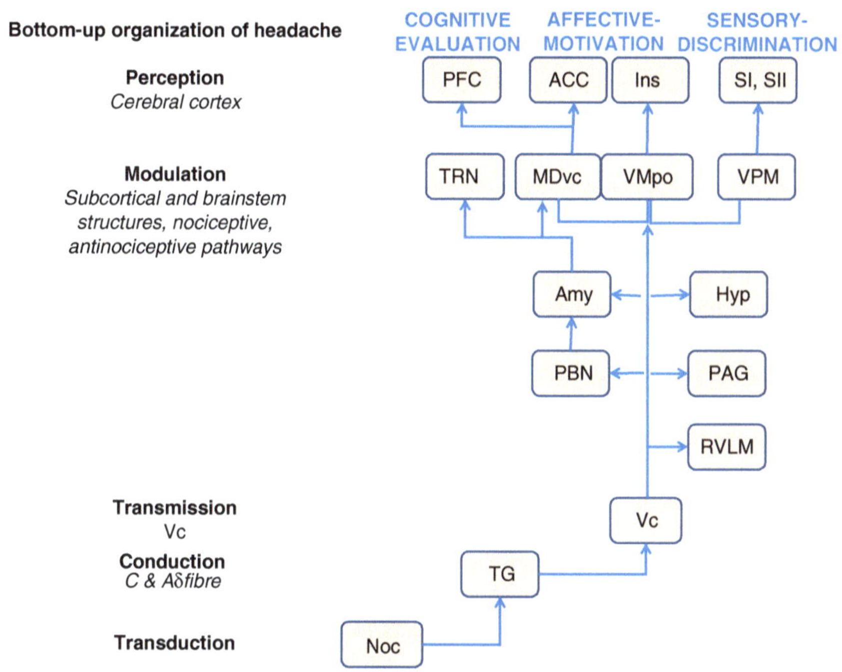

Fig. 1.3 Schematic representation of ascending nociceptive pathways in perception of headache. Bipolar neurons in the trigeminal ganglion (*TG*) conduct noxious signals from peripheral nociceptors of perivascular trigeminal afferents to nociceptive laminae in the spinal trigeminal nucleus (*Vc*) through Aδ and C fibers. Transmission of nociceptive impulses to second-order neurons in the dorsal Vc is a critical step in pain transmission. Trigeminal projections from the Vc to third-order neurons in the thalamic ventroposteromedial (*VPM*) nuclei are relayed to the primary and secondary somatosensory cortex. The latter pathway is also known as a lateral pain system and related to the sensory discriminative aspects of headache perception. The Vc also projects to a large number of the brainstem and diencephalic structures including the rostral ventrolateral medulla (*RVLM*), parabrachial nucleus (*PBN*), periaqueductal gray matter (*PAG*), amygdala, and hypothalamus. The pathway from Vc-PBN-amygdala-medial thalamus to the anterior cingulate cortex (*ACC*) and insula is known as medial pain pathway and is involved in affective aspects of headache. The prefrontal cortex plays a role in cognitive evaluation of headache perception. The thalamic reticular nucleus (*TRN*) seems to be involved in attention and emotional components of pain perception. Reciprocal connections between subcortical structures and cerebral cortex are not shown for clarity. *ACC* anterior cingulate cortex, *Amyg* amygdala, *Ins* insular cortex, *Noc* nociceptor, *MDvc* thalamic mediodorsal ventrocaudal nucleus, *PAG* periaquaductal gray matter, *PBN* parabrachial nucleus, *PFC* prefrontal cortex, *Po* posterior group, *RVLM* rostroventrolateral medulla, *SI and SII* primary and secondary somatosensory cortices, *Vc* caudal trigeminal brainstem nucleus, *VMpo* thalamic ventromedial posterior nucleus, *VPM* ventroposteromedial thalamic nucleus, *TG* trigeminal ganglia, *TRN* thalamic reticular nucleus

VGLUT1 [121]. Trigeminal projections also target the lateral reticular formation, mainly the rostral ventrolateral medullary reticular formation (RVLM), which participates in viscero-sympathetic reflexes. It was shown that trigemino-RVLM axons can be CGRP immunopositive [122].

The parabrachial nuclei (PBN), particularly their lateral parts, are an important projection site for trigeminal nociceptive information. The majority of nociceptive neurons in lamina I mainly in the Vc and Vi send axons to the PBN where many neurons respond preferably to noxious stimuli [90, 123, 124]. When retrograde tracer was injected into the PBN, projection neurons were detected ipsilaterally in lamina I of the Vc [125] though bilateral projection from the Vc to parabrachial nuclei has been identified [124]. Many of the neurons in the lateral parabrachial area project to the ventromedial hypothalamus and central nuclei of amygdala. Parabrachial-projecting neurons in the Vc have a topographic distribution [126]. Corneal afferents rather synapse with parabrachial-projecting neurons in the trigeminal nucleus and barely target neurons projecting to the thalamus. The trigeminal projections to the PBN are related with autonomic emotional responses to pain [127–129]. It is noteworthy that the PBN has reciprocal connections with forebrain areas and the insular cortex [130] and receives input from the amygdala [131]. The PBN was activated during CSD in rodents [132].

The amygdala, as a part of the trigemino–parabrachio–amygdaloid pathway, is another important subcortical nociceptive relay station. The central nucleus of the amygdala receives noxious information indirectly from the superficial lamina of the Vc through the lateral parabrachial area [133]. The amygdala, particularly the basolateral amygdaloid (BLA) nucleus, projects to the perirhinal area, the agranular insular area, and the mediodorsal thalamic nucleus. A direct amygdalofugal pathway to the trigeminal nuclear complex also exists in rodents. Particularly, the central amygdaloid nucleus sends extensive unilateral projections to all the trigeminal sensory nuclei, in addition to relatively light projections to the contralateral Vc [134]. Activation of amygdaloid nuclei along with the ipsilateral Vc was identified following CSD in awake freely moving rodents [135].

The perception of trigeminal pain is significantly modulated by the hypothalamus in the diencephalon. A considerable number of Vc neurons directly send their axons to hypothalamic regions [102]. Vc neurons projecting to the hypothalamus respond exclusively to noxious stimulation of the dura mater. In turn, the paraventricular nucleus, the lateral hypothalamic area, the perifornical hypothalamic area, the A11 nucleus, and the retrochiasmatic area send projections to the Vc. Hypothalamic projections are preferentially involved in the processing of meningeal and cutaneous inputs from the ophthalmic branch of the trigeminal nerve [136]. The latter finding indicates a somatotopic modulation of the Vc neurons by the hypothalamus. Descending hypothalamic projections to the Vc are bilateral, except those from the paraventricular nucleus that exhibit an ipsilateral predominance.

1.3.1.3 Descending Antinociceptive Systems

The transmission of nociceptive information to second-order trigeminal neuron is controlled by the inhibitory pathway descending from the periaqueductal gray matter (PAG) and from the rostral ventromedial medulla (RVM) (Figs. 1.3 and 1.4). The PAG located in the mesencephalon around the Sylvius aqueduct is the key structure

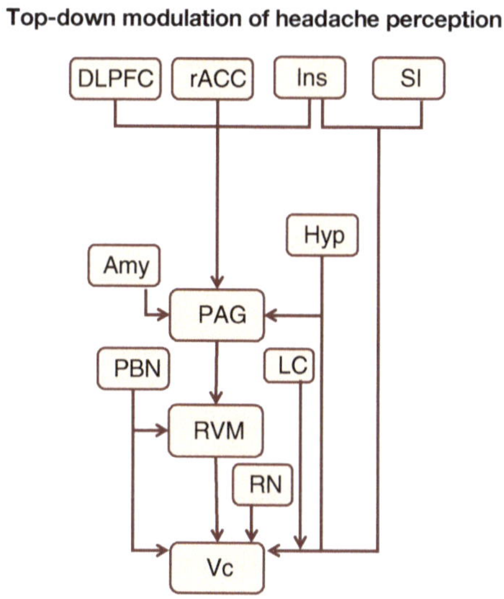

Fig. 1.4 Schematic representation of top-down, descending modulation of headache. The transmission of nociceptive information to second-order trigeminal neurons is controlled by inhibitory pathways descending from the periaqueductal gray (*PAG*) and from the rostral ventromedial medulla (*RVM*). Cortical inputs from pain-processing cortical areas such as the somatosensory areas, the insular cortex, the prefrontal cortex, and the anterior cingulate cortex project to the PAG. The hypothalamus, amygdala, PBN, locus coeruleus, and Raphé nuclei send projections to the descending pain inhibitory system. The dorsolateral prefrontal cortex (*DLPFC*) is related to attentional modulation of pain, the rostral anterior cingulate cortex (*rACC*) is implicated in placebo response, and morphine, noradrenaline, and serotonin are essential players in that system. *rACC* rostral anterior cingulate cortex, *Amy* amygdala, *Ins* insular cortex, *LC* locus coeruleus, *PAG* periaquaductal gray matter, *PBN* parabrachial nucleus, *DLPFC* dorsolateral prefrontal cortex, *RN* Raphé nucleus, *RVM* rostroventromedial medulla, *SI* primary somatosensory cortex, *Vc* caudal trigeminal brainstem nucleus

in descending pain modulation with its powerful inhibitory properties on pain perception. Higher cerebral cortical structures, such as the hypothalamus and amygdala, have been also implicated in the descending modulation of nociceptive activity [137–139]. Cortical inputs from pain-processing cortical areas such as the somatosensory areas, the insular cortex, the prefrontal cortex, and the anterior cingulate cortex to the PAG were demonstrated in several species [140, 141]. It is suggested that the RVM is the final relay station for descending antinociceptive information from the forebrain [142], as inputs from higher brain centers converge on the PAG and the RVM to exert pain-suppressive effects. Stimulation of the RVM as well as PAG stimulation has been shown to suppress nociceptive responses [143].

There are distinct neuronal subpopulations within the RVM that project caudally. Depending on the features of nociceptive stimuli (such as strength and duration), RVM neurons could yield either excitatory (on cells) or inhibitory (off cells) response to a noxious stimulus. Both neuronal subtypes are activated by electrical stimulation of the PAG. Morphine applied into the PAG or given systemically suppresses on-cell activity and increases off-cell activity [144]. Supraspinal opioid receptors play a key role in descending inhibitory controls relaying through the PAG and RVM.

In addition, there are serotonergic RVM cells projecting to the spinal cord [145] which contribute to descending antinociceptive inhibition by stimulation of the RVM or PAG [138, 146]. Descending projections from the noradrenergic neuronal cell groups in the locus coeruleus, subcoeruleus, A5, and A7 have a significant antinociceptive influence through spinal $\alpha 2$-adrenoceptors [146, 147]. It is notable that the locus coeruleus has direct inputs from the central nucleus of the amygdala, preoptic area, and paraventricular nucleus of the hypothalamus. GABAergic and glycinergic interneurons within the nociceptive laminae of the Vc mediate inhibitory effect on the transduction of nociceptive impulses to second-order neurons [148].

The amygdala plays a key role in emotional behavior, as inputs from the trigemino–parabrachio–amygdaloid pathway contribute to pain-induced changes in affective behavior and direct amygdalofugal projections to the PAG–RVM system provide feedback modulation of emotions on pain [137]. Through the latter pathway, application of opioids into the amygdala has been shown to induce antinociceptive effects [137] (Fig. 1.4). The parabrachial nuclei have direct projections to the trigeminocervical complex [149], and their electrical stimulation also exerted inhibitory effects on the activity of nociceptive neurons in the Vc.

Neurons of the paraventricular nucleus of the hypothalamus (PVN) send descending projections to laminae I and II of the Vc as well as the superior salivatory nucleus (Fig. 1.4). The latter nucleus gives rise to parasympathetic outflow to the cephalic vasculature, particularly in response to trigeminal activation, and could modulate neurogenic inflammation in the meninges [56, 150, 151]. Stimulation of the hypothalamic A11 nucleus has been shown to decrease the dural stimulation-evoked responses of Vc neurons. As a whole, those findings support the top-down modulation of the hypothalamus on Vc activities particularly driven by meningeal nociceptors.

The hypothalamic nuclei, particularly the paraventricular nucleus and/or arcuate nucleus, are involved in stress-induced analgesia [152]. Electrical stimulation of the hypothalamus results in antinociception. The hypothalamic PVN, PAG, and central nuclei of the amygdala take part in the stress-induced analgesic system [153]. It is notable that the rostral ACC, which is rich in opiate receptors, also participates in the endogenous analgesia [154]. During the analgesia induced by opioids or placebo, functional connectivity between the rACC and PAG has been found [155].

1.3.2 *Cortical Areas Associated with Discriminative and Affective Aspects of Nociception*

Perception of headache is a complex function of the cerebral cortex and involves distinct parts of the brain, which processes sensory discriminative, affective–emotional, and metacognitive aspects of nociception. Pain studies demonstrated that the activation of a cortical network of brain structures involving the somatosensory cortices SI and SII, the insular cortex, the anterior cingulate cortex (ACC), and the frontal cortex (DLPFC, orbitofrontal) is associated with nociceptive experience.

1.3.2.1 Cortical Structures of Discriminative Head Pain

The sensory discriminative and affective–motivational aspects of pain are transmitted through different systems and encoded in distinct cortical regions. The bottom-up organization of pain perception occurs through at least two major ascending pathways, the lateral and medial nociceptive systems [156]. The lateral nociceptive system comprises lamina I neurons projecting to the primary and secondary somatosensory cortex via the lateral thalamus and is involved in sensory discrimination of nociception [157, 158] (Fig. 1.3). Accordingly, it was demonstrated that manipulation of pain intensity was associated with changes mainly in the SI cortex, while the subjective ratings of pain unpleasantness were correlated with activity in the ACC [159].

Nociceptive stimulation of the ophthalmic branch of the trigeminal nerve, which preferably provides nociceptive information mediating headache, also activates similar cerebral cortical regions such as the somatosensory cortex, the insula, and the anterior cingulate cortex [160]. Descending cortical projections from the cerebral cortex to the trigeminal nucleus caudalis (Vc) have been demonstrated. Cortical projections originate contralaterally from insular (Ins) and primary somatosensory (SI) cortices. Projections from the primary sensory cortex terminate in deeper lamina, while projections from the insular cortex target solely superficial nociceptive laminae (laminae I and II) [161, 162] and inhibit trigeminal nociception, and meningeal-driven nociceptive inputs onto Vc were shown to be facilitated and inhibited by projections from the insula and SI, respectively [162].

1.3.2.2 Cortical Structures of Affective Pain Modulation

The medial nociceptive system that is directed toward the anterior cingulate cortex (ACC) through the parabrachial nucleus (PBN), parafascicular nucleus, and amygdala is involved in the affective–emotional aspects of nociception [157, 163, 164] (Fig. 1.3). Furthermore nociceptive information may be transmitted to the forebrain from the PBN and amygdala. The prefrontal cortex and the orbitofrontal cortex are implicated in the evaluation of affective experiences [165]. These pathways may contribute to the emotional aspects of pain and to the interactions between hedonic and cognitive processes of pain.

The majority of trigeminal ganglion neurons project multisynaptically to the anterior cingulate cortex (ACC). Iwata et al. [164] demonstrated by employing extracellular unit recordings that the ascending somatosensory pathways to the ACC from the trigeminal primary afferents arise mainly from Aδ and Aβ fibers but not from C fibers. The ACC predominantly receives projections from the lateral parabrachial nucleus. The parabrachial nuclei convey incoming nociceptive information from the Vc to the basolateral amygdaloid BLA nucleus [133]. The anatomical connections from Vc-parabrachial-BLA-ACC are thought to be involved in emotional and autonomic functions during trigeminal nociception.

The insula and amygdala, as components of the medial nociceptive pathway, have been implicated in evaluative and affective processes. The insular cortex sends projections to trigeminal nucleus caudalis. Tracing studies identified that many neurons in the granular and dysgranular insular cortex project to the laminae I/II of Vc bilaterally with a contralateral predominance. It is important to note that the direct projections from the insula only target to laminae I/II while sparing the lamina V of Vc [166]. Strong projections from the insular cortex to the bilateral rostral ventromedial medulla (RVM) and the nucleus of the solitary tract were detected. Bilateral insular projections with an ipsilateral predominance to the parabrachial nucleus were shown [161]. The insula is one of the important cortical pain-associated centers, and nociceptive processing of Vc neurons may be directly modulated through the insula or indirectly through brainstem nuclei such as PAG, PBN, and RVM. The insular cortex is pivotal in interoception and homeostatic functions [167]. Lesions of the insula are often associated with increased tolerance of pain.

Pain perception was also modulated by expectations and attention, and studies implicated the role of the dorsolateral prefrontal cortex (DLPFC) and the orbitofrontal cortex (OFC) during distraction and anticipation. The DLPFC may have a "top-down" mode of inhibition on the ascending nociceptive systems and is related with attentional modulation of pain (Figs. 1.3 and 1.4). In support of the pain modulatory function of the DLPFC, the anatomical connections between prefrontal cortices and the midbrain structure periaqueductal gray (PAG) were demonstrated by using diffuse tensor imaging [168]. Increased DLPFC activity was correlated with reduction in the affective component of nociceptive pain [169] and increased activity in the anterior cingulate cortex (ACC) along with increased activity in the PAG. Since the DLPFC activation was associated with decrease of nociception, repetitive transcranial magnetic stimulation (TMS) application to DLPFC was used for chronic migraine and fibromyalgia management [169, 170].

1.3.2.3 Functional Connectivity and Cortical Networks

Acute pain is a complex experience that is associated with activation of many structures in the brain that is often called pain matrix. The main components of the pain network are cortical structures of SI, SII, IC, ACC and PFC and the thalamus, which is the gate to sensorial input to the cerebral cortex and pacemaker for thalamocortical oscillations. The functional connectivity of such a network is important, and recent imaging studies have been focused on the alterations of functional connectivity during resting state (default mode network) or task performance. Chronic headache disorders have often been reported to be associated with changes in the brain networks during resting state and/or in response to stimuli.

Resting-state abnormalities were found in brain regions associated with pain processing and cognition in migraine patients [171–173]. Functional connectivity studies in patients suffering from temporomandibular disorders (TMD) revealed an increased connectivity of the anterior insula and anterior cingulate cortex. The latter finding was suggested to indicate an adaptation of the pain modulatory system early in the chronification process [174].

The resting-state functional connectivity of the hypothalamus was increased with parts of the frontal, parietal, and temporal cortex interictally in cluster headache patients [175]. However, the increased resting-state functional connectivity of the hypothalamus with the ACC and the posterior cingulate cortex (PCC) was detected during acute spontaneous cluster headache attacks. Functional MR studies revealed a diffuse abnormality of brain functional connectivity in cluster headache patients, which extends beyond the pain matrix primarily to cerebellar, frontal, and occipital areas [176, 177]. A recent case study demonstrated that cerebral activation of the ipsilateral trigeminal root entry zone, ventral pons, red nucleus, basal ganglia, cerebellum, prefrontal cortex, insula, and cingulate cortex was associated with ipsilateral hypothalamic activation during the cluster headache attack [178].

Structural MR studies in medication overuse headache (MOH) patients demonstrated increased gray matter thickness in the midbrain including periaqueductal gray matter and nucleus cuneiformis, which was partially reversed by the treatment in parallel to clinical improvement. It was proposed that the decreased gray matter in the orbitofrontal cortex could be predictive of poor response to treatment in MOH [179].

1.4 Gender Differences in Headache Anatomy

Gender differences in the epidemiology of headache disorders are well known; however, experimental headache studies conducted on females are significantly scarce [180, 181]. After reviewing the literature on headache and gender differences, it can be concluded that the medial pain pathway related to the affective–motivational aspects of headache seems to be more involved in women, which is briefly presented in the following section: (a) Capsaicin-induced trigeminal sensitization as detected by the visual flair and allodynic areas was greater in women particularly during menstruation phase compared to men [182]. (b) In women the resting cerebral metabolic glucose utilization in the orbitofrontal area was greater than in men [183]. (c) During negative affects such as anxiety and anger, the cerebellum, midbrain, thalamus, and ACC were more activated in women [184]. (d) In default mode network, there is a difference in insular processing between men and women's brain [185]. (e) In migraineurs, the posterior insular and precuneus cortices are thicker in women [186]. (f) Heat pain induced greater activation in the PFC, ACC, insula, and thalamus in women [187–189].

Experimental animal studies are in line with human data: (a) In female mice, the functional connectivity between nodes of descending antinociception and affective system was stronger compared to male mice [63]. (b) Female rats exhibited significant sex differences in activation pattern of temporomandibular joint-responding neurons in the trigeminal brainstem and their projections on subcortical structures

[190]. (c) In female rats, the density of dural mast cells is higher than in males, and estradiol promotes an increase in mast cell numbers along with a change in the phenotype [191]. (d) There are estrogen receptors and terminals in PBN subregions that are related to pain modulation [192]. (e) Estrogens modify nitroglycerin-induced c-fos expression in PVH, SON, SPVC, and CGRP expression in female rat brain [193]. (f) Cortical spreading depression susceptibility of FHM-1 knock-in mice was increased in female sex which was influenced by gonadal hormones [194, 180].

References

1. Olesen J, Burstein R, Ashina M, Tfelt-Hansen P (2009) Origin of pain in migraine: evidence for peripheral sensitisation. Lancet Neurol 8:679–690. doi:10.1016/S1474-4422(09)70090-0
2. Pietrobon D, Striessnig J (2003) Neurobiology of migraine. Nat Rev Neurosci 4:386–398. doi:10.1038/nrn1102
3. Ray BS, Wolff HG (1940) Experimental studies on headache: pain sensitive structures of the head and their significance in headache. Arch Surg 1:813–856
4. Feindel W, Penfield W, Mcnaughton F (1960) The tentorial nerves and localization of intracranial pain in man. Neurology 10:555–563
5. Penfield W, McNaughton M (1940) Dural headache and innervation of the dura mater. Arch Neurol Psychiatry 44:43–75
6. Bove GM, Moskowitz MA (1997) Primary afferent neurons innervating guinea pig dura. J Neurophysiol 77:299–308
7. Strassman AM, Raymond SA, Burstein R (1996) Sensitization of meningeal sensory neurons and the origin of headaches. Nature 384:560–564. doi:10.1038/384560a0
8. Strassman A, Mason P, Moskowitz M, Maciewicz R (1986) Response of brainstem trigeminal neurons to electrical stimulation of the dura. Brain Res 379:242–250
9. Schepelmann K, Ebersberger A, Pawlak M et al (1999) Response properties of trigeminal brain stem neurons with input from dura mater encephali in the rat. Neuroscience 90: 543–554
10. Davis KD, Dostrovsky JO (1988) Properties of feline thalamic neurons activated by stimulation of the middle meningeal artery and sagittal sinus. Brain Res 454:89–100
11. Burstein R, Jakubowski M, Garcia-Nicas E et al (2010) Thalamic sensitization transforms localized pain into widespread allodynia. Ann Neurol 68:81–91. doi:10.1002/ana.21994
12. Mayberg MR, Zervas NT, Moskowitz MA (1984) Trigeminal projections to supratentorial pial and dural blood vessels in cats demonstrated by horseradish peroxidase histochemistry. J Comp Neurol 223:46–56. doi:10.1002/cne.902230105
13. Steiger HJ, Meakin CJ (1984) The meningeal representation in the trigeminal ganglion–an experimental study in the cat. Headache 24:305–309
14. Schueler M, Neuhuber WL, De Col R, Messlinger K (2014) Innervation of rat and human dura mater and pericranial tissues in the parieto-temporal region by meningeal afferents. Headache 54:996–1009. doi:10.1111/head.12371
15. McNaughton M (1938) The innervation of the intracranial blood vessels and dural sinuses. Assoc Res Nerv Ment Dis 18:178–200
16. Liu Y, Broman J, Edvinsson L (2004) Central projections of sensory innervation of the rat superior sagittal sinus. Neuroscience 129:431–437. doi:10.1016/j.neuroscience.2004.07.045
17. Arnold F (1831) Der Kopfteil des vegetativen Nervensystems beim Menschen. K. Groos, Heidelberg
18. Luschka H (1856) Die Nerven der harten Hirnhaut. H. Laupp, Tübingen
19. O'Connor TP, van der Kooy D (1986) Pattern of intracranial and extracranial projections of trigeminal ganglion cells. J Neurosci Off J Soc Neurosci 6:2200–2207

20. Strassman AM, Weissner W, Williams M et al (2004) Axon diameters and intradural trajectories of the dural innervation in the rat. J Comp Neurol 473:364–376. doi:10.1002/cne.20106
21. Edvinsson L, Uddman R (1981) Adrenergic, cholinergic and peptidergic nerve fibres in dura mater–involvement in headache? Cephalalgia Int J Headache 1:175–179
22. Keller JT, Marfurt CF, Dimlich RV, Tierney BE (1989) Sympathetic innervation of the supratentorial dura mater of the rat. J Comp Neurol 290:310–321. doi:10.1002/cne.902900210
23. Amenta F, Sancesario G, Ferrante F, Cavallotti C (1980) Acetylcholinesterase-containing nerve fibers in the dura mater of guinea pig, mouse, and rat. J Neural Transm 47:237–242
24. Edvinsson L, Hara H, Uddman R (1989) Retrograde tracing of nerve fibers to the rat middle cerebral artery with true blue: colocalization with different peptides. J Cereb Blood Flow Metab Off J Int Soc Cereb Blood Flow Metab 9:212–218. doi:10.1038/jcbfm.1989.31
25. Hardebo JE, Arbab M, Suzuki N, Svendgaard NA (1991) Pathways of parasympathetic and sensory cerebrovascular nerves in monkeys. Stroke J Cereb Circ 22:331–342
26. Suzuki N, Hardebo JE (1991) The pathway of parasympathetic nerve fibers to cerebral vessels from the otic ganglion in the rat. J Auton Nerv Syst 36:39–46
27. Von Düring M, Bauersachs M, Böhmer B et al (1990) Neuropeptide Y- and substance P-like immunoreactive nerve fibers in the rat dura mater encephali. Anat Embryol (Berl) 182:363–373
28. Keller JT, Marfurt CF (1991) Peptidergic and serotoninergic innervation of the rat dura mater. J Comp Neurol 309:515–534. doi:10.1002/cne.903090408
29. Messlinger K, Hanesch U, Baumgärtel M et al (1993) Innervation of the dura mater encephali of cat and rat: ultrastructure and calcitonin gene-related peptide-like and substance P-like immunoreactivity. Anat Embryol (Berl) 188:219–237
30. Edvinsson L, Brodin E, Jansen I, Uddman R (1988) Neurokinin A in cerebral vessels: characterization, localization and effects in vitro. Regul Pept 20:181–197
31. You J, Gulbenkian S, Jansen Olesen I et al (1995) Peptidergic innervation of guinea-pig brain vessels: comparison with immunohistochemistry and in vitro pharmacology in rostrally and caudally located arteries. J Auton Nerv Syst 55:179–188
32. Huang D, Li S, Dhaka A et al (2012) Expression of the transient receptor potential channels TRPV1, TRPA1 and TRPM8 in mouse trigeminal primary afferent neurons innervating the dura. Mol Pain 8:66. doi:10.1186/1744-8069-8-66
33. Staikopoulos V, Sessle BJ, Furness JB, Jennings EA (2007) Localization of P2X2 and P2X3 receptors in rat trigeminal ganglion neurons. Neuroscience 144:208–216. doi:10.1016/j.neuroscience.2006.09.035
34. Andres KH, von Düring M, Muszynski K, Schmidt RF (1987) Nerve fibres and their terminals of the dura mater encephali of the rat. Anat Embryol (Berl) 175:289–301
35. Fricke B, Andres KH, Von Düring M (2001) Nerve fibers innervating the cranial and spinal meninges: morphology of nerve fiber terminals and their structural integration. Microsc Res Tech 53:96–105. doi:10.1002/jemt.1074
36. Messlinger K (1996) Functional morphology of nociceptive and other fine sensory endings (free nerve endings) in different tissues. In: The polymodal receptor: a gateway to pathological pain. Elsevier, Amsterdam, pp 273–298
37. Von Düring M, Andres KH (1991) Sensory nerve fiber terminals in the arachnoid granulations of non-human primates. Neurosci Lett 127:121–124
38. Kosaras B, Jakubowski M, Kainz V, Burstein R (2009) Sensory innervation of the calvarial bones of the mouse. J Comp Neurol 515:331–348. doi:10.1002/cne.22049
39. Calhoun AH, Ford S, Millen C et al (2010) The prevalence of neck pain in migraine. Headache 50:1273–1277. doi:10.1111/j.1526-4610.2009.01608.x
40. Svensson P, Ashina M (2006) Human studies of experimental pain from muscle. In: The headaches. Lippincott Williams & Wilkins, Philadelphia, pp 627–635
41. Malick A, Burstein R (2000) Peripheral and central sensitization during migraine. Funct Neurol 15(Suppl 3):28–35
42. Schueler M, Messlinger K, Dux M et al (2013) Extracranial projections of meningeal afferents and their impact on meningeal nociception and headache. Pain 154:1622–1631. doi:10.1016/j.pain.2013.04.040

43. Moskowitz MA, Buzzi MG (1991) Neuroeffector functions of sensory fibres: implications for headache mechanisms and drug actions. J Neurol 238(Suppl 1):S18–S22
44. Moskowitz MA (1993) Neurogenic inflammation in the pathophysiology and treatment of migraine. Neurology 43:S16–S20
45. Goadsby PJ (2007) Recent advances in understanding migraine mechanisms, molecules and therapeutics. Trends Mol Med 13:39–44. doi:10.1016/j.molmed.2006.11.005
46. Sicuteri F, Fusco BM, Marabini S et al (1989) Beneficial effect of capsaicin application to the nasal mucosa in cluster headache. Clin J Pain 5:49–53
47. Marks DR, Rapoport A, Padla D et al (1993) A double-blind placebo-controlled trial of intranasal capsaicin for cluster headache. Cephalalgia Int J Headache 13:114–116
48. Hou M, Uddman R, Tajti J et al (2002) Capsaicin receptor immunoreactivity in the human trigeminal ganglion. Neurosci Lett 330:223–226
49. Belvisi MG, Dubuis E, Birrell MA (2011) Transient receptor potential A1 channels: insights into cough and airway inflammatory disease. Chest 140:1040–1047. doi:10.1378/chest. 10-3327
50. Jordt S-E, Bautista DM, Chuang H-H et al (2004) Mustard oils and cannabinoids excite sensory nerve fibres through the TRP channel ANKTM1. Nature 427:260–265. doi:10.1038/nature02282
51. Nassini R, Materazzi S, Vriens J et al (2012) The "headache tree" via umbellulone and TRPA1 activates the trigeminovascular system. Brain J Neurol 135:376–390. doi:10.1093/brain/awr272
52. Eberhardt M, Dux M, Namer B et al (2014) H2S and NO cooperatively regulate vascular tone by activating a neuroendocrine HNO-TRPA1-CGRP signalling pathway. Nat Commun 5:4381. doi:10.1038/ncomms5381
53. Salas MM, Hargreaves KM, Akopian AN (2009) TRPA1-mediated responses in trigeminal sensory neurons: interaction between TRPA1 and TRPV1. Eur J Neurosci 29:1568–1578. doi:10.1111/j.1460-9568.2009.06702.x
54. Yan J, Edelmayer RM, Wei X et al (2011) Dural afferents express acid-sensing ion channels: a role for decreased meningeal pH in migraine headache. Pain 152:106–113. doi:10.1016/j.pain.2010.09.036
55. Wemmie JA, Price MP, Welsh MJ (2006) Acid-sensing ion channels: advances, questions and therapeutic opportunities. Trends Neurosci 29:578–586. doi:10.1016/j.tins.2006.06.014
56. Bolay H, Reuter U, Dunn AK et al (2002) Intrinsic brain activity triggers trigeminal meningeal afferents in a migraine model. Nat Med 8:136–142. doi:10.1038/nm0202-136
57. Lambert GA, Michalicek J (1994) Cortical spreading depression reduces dural blood flow–a possible mechanism for migraine pain? Cephalalgia Int J Headache 14:430–436; discussion 393–394
58. Rothwell NJ, Hopkins SJ (1995) Cytokines and the nervous system II: actions and mechanisms of action. Trends Neurosci 18:130–136
59. Vitkovic L, Bockaert J, Jacque C (2000) "Inflammatory" cytokines: neuromodulators in normal brain? J Neurochem 74:457–471
60. Magni G, Ceruti S (2013) P2Y purinergic receptors: new targets for analgesic and antimigraine drugs. Biochem Pharmacol 85:466–477. doi:10.1016/j.bcp.2012.10.027
61. Amrutkar DV, Ploug KB, Hay-Schmidt A et al (2012) mRNA expression of 5-hydroxytryptamine 1B, 1D, and 1F receptors and their role in controlling the release of calcitonin gene-related peptide in the rat trigeminovascular system. Pain 153:830–838. doi:10.1016/j.pain.2012.01.005
62. Buzzi MG, Moskowitz MA (1991) Evidence for 5-HT1B/1D receptors mediating the antimigraine effect of sumatriptan and dihydroergotamine. Cephalalgia Int J Headache 11:165–168
63. Wang X, Fang Y, Liang J et al (2014) 5-HT7 receptors are involved in neurogenic dural vasodilatation in an experimental model of migraine. J Mol Neurosci MN. doi:10.1007/s12031-014-0268-9
64. Tao J, Liu P, Xiao Z et al (2012) Effects of familial hemiplegic migraine type 1 mutation T666M on voltage-gated calcium channel activities in trigeminal ganglion neurons. J Neurophysiol 107:1666–1680. doi:10.1152/jn.00551.2011

65. Cao Y-Q, Tsien RW (2005) Effects of familial hemiplegic migraine type 1 mutations on neuronal P/Q-type Ca2+ channel activity and inhibitory synaptic transmission. Proc Natl Acad Sci U S A 102:2590–2595. doi:10.1073/pnas.0409896102
66. Westenbroek RE, Hoskins L, Catterall WA (1998) Localization of Ca2+ channel subtypes on rat spinal motor neurons, interneurons, and nerve terminals. J Neurosci Off J Soc Neurosci 18:6319–6330
67. Ripsch MS, Ballard CJ, Khanna M et al (2012) A peptide uncoupling CRMP-2 from the presynaptic Ca(2+) channel complex demonstrates efficacy in animal models of migraine and aids therapy-induced neuropathy. Transl Neurosci 3:1–8. doi:10.2478/s13380-012-0002-4
68. Lampert A, O'Reilly AO, Reeh P, Leffler A (2010) Sodium channelopathies and pain. Pflugers Arch Eur J Physiol 460:249–263. doi:10.1007/s00424-009-0779-3
69. Liang J, Liu X, Pan M et al (2014) Blockade of Nav1.8 currents in nociceptive trigeminal neurons contributes to anti-trigeminovascular nociceptive effect of amitriptyline. Neuromolecular Med 16:308–321. doi:10.1007/s12017-013-8282-6
70. Devulder J (2010) Flupirtine in pain management: pharmacological properties and clinical use. CNS Drugs 24:867–881. doi:10.2165/11536230-000000000-00000
71. Mastronardi P, D'Onofrio M, Scanni E et al (1988) Analgesic activity of flupirtine maleate: a controlled double-blind study with diclofenac sodium in orthopaedics. J Int Med Res 16:338–348
72. Peretz A, Degani N, Nachman R et al (2005) Meclofenamic acid and diclofenac, novel templates of KCNQ2/Q3 potassium channel openers, depress cortical neuron activity and exhibit anticonvulsant properties. Mol Pharmacol 67:1053–1066. doi:10.1124/mol.104.007112
73. Ooi L, Gigout S, Pettinger L, Gamper N (2013) Triple cysteine module within M-type K+ channels mediates reciprocal channel modulation by nitric oxide and reactive oxygen species. J Neurosci Off J Soc Neurosci 33:6041–6046. doi:10.1523/JNEUROSCI. 4275-12.2013
74. Ottosson A, Edvinsson L (1997) Release of histamine from dural mast cells by substance P and calcitonin gene-related peptide. Cephalalgia Int J Headache 17:166–174
75. Krabbe AA, Olesen J (1980) Headache provocation by continuous intravenous infusion of histamine. Clinical results and receptor mechanisms. Pain 8:253–259
76. Lassen LH, Thomsen LL, Olesen J (1995) Histamine induces migraine via the H1-receptor. Support for the NO hypothesis of migraine. Neuroreport 6:1475–1479
77. Varatharaj A, Mack J, Davidson JR et al (2012) Mast cells in the human dura: effects of age and dural bleeding. Childs Nerv Syst ChNS Off J Int Soc Pediatr Neurosurg 28:541–545. doi:10.1007/s00381-012-1699-7
78. Dimlich RV, Keller JT, Strauss TA, Fritts MJ (1991) Linear arrays of homogeneous mast cells in the dura mater of the rat. J Neurocytol 20:485–503
79. Dux M, Sántha P, Jancsó G (2012) The role of chemosensitive afferent nerves and TRP ion channels in the pathomechanism of headaches. Pflugers Arch Eur J Physiol 464:239–248. doi:10.1007/s00424-012-1142-7
80. Dux M, Schwenger N, Messlinger K (2002) Possible role of histamine (H1- and H2-) receptors in the regulation of meningeal blood flow. Br J Pharmacol 137:874–880. doi:10.1038/sj.bjp.0704946
81. Metcalfe DD, Baram D, Mekori YA (1997) Mast cells. Physiol Rev 77:1033–1079
82. Dux M, Rosta J, Sántha P, Jancsó G (2009) Involvement of capsaicin-sensitive afferent nerves in the proteinase-activated receptor 2-mediated vasodilatation in the rat dura mater. Neuroscience 161:887–894. doi:10.1016/j.neuroscience.2009.04.010
83. Reuter U, Bolay H, Jansen-Olesen I et al (2001) Delayed inflammation in rat meninges: implications for migraine pathophysiology. Brain J Neurol 124:2490–2502
84. Reuter U, Chiarugi A, Bolay H, Moskowitz MA (2002) Nuclear factor-kappaB as a molecular target for migraine therapy. Ann Neurol 51:507–516
85. Olszewski J (1950) On the anatomical and functional organization of the spinal trigeminal nucleus. J Comp Neurol 92:401–413
86. Gobel S, Falls WM, Hockfield S (1977) The division of the dorsal and ventral horns of the mammalian caudal medulla into eight layers using anatomical criteria. In: Pain in the trigeminal region. Elsevier/North-Holland Biomedical Press, Amsterdam/New York, pp 443–453

87. Rexed B (1952) The cytoarchitectonic organization of the spinal cord in the cat. J Comp Neurol 96:414–495
88. Nord SG, Kyler HJ (1968) A single unit analysis of trigeminal projections to bulbar reticular nuclei of the rat. J Comp Neurol 134:485–494. doi:10.1002/cne.901340407
89. Phelan KD, Falls WM (1989) The interstitial system of the spinal trigeminal tract in the rat: anatomical evidence for morphological and functional heterogeneity. Somatosens Mot Res 6:367–399
90. Hayashi H, Tabata T (1989) Physiological properties of sensory trigeminal neurons projecting to mesencephalic parabrachial area in the cat. J Neurophysiol 61:1153–1160
91. Hayashi H, Sumino R, Sessle BJ (1984) Functional organization of trigeminal subnucleus interpolaris: nociceptive and innocuous afferent inputs, projections to thalamus, cerebellum, and spinal cord, and descending modulation from periaqueductal gray. J Neurophysiol 51:890–905
92. Strassman AM, Vos BP (1993) Somatotopic and laminar organization of fos-like immunoreactivity in the medullary and upper cervical dorsal horn induced by noxious facial stimulation in the rat. J Comp Neurol 331:495–516. doi:10.1002/cne.903310406
93. Yokota T, Nishikawa N (1980) Reappraisal of somatotopic tactile representation within trigeminal subnucleus caudalis. J Neurophysiol 43:700–712
94. Torvik A (1956) Afferent connections to the sensory trigeminal nuclei, the nucleus of the solitary tract and adjacent structures; an experimental study in the rat. J Comp Neurol 106:51–141
95. Kruger L, Siminoff R, Witkovsky P (1961) Single neuron analysis of dorsal column nuclei and spinal nucleus of trigeminal in cat. J Neurophysiol 24:333–349
96. Jacquin MF, Barcia M, Rhoades RW (1989) Structure-function relationships in rat brainstem subnucleus interpolaris: IV. Projection neurons. J Comp Neurol 282:45–62. doi:10.1002/cne.902820105
97. Liu Y, Zhang M, Broman J, Edvinsson L (2003) Central projections of sensory innervation of the rat superficial temporal artery. Brain Res 966:126–133
98. Lisney SJ (1983) Some current topics of interest in the physiology of trigeminal pain: a review. J R Soc Med 76:292–296
99. Sjoqvist O (1938) Studies on pain conduction in the trigeminal nerve. A contribution to the surgical treatment of facial pain. Acta Psychiatry Scand 17(Suppl):1–139
100. Young RF (1982) Effect of trigeminal tractotomy on dental sensation in humans. J Neurosurg 56:812–818. doi:10.3171/jns.1982.56.6.0812
101. Broton JG, Rosenfeld JP (1986) Cutting rostral trigeminal nuclear complex projections preferentially affects perioral nociception in the rat. Brain Res 397:1–8
102. Malick A, Strassman RM, Burstein R (2000) Trigeminohypothalamic and reticulohypothalamic tract neurons in the upper cervical spinal cord and caudal medulla of the rat. J Neurophysiol 84:2078–2112
103. Hayashi H (1985) Morphology of terminations of small and large myelinated trigeminal primary afferent fibers in the cat. J Comp Neurol 240:71–89. doi:10.1002/cne.902400106
104. Panneton WM, Burton H (1981) Corneal and periocular representation within the trigeminal sensory complex in the cat studied with transganglionic transport of horseradish peroxidase. J Comp Neurol 199:327–344. doi:10.1002/cne.901990303
105. Pearson JC, Jennes L (1988) Localization of serotonin- and substance P-like immunofluorescence in the caudal spinal trigeminal nucleus of the rat. Neurosci Lett 88:151–156
106. Boissonade FM, Sharkey KA, Lucier GE (1993) Trigeminal nuclear complex of the ferret: anatomical and immunohistochemical studies. J Comp Neurol 329:291–312. doi:10.1002/cne.903290302
107. Bae YC, Oh JM, Hwang SJ et al (2004) Expression of vanilloid receptor TRPV1 in the rat trigeminal sensory nuclei. J Comp Neurol 478:62–71. doi:10.1002/cne.20272
108. Henry MA, Johnson LR, Nousek-Goebl N, Westrum LE (1996) Light microscopic localization of calcitonin gene-related peptide in the normal feline trigeminal system and following retrogasserian rhizotomy. J Comp Neurol 365:526–540. doi:10.1002/(SICI)1096-9861(19960219)365:4<526::AID-CNE2>3.0.CO;2-6

109. Tashiro T, Takahashi O, Satoda T et al (1991) Distribution of axons showing calcitonin gene-related peptide- and/or substance P-like immunoreactivity in the sensory trigeminal nuclei of the cat. Neurosci Res 11:119–133
110. Henry MA, Nousek-Goebl NA, Westrum LE (1993) Light and electron microscopic localization of calcitonin gene-related peptide immunoreactivity in lamina II of the feline trigeminal pars caudalis/medullary dorsal horn: a qualitative study. Synap N Y N 13:99–107. doi:10.1002/syn.890130202
111. Lennerz JK, Rühle V, Ceppa EP et al (2008) Calcitonin receptor-like receptor (CLR), receptor activity-modifying protein 1 (RAMP1), and calcitonin gene-related peptide (CGRP) immunoreactivity in the rat trigeminovascular system: differences between peripheral and central CGRP receptor distribution. J Comp Neurol 507:1277–1299. doi:10.1002/cne.21607
112. Eftekhari S, Warfvinge K, Blixt FW, Edvinsson L (2013) Differentiation of nerve fibers storing CGRP and CGRP receptors in the peripheral trigeminovascular system. J Pain Off J Am Pain Soc 14:1289–1303. doi:10.1016/j.jpain.2013.03.010
113. Guy N, Chalus M, Dallel R, Voisin DL (2005) Both oral and caudal parts of the spinal trigeminal nucleus project to the somatosensory thalamus in the rat. Eur J Neurosci 21: 741–754. doi:10.1111/j.1460-9568.2005.03918.x
114. Mantle-St John LA, Tracey DJ (1987) Somatosensory nuclei in the brainstem of the rat: independent projections to the thalamus and cerebellum. J Comp Neurol 255:259–271. doi:10.1002/cne.902550209
115. Ring G, Ganchrow D (1983) Projections of nucleus caudalis and spinal cord to brainstem and diencephalon in the hedgehog (Erinaceus europaeus and Paraechinus aethiopicus): a degeneration study. J Comp Neurol 216:132–151. doi:10.1002/cne.902160203
116. Craig AD (2004) Distribution of trigeminothalamic and spinothalamic lamina I terminations in the macaque monkey. J Comp Neurol 477:119–148. doi:10.1002/cne.20240
117. Bolay H, Tepe N, Filiz A et al (2013) The thalamic reticular nucleus is activated by cortical spreading depression in freely moving rats: prevention by acute valproate. Cephalalgia 33:5–6
118. Tepe N, Filiz A, Akcali D et al (2015) The thalamic reticular nucleus is activated by cortical spreading depression in freely moving rats: prevention by acute valproate administration. Eur J Neurosci. 41(1):120–8.
119. Zikopoulos B, Barbas H (2012) Pathways for emotions and attention converge on the thalamic reticular nucleus in primates. J Neurosci Off J Soc Neurosci 32:5338–5350. doi:10.1523/JNEUROSCI. 4793-11.2012
120. Bernard JF, Peschanski M, Besson JM (1989) A possible spino (trigemino)-ponto-amygdaloid pathway for pain. Neurosci Lett 100:83–88
121. Ge S-N, Li Z-H, Tang J et al (2014) Differential expression of VGLUT1 or VGLUT2 in the trigeminothalamic or trigeminocerebellar projection neurons in the rat. Brain Struct Funct 219:211–229. doi:10.1007/s00429-012-0495-1
122. Panneton WM, Gan Q (2014) Direct reticular projections of trigeminal sensory fibers immunoreactive to CGRP: potential monosynaptic somatoautonomic projections. Front Neurosci 8:136. doi:10.3389/fnins.2014.00136
123. Barnett EM, Evans GD, Sun N et al (1995) Anterograde tracing of trigeminal afferent pathways from the murine tooth pulp to cortex using herpes simplex virus type 1. J Neurosci Off J Soc Neurosci 15:2972–2984
124. Feil K, Herbert H (1995) Topographic organization of spinal and trigeminal somatosensory pathways to the rat parabrachial and Kölliker-Fuse nuclei. J Comp Neurol 353:506–528. doi:10.1002/cne.903530404
125. Mitchell JL, Silverman MB, Aicher SA (2004) Rat trigeminal lamina I neurons that project to thalamic or parabrachial nuclei contain the mu-opioid receptor. Neuroscience 128:571–582. doi:10.1016/j.neuroscience.2004.07.026
126. Aicher SA, Hermes SM, Hegarty DM (2012) Corneal afferents differentially target thalamic- and parabrachial-projecting neurons in spinal trigeminal nucleus caudalis. Neuroscience. doi:10.1016/j.neuroscience.2012.11.033

127. Saper CB (1995) The spinoparabrachial pathway: shedding new light on an old path. J Comp Neurol 353:477–479. doi:10.1002/cne.903530402
128. Sessle BJ (1999) Neural mechanisms and pathways in craniofacial pain. Can J Neurol Sci J Can Sci Neurol 26(Suppl 3):S7–S11
129. Gauriau C, Bernard J-F (2002) Pain pathways and parabrachial circuits in the rat. Exp Physiol 87:251–258
130. Yasui Y, Takada M, Mitani A et al (1985) Direct cortical projections to the parabrachial nucleus in the cat. J Comp Neurol 234:77–86
131. Tokita K, Inoue T, Boughter JD (2009) Afferent connections of the parabrachial nucleus in C57BL/6J mice. Neuroscience 161:475–488. doi:10.1016/j.neuroscience.2009.03.046
132. Moskowitz MA, Nozaki K, Kraig RP (1993) Neocortical spreading depression provokes the expression of c-fos protein-like immunoreactivity within trigeminal nucleus caudalis via trigeminovascular mechanisms. J Neurosci Off J Soc Neurosci 13:1167–1177
133. Jasmin L, Burkey AR, Card JP, Basbaum AI (1997) Transneuronal labeling of a nociceptive pathway, the spino-(trigemino-)parabrachio-amygdaloid, in the rat. J Neurosci Off J Soc Neurosci 17:3751–3765
134. Lazarov NE, Usunoff KG, Schmitt O et al (2011) Amygdalotrigeminal projection in the rat: an anterograde tracing study. Ann Anat Anat Anz Off Organ Anat Ges 193:118–126. doi:10.1016/j.aanat.2010.12.004
135. Akcali D, Sayin A, Sara Y, Bolay H (2010) Does single cortical spreading depression elicit pain behaviour in freely moving rats? Cephalalgia Int J Headache 30:1195–1206. doi:10.1177/0333102409360828
136. Abdallah K, Artola A, Monconduit L et al (2013) Bilateral descending hypothalamic projections to the spinal trigeminal nucleus caudalis in rats. PLoS One 8:e73022. doi:10.1371/journal.pone.0073022
137. Helmstetter FJ, Tershner SA, Poore LH, Bellgowan PS (1998) Antinociception following opioid stimulation of the basolateral amygdala is expressed through the periaqueductal gray and rostral ventromedial medulla. Brain Res 779:104–118
138. Aimone LD, Gebhart GF (1988) Serotonin and/or an excitatory amino acid in the medial medulla mediates stimulation-produced antinociception from the lateral hypothalamus in the rat. Brain Res 450:170–180
139. Messlinger K, Dostrovsky JO, Strassman A (2006) Anatomy and physiology of head pain. In: The headaches, 3rd edn. Lippincott Williams & Wilkins, Philadelphia, pp 95–109
140. An X, Bandler R, Ongür D, Price JL (1998) Prefrontal cortical projections to longitudinal columns in the midbrain periaqueductal gray in macaque monkeys. J Comp Neurol 401:455–479
141. Mantyh PW (1982) Forebrain projections to the periaqueductal gray in the monkey, with observations in the cat and rat. J Comp Neurol 206:146–158. doi:10.1002/cne.902060205
142. Gebhart GF (2004) Descending modulation of pain. Neurosci Biobehav Rev 27:729–737. doi:10.1016/j.neubiorev.2003.11.008
143. Morgan MM, Heinricher MM, Fields HL (1992) Circuitry linking opioid-sensitive nociceptive modulatory systems in periaqueductal gray and spinal cord with rostral ventromedial medulla. Neuroscience 47:863–871
144. Pertovaara A, Almeide A (2006) Endogenous pain modulation, chapter 13, descending inhibitory systems. Handb Clin Neurol 81(3rd series vol. 3), 179–192
145. Lakos S, Basbaum AI (1988) An ultrastructural study of the projections from the midbrain periaqueductal gray to spinally projecting, serotonin-immunoreactive neurons of the medullary nucleus raphe magnus in the rat. Brain Res 443:383–388
146. Kwiat GC, Basbaum AI (1992) The origin of brainstem noradrenergic and serotonergic projections to the spinal cord dorsal horn in the rat. Somatosens Mot Res 9:157–173
147. Yaksh TL (1985) Pharmacology of spinal adrenergic systems which modulate spinal nociceptive processing. Pharmacol Biochem Behav 22:845–858
148. Iliakis B, Anderson NL, Irish PS et al (1996) Electron microscopy of immunoreactivity patterns for glutamate and gamma-aminobutyric acid in synaptic glomeruli of the feline spinal trigeminal nucleus (Subnucleus Caudalis). J Comp Neurol 366:465–477. doi:10.1002/(SICI)1096-9861(19960311)366:3<465::AID-CNE7>3.0.CO;2-2

149. Yoshida A, Chen K, Moritani M et al (1997) Organization of the descending projections from the parabrachial nucleus to the trigeminal sensory nuclear complex and spinal dorsal horn in the rat. J Comp Neurol 383:94–111
150. Delépine L, Aubineau P (1997) Plasma protein extravasation induced in the rat dura mater by stimulation of the parasympathetic sphenopalatine ganglion. Exp Neurol 147:389–400. doi:10.1006/exnr.1997.6614
151. Johnson K, Bolay H (2006) Neurogenic inflammatory mechanisms in migraine. In: The headaches, 3rd edn. Lippincott Williams & Wilkins, Philadelphia, pp 309–319
152. Millan MJ, Gramsch C, Przewłocki R et al (1980) Lesions of the hypothalamic arcuate nucleus produce a temporary hyperalgesia and attenuate stress-evoked analgesia. Life Sci 27:1513–1523
153. Butler RK, Finn DP (2009) Stress-induced analgesia. Prog Neurobiol 88:184–202. doi:10.1016/j.pneurobio.2009.04.003
154. Frost JJ, Mayberg HS, Sadzot B et al (1990) Comparison of [11C]diprenorphine and [11C] carfentanil binding to opiate receptors in humans by positron emission tomography. J Cereb Blood Flow Metab Off J Int Soc Cereb Blood Flow Metab 10:484–492. doi:10.1038/jcbfm.1990.90
155. Valet M, Sprenger T, Boecker H et al (2004) Distraction modulates connectivity of the cingulo-frontal cortex and the midbrain during pain–an fMRI analysis. Pain 109:399–408. doi:10.1016/j.pain.2004.02.033
156. Treede RD, Kenshalo DR, Gracely RH, Jones AK (1999) The cortical representation of pain. Pain 79:105–111
157. Gauriau C, Bernard J-F (2004) A comparative reappraisal of projections from the superficial laminae of the dorsal horn in the rat: the forebrain. J Comp Neurol 468:24–56. doi:10.1002/cne.10873
158. Gauriau C, Bernard J-F (2004) Posterior triangular thalamic neurons convey nociceptive messages to the secondary somatosensory and insular cortices in the rat. J Neurosci Off J Soc Neurosci 24:752–761. doi:10.1523/JNEUROSCI. 3272-03.2004
159. Rainville P, Duncan GH, Price DD et al (1997) Pain affect encoded in human anterior cingulate but not somatosensory cortex. Science 277:968–971
160. May A, Kaube H, Büchel C et al (1998) Experimental cranial pain elicited by capsaicin: a PET study. Pain 74:61–66
161. Sato F, Akhter F, Haque T et al (2013) Projections from the insular cortex to pain-receptive trigeminal caudal subnucleus (medullary dorsal horn) and other lower brainstem areas in rats. Neuroscience 233:9–27. doi:10.1016/j.neuroscience.2012.12.024
162. Noseda R, Constandil L, Bourgeais L et al (2010) Changes of meningeal excitability mediated by corticotrigeminal networks: a link for the endogenous modulation of migraine pain. J Neurosci Off J Soc Neurosci 30:14420–14429. doi:10.1523/JNEUROSCI. 3025-10.2010
163. Price DD, Dubner R (1977) Neurons that subserve the sensory-discriminative aspects of pain. Pain 3:307–338
164. Iwata K, Miyachi S, Imanishi M et al (2011) Ascending multisynaptic pathways from the trigeminal ganglion to the anterior cingulate cortex. Exp Neurol 227:69–78. doi:10.1016/j.expneurol.2010.09.013
165. Lorenz J, Cross DJ, Minoshima S et al (2002) A unique representation of heat allodynia in the human brain. Neuron 35:383–393
166. Dubner R, Bennett GJ (1983) Spinal and trigeminal mechanisms of nociception. Annu Rev Neurosci 6:381–418. doi:10.1146/annurev.ne.06.030183.002121
167. Craig AD (2003) A new view of pain as a homeostatic emotion. Trends Neurosci 26:303–307
168. Hadjipavlou G, Dunckley P, Behrens TE, Tracey I (2006) Determining anatomical connectivities between cortical and brainstem pain processing regions in humans: a diffusion tensor imaging study in healthy controls. Pain 123:169–178. doi:10.1016/j.pain.2006.02.027
169. Lorenz J, Minoshima S, Casey KL (2003) Keeping pain out of mind: the role of the dorsolateral prefrontal cortex in pain modulation. Brain J Neurol 126:1079–1091

170. Brighina F, Piazza A, Vitello G et al (2004) rTMS of the prefrontal cortex in the treatment of chronic migraine: a pilot study. J Neurol Sci 227:67–71. doi:10.1016/j.jns.2004.08.008
171. Zhao L, Liu J, Dong X et al (2013) Alterations in regional homogeneity assessed by fMRI in patients with migraine without aura stratified by disease duration. J Headache Pain 14:85. doi:10.1186/1129-2377-14-85
172. Tessitore A, Russo A, Giordano A et al (2013) Disrupted default mode network connectivity in migraine without aura. J Headache Pain 14:89. doi:10.1186/1129-2377-14-89
173. Schwedt TJ, Schlaggar BL, Mar S et al (2013) Atypical resting-state functional connectivity of affective pain regions in chronic migraine. Headache 53:737–751. doi:10.1111/head.12081
174. Ichesco E, Quintero A, Clauw DJ et al (2012) Altered functional connectivity between the insula and the cingulate cortex in patients with temporomandibular disorder: a pilot study. Headache 52:441–454. doi:10.1111/j.1526-4610.2011.01998.x
175. Qiu E, Wang Y, Ma L et al (2013) Abnormal brain functional connectivity of the hypothalamus in cluster headaches. PLoS One 8:e57896. doi:10.1371/journal.pone.0057896
176. Rocca MA, Valsasina P, Absinta M et al (2010) Central nervous system dysregulation extends beyond the pain-matrix network in cluster headache. Cephalalgia Int J Headache 30:1383–1391. doi:10.1177/0333102410365164
177. Yang P-F, Qi H-X, Kaas JH, Chen LM (2014) Parallel functional reorganizations of somatosensory areas 3b and 1, and S2 following spinal cord injury in squirrel monkeys. J Neurosci Off J Soc Neurosci 34:9351–9363. doi:10.1523/JNEUROSCI. 0537-14.2014
178. Morelli N, Rota E, Gori S et al (2013) Brainstem activation in cluster headache: an adaptive behavioural response? Cephalalgia Int J Headache 33:416–420. doi:10.1177/0333102412474505
179. Riederer F, Gantenbein AR, Marti M et al (2013) Decrease of gray matter volume in the midbrain is associated with treatment response in medication-overuse headache: possible influence of orbitofrontal cortex. J Neurosci Off J Soc Neurosci 33:15343–15349. doi:10.1523/JNEUROSCI. 3804-12.2013
180. Bolay H, Berman NEJ, Akcali D (2011) Sex-related differences in animal models of migraine headache. Headache 51:891–904. doi:10.1111/j.1526-4610.2011.01903.x
181. Peterlin BL, Gupta S, Ward TN, Macgregor A (2011) Sex matters: evaluating sex and gender in migraine and headache research. Headache 51:839–842. doi:10.1111/j.1526-4610.2011.01900.x
182. Gazerani P, Andersen OK, Arendt-Nielsen L (2005) A human experimental capsaicin model for trigeminal sensitization. Gender-specific differences. Pain 118:155–163. doi:10.1016/j.pain.2005.08.009
183. Andreason PJ, Zametkin AJ, Guo AC et al (1994) Gender-related differences in regional cerebral glucose metabolism in normal volunteers. Psychiatry Res 51:175–183
184. Wager TD, Phan KL, Liberzon I, Taylor SF (2003) Valence, gender, and lateralization of functional brain anatomy in emotion: a meta-analysis of findings from neuroimaging. Neuroimage 19:513–531
185. Lopez-Larson MP, Anderson JS, Ferguson MA, Yurgelun-Todd D (2011) Local brain connectivity and associations with gender and age. Dev Cogn Neurosci 1:187–197. doi:10.1016/j.dcn.2010.10.001
186. Maleki N, Becerra L, Brawn J et al (2012) Concurrent functional and structural cortical alterations in migraine. Cephalalgia Int J Headache 32:607–620. doi:10.1177/0333102412445622
187. Paulson PE, Minoshima S, Morrow TJ, Casey KL (1998) Gender differences in pain perception and patterns of cerebral activation during noxious heat stimulation in humans. Pain 76:223–229
188. Derbyshire SWG, Nichols TE, Firestone L et al (2002) Gender differences in patterns of cerebral activation during equal experience of painful laser stimulation. J Pain Off J Am Pain Soc 3:401–411
189. De Leeuw R, Albuquerque RJC, Andersen AH, Carlson CR (2006) Influence of estrogen on brain activation during stimulation with painful heat. J Oral Maxillofac Surg Off J Am Assoc Oral Maxillofac Surg 64:158–166. doi:10.1016/j.joms.2005.10.006
190. Chang Z, Okamoto K, Bereiter DA (2012) Differential ascending projections of temporomandibular joint-responsive brainstem neurons to periaqueductal gray and posterior thalamus of male and female rats. Neuroscience 203:230–243. doi:10.1016/j.neuroscience.2011.11.042

191. Boes T, Levy D (2012) Influence of sex, estrous cycle, and estrogen on intracranial dural mast cells. Cephalalgia Int J Headache 32:924–931. doi:10.1177/0333102412454947
192. Saleh TM, Connell BJ, McQuaid T, Cribb AE (2003) Estrogen-induced neurochemical and electrophysiological changes in the parabrachial nucleus of the male rat. Brain Res 990:58–65
193. Greco R, Mangione A, Siani F et al (2013) Effects of CGRP receptor antagonism in nitroglycerin-induced hyperalgesia. Cephalalgia Int J Headache 34:594–604. doi:10.1177/0333102413517776
194. Eikermann-Haerter K, Baum MJ, Ferrari MD et al (2009) Androgenic suppression of spreading depression in familial hemiplegic migraine type 1 mutant mice. Ann Neurol 66:564–568. doi:10.1002/ana.21779

Chapter 2
Animal Models of Migraine

Anna P. Andreou and Michael L. Oshinsky

The headache research field is privileged to have in its preclinical laboratories well-established animal models that significantly facilitate and improve our understanding of headache mechanisms, in particular in terms of the molecular signalling and brain networks involved. A variety of pharmacological screening approaches for novel therapeutics and for the improvement of advanced pharmacological agents can be achieved in translational research utilising these models. The available migraine models have been developed based on our understanding of migraine from clinical, migraine patient-specific evidence. These clinical phenotypes have been successfully employed to model features of the disease physiology in animals and to provide reproducible meaningful physiological measures in the laboratory.

2.1 Animal Models of Migraine

2.1.1 What Defines an Animal Model?

Any disease model, in humans or animals, needs to fulfil three essential criteria: (1) provide trustful replication of disease physiology, (2) demonstrate good efficacy of known disease treatments and (3) demonstrate a lack of efficacy of clinically known unsuccessful disease treatments. The end goals of the development and use of

A.P. Andreou (✉)
Headache Research – Section of Anaesthetics, Pain Medicine and Intensive Care, Faculty of Medicine, Imperial College London, Chelsea and Westminster Hospital Campus, 369 Fulham Road, London SW10 9NH, UK
e-mail: anna.andreou@headache-research.com

M.L. Oshinsky
National Institute of Neurological Disorders and Stroke, National Institutes of Health, 6001 Executive Blvd, Room 2116, MSC 9521, Bethesda, MD 20892-9521, USA

M. Ashina, P. Geppetti (eds.), *Pathophysiology of Headaches: From Molecule to Man*, Headache, DOI 10.1007/978-3-319-15621-7_2

animal models are to enhance our understanding of disease mechanisms and to aid the discovery of novel anti-migraine treatments.

Migraine is the most well-studied type of the primary headache, and the animal models used in migraine research have proven their modelling potentials in terms of screening clinically effective and ineffective treatments. Their major disadvantage, however, lies in their ability to replicate a complete disease phenotype, although this does not necessarily hamper their use in drug discovery or pathophysiology studies. Migraine is a complex neurological disorder that involves specific characteristics, spontaneous or biologically (stress, lack of sleep and missing meals are few of the well-recognised migraine triggers) triggered episodic attacks, characterised by intense headache that becomes worst with movement, accompanied by nausea, photophobia, phonophobia or even osmophobia and occasionally aura [1]. Outside the headache and aura phase, the attack onset appears to occur earlier during the premonitory phase [2]. Even in the interictal phase, migraine sufferers may have a differential processing of sensory information, as indicated by their lack of habituation to normal stimuli [3]. The migraine animal models are somewhat limited as they model aspects of the migraine syndrome and not the entire spectrum of symptoms. Currently, no animal model exists that replicates all components of migraine, particularly the sensory disturbances seen during the attack, as well as the premonitory phase events and the lack of habituation. This disadvantage largely reflects the lack of understanding we have on the migraine pathophysiology itself, and it is partly compromising the human models of migraine too. However, despite their inability to model the full spectrum of migraine, the migraine animal models are considered among the most successful neurological disease models, as they do model aspects of the disease and are indeed reliable tools for pharmacological investigations. As a comparison, the widely used cerebral infraction model (middle cerebral artery occlusion model) in the field of stroke research fails to prove the efficacy of the sole clinically available treatment, the tissue plasminogen activator [4].

Animal models of migraine are mostly acute system activation models. Acute migraine models are more widely used and are divided into three large categories with regard to the aspect of migraine pathophysiology they model: (A) Models of activation of the involved pain pathway, the trigeminovascular system – the trigeminal nerve innervation of dural structures, mainly blood vessels, and the trigeminal ganglion. This model mainly reproduces the process of events that are thought to at least replicate components of the pain perception that occurs during the headache phase of a migraine attack. (B) Models of cortical spreading depression that reproduce the possible events occurring during aura. (C) Models of nitric oxide (NO) signalling (provocation models). This model is based on clinical studies establishing that NO donors can trigger a migraine attack in sufferers after a delay of hours [5] and even reproduce premonitory symptoms and nausea in some patients [6, 7]. Neither in humans nor in animals NO donors induce a migraine aura or attenuate cortical excitability. Additionally, the identification of monogenic mutations as the cause of rare types of migraine has further allowed the development of genetic models with knock-in mutations in their genome [8]. A combination of the above modelling assays in these genetically modified animals has been further employed to answer the enigma of migraine neurobiology. In the past few years, some attempts have been made to develop chronic, conscious migraine models; however, the characterisation of the produced phenotype, as well as their efficacy to migraine treatments, is yet to be validated among different laboratories.

2.1.2 Ethical Considerations

The use of animals in experimental research is under strict regulatory control by different authorities. The considerations around the use of animals in medical research evolve around three directions, known as the 3Rs: reduction, measures to ensure that the minimum number of animals will be used; refinement, how to achieve objectives with minimum animal suffering; and replacement, how to achieve the same objectives without using animals. Any research design should justify the 3Rs. Migraine is a complex disease, and the determination of neuronal changes not only requires the presence of neurons at a state of nociceptive condition but also requires intact brain pathways that influence each other. As such, pathway investigations cannot be ethically conducted in humans; there is no feasible alternative that would entirely replace the use of living animals that would allow the objectives to be met. Good laboratory practices should be used, however, to minimise animal suffering and to reduce the number of animals used. The majority of migraine models have been thus developed in anaesthetised animals, in which suffering is considered minimum. In preclinical research in general, however, there is growing concern that poor experimental design and lack of transparent reporting contribute to the frequent failure of preclinical animal studies to translate into treatments for human disease. In 2010, the Animal Research Reporting of In Vivo Experiments (ARRIVE) guidelines were introduced [9–11], and adapted by many regulatory authorities and medical journals, to help improve reporting standards. The guidelines refer to common good laboratory practice and should be seen in the same perspective as the guidelines established for clinical trials in humans.

2.1.3 Migraine Pathophysiology

Migraine pathophysiology is extensively analysed elsewhere in this book; however, for the purposes of better understanding the neurobiology utilised in animal models, a brief description is given below.

The migraine headache is perceived to be felt on intracranial structures, such as the dura mater and intracranial vasculature [12]. The sensory innervation of these structures arises from the trigeminal nerve, mainly from unmyelinated C-, and thinly myelinated Aδ-fibres, which have their cell bodies in the trigeminal ganglion. Nociceptive activation of these trigeminal fibres is referred to as "trigeminovascular activation" [13]. The trigeminal fibres that transmit sensory information from intracranial structures synapse on second-order neurons within the trigeminocervical complex (TCC; trigeminal nucleus caudalis, C1 and C2 spinal levels). These neurons give rise to the main ascending trigeminothalamic pathway that relays sensory information to third-order neurons in the contralateral thalamus. The thalamus, mainly the ventroposteromedial thalamic nucleus (VPM) and the posterior thalamic nucleus (Po), is acting as the last gate before sensory information is transmitted to cortical areas involved in the processing of pain perception. A complex of descending networks from multiple brainstem, midbrain and cortical nuclei modulate the excitability of the ascending trigeminothalamic pathway [14]. In the absence of any evidence of malfunction in the trigeminovascular system [13, 15], a disruption of normal endogenous descending modulatory tone in the brain may play a critical role in migraine. However, what really alters the excitability of the ascending trigeminothalamic pathway, in a manner that a migraine attack may develop in susceptible individuals, remains to be revealed.

The migraine aura is now believed to result from the neurophysiological event called cortical spreading depression (CSD) [16]. CSD is a wave of cortical neuronal depolarisation, followed by depressed activity and associated with blood flow changes [17]. In migraine patients, CSD is believed to spread out from the occipital cortex, but it remains enigmatic how CSD is triggered in patients during migraine aura.

Accumulating evidence exists as to why the trigger of migraine attacks should be sought in the hypothalamus [18]. The strongest, direct evidence for hypothalamic activation in migraine patients arises from brain imaging studies. These studies demonstrated, using positron emission tomography (PET), increased blood flow in the posterior region of the hypothalamus during the very early stages of spontaneous migraine attacks [19] and during the premonitory phase of the NO donor, nitroglycerin (NTG)-induced migraine attacks [20].

2.2 Models of the Peripheral Trigeminovascular System

Animal models investigating changes in the peripheral branch of the trigeminovascular system, i.e. the dural environment and vasculature and the trigeminal fibres innervating these, are believed to model peripheral events that are likely to occur during a migraine attack. As our understanding of migraine has progressed over the years, some of these models are now considered redundant, particularly those demonstrating a pure vascular site of action. Nevertheless, their utilisation in the triptan era, as well as the lessons learned from them, makes knowledge around these models an integrated facet of the scientific progression in the field.

2.2.1 Vascular Models

Vascular animal models had been developed during the first epoch phase of scientific exploration in migraine models and were based on the view that extracerebral vasodilation could occur during a migraine attack. Indeed, successful migraine treatments include drugs which do not cross the blood-brain barrier (BBB) and evoke vasoconstriction, such as sumatriptan and ergot alkaloids [21]. Identification of elevated levels of the vasodilatory peptide calcitonin gene-related peptide (CGRP) in patients during a migraine attack [22] further reinforced, at the time, the theory of an integrated vascular involvement in migraine pathophysiology. However, a role for cephalic vessels in the development of migraine as a syndrome has been criticised over the years. Magnetic resonance angiography data during NTG-triggered migraine attacks suggests no association with vasodilation of cerebral or meningeal vessels [23]. It is now well established that not all vasodilatory peptides trigger a migraine attack [24], while following intravenous infusion of other vasodilatory peptides, a migraine attack is triggered hours after the cease of the vasodilatory effects [25]. Vasodilation alone may be an epiphenomenon of migraine attacks, which is not sufficient to induce pain. Additionally, not all vasoconstrictive therapies alleviate migraine symptoms, while many of the clinically effective anti-migraine drugs do not have a vasoactive action [26]. Nevertheless, the use of vascular models in migraine research made it clear that it is not the vasodilation, as such, that is important in migraine pathophysiology but the induction of second messenger signalling pathways that vasoactive substances induce through their interactions with G-protein-coupled receptors. Vascular models have thus allowed the collection of reliable information for the vasodilatory role of neuropeptides found in the trigeminal nerve endings, measurement of intracellular calcium changes and second messengers' concentration.

Nevertheless, the vascular models of migraine are worth referring to, not only due to their use during the triptan era but also for their usefulness in evaluating the vasoconstrictive profile of potential anti-migraine therapeutics, which can be contraindicated for some patients. In combination with immunocytochemical methods on isolated cerebral vessels, the anatomical localisation of receptors, such as of 5-HT_{1B} and CGRP receptors, had become more clear [27, 28]. Additionally, with the new development of CGRP antibodies [29–31], the models may be found once again useful in identifying the long-term effects of vasodilation blockade (due to CGRP) on the actual vascular bed.

2.2.1.1 Constriction of the Carotid Arteriovenous Anastomoses

This model was used in the early phases of triptans' validation as potential therapeutic agents. The model aimed to replicate clinical evidence showing that the anterior jugular oxygen saturation is reduced to the ipsilateral side of the headache during a migraine attack [32], probably due to dilation of the carotid arteriovenous anastomoses, which will reduce the available oxygenated blood through its thrust into the veins [33]. This was further thought to explain clinical observations in some patients, such as facial paleness, swelling and reduced facial temperature of the frontal vein ipsilateral to the headache. The model did not gain much interest in the field, not only due to its technical difficulties but also due to that fact that arteriovenous anastomoses do not appear dysfunctional in humans, particularly due to the intact sympathetic nervous system. Decreased oxygenation of blood has been also supported to have a causative role in a minority of migraine patients with right-to-left shunts; however, studies support an unlikely significant role in migraine triggering or chronification [34].

The model used primarily anaesthetised pigs, in which the strong sympathetic influence is suppressed, allowing ~80 % of the total carotid blood flow to be shunted via arteriovenous anastomoses into the jugular venous circulation [35]. Alternatively, vasoconstriction of potential drugs in vasosympathectomised dogs was also employed [36]. Sumatriptan, ergot alkaloids and α-adrenoceptor agonists were shown to reduce carotid arteriovenous anastomotic shunting [21, 37]. The model has thus been used to predict a clear vascular site of action of potential treatments. Such a complicated and demanding model these days can be easily replaced by in vitro vascular preparations.

2.2.1.2 Constriction of Cephalic Blood Vessels

More direct vascular models that investigate the pure vasoconstriction properties of therapeutics were also developed both as in vivo and in vitro setups. In vitro vascular models use isolated cranial vessels (including human arteries) and isometric measures of vessel diameter in order to study the contraction or relaxation of vascular segments mounted in organ baths during application of potential anti-migraine drugs. This model has been successfully used to evaluate the vascular action of triptans and demonstrated the 5-HT_{1B} receptor efficacy of sumatriptan [38]. Using this model, Müller-Schhweinitzer and Weidmann suggested as early as in 1977 [39] that the anti-migraine efficacy of ergotamine was due to a pure vasoconstrictive action.

A variety of specific acute anti-migraine drugs, including sumatriptan and ergot alkaloids, have been shown to produce selective vasoconstriction of cephalic blood vessels in an in vivo vascular model [21], utilising initially dogs and rabbits, and rats and guinea pigs later. The vasomotor role of endogenous neuropeptides of the perivascular trigeminal nerve endings has been further studied in this model by local luminal and abluminal applications [40, 41]. Using intravital microscopy over a cranial window that allows direct measurements of the dural blood vessel diameter or blood flow, it was shown that topical or intravenous administration of CGRP induces vasodilation, which is blocked by the CGRP antagonist CGRP_{8-37} [42, 43]. Exogenous CGRP acts directly on CGRP receptors on the smooth muscle of dural arteries and compounds that inhibit CGRP-induced dilation demonstrate at least their partial action on the smooth muscles of blood vessels [44]. Similarly, systemic administration of NO donors causes reproducible dural blood vessel dilation [45, 46]. A number of compounds including triptans, CGRP antagonists, cannabinoid receptor 1 (CB1) agonists and nitric oxide synthase (NOS) inhibitors have been found to attenuate this chemically induced vasodilation [45, 47].

A major limitation of the above model when studying the effects of vasodilatory peptides, such as CGRP, is the resulting hypotension and potential activation of autoregulatory mechanisms [41]. The subsequent vasodilation of the cranial vasculature makes the involvement of an actual vasomotor pharmacological interaction or of an autoregulation mechanism activated in response to hypotension difficult to interpret [48]. An alternative to the potent hypotension induced by intravenous administration of vasodilators was suggested by Gupta and colleagues, who demonstrated that intracarotid administration of CGRP induces maximum middle meningeal artery dilation with minimum blood pressure effects [49]. This route of administration can be successfully adopted for other vasodilatory substances, including NO donors.

2.2.2 *Neurovascular Models for Peripheral Investigations of the Trigeminovascular System*

Neurovascular animal models aim to reflect more appropriately the involvement of the peripheral nerve fibres in the modulation of the dural vascular tone.

2.2.2.1 Neurogenic Dural Vasodilation and Blood Flow Model

The involvement of CGRP in migraine pathophysiology has been crucial in the development of experimental animal models of migraine. Migraine patients appear to have elevated levels of CGRP in the cerebral circulation during a migraine attack [22], although these findings have been criticised in other studies [50]. Using animals, it was later shown that stimulation of trigeminal nerve fibres innervating the dura mater induces the release of CGRP [51], which results in dural blood vessel dilation via CGRP receptors located on the vascular smooth muscle [46]. Further to the vasodilation, an output of trigeminovascular activation is a neurogenic, CGRP-driven, reproducible increase in meningeal blood flow [52, 53]. How CGRP release may be triggered in migraine patients is not clear, although an antidromic activation of the trigeminal system as an epiphenomenon of central mechanisms has been suggested, but not fully supported [54]. Nevertheless, the model directly activates the nociceptive pathway thought to be involved during the migraine headache phase. Thus, the vasodilatory reaction and blood flow increase of dural vessels following stimulation of the trigeminal fibres is used as an indirect indication of trigeminal system activation, and it models peripheral aspects of the migraine attack. It is important to note the correct interpretation of the model that it is not the CGRP-induced vasodilation or blood flow change as such that should be aimed to be blocked but the activation of the peripheral trigeminal fibres. CGRP itself is not known to sensitise trigeminal fibres either and thus does not contribute to nociceptive activation [55]. In this model, as vasodilation is a result of neuronal fibre activation, the model is known as neurogenic dural vasodilation (NDV) and permits the study of the peripheral branch of the trigeminovascular system.

In the NDV model, trigeminal fibre stimulation is mostly achieved through application of electrical stimulation from bipolar electrodes positioned near dural arteries, such as the middle meningeal artery, on a closed cranial window. The model utilises mostly intravital microscopy which permits the direct study of cranial blood vessels' diameter or laser Doppler flowmetry for detection of dural blood flow changes. Electrical stimulation of the closed cranial window causes a neurogenic reproducible dilation and increase of blood flow of the underlying dural vessels, via activation of the trigeminal nerve fibres. Vessel dilation is due to CGRP release from pre-junctional trigeminal nerve endings innervating the dural vessels [46, 52, 53, 56], which binds to CGRP receptors on the smooth muscle of dural vessels resulting in vasodilation. The CGRP antagonist $CGRP_{8-37}$ is able to completely inhibit NDV, further indicating the importance of this peptide in NDV and the usefulness of modelling, at least partly, the pharmacology of the trigeminovascular system. The model is performed in anaesthetised rodents, and good laboratory practice that can assess the depth of anaesthesia, changes in blood pressure and temperature must be employed during its use, to allow for reliable outcomes to be delivered.

Beyond perivascular electrical stimulation, it was further shown that employment of chemical stimulation of the trigeminal fibres, through capsaicin, for example, could be also used to study neurogenic vasodilation. Capsaicin-induced vasodilation is elicited by the release of CGRP, as it can be blocked by a CGRP antagonist [57]. Capsaicin binds on the transient receptor potential vanilloid type-1 (TRPV1) found mostly on small diameter sensory fibres and depolarises them [58]. The induced vasodilation occurs due to the release of, among other peptides, CGRP. However, TRPV1 antagonism does not block NDV induced by perivascular dural electrical stimulation [59], indicating that electrical stimulation of the trigeminal fibres does not activate TRPV1 channels. This may further suggest that TRPV1 receptors do not play a significant role in, at least the antidromic, activation of the peripheral side of the trigeminovascular system. Additionally, although NO can itself act as a smooth muscle relaxant [60], there is evidence suggesting that NO activates trigeminal neurons by inducing CGRP release [61]. A synergistic relationship between CGRP and NO may exist, as CGRP receptor activation can increase the expression of inducible nitric oxide synthase (NOS) and stimulate NO release from glial cells in the trigeminal ganglion [62, 63]. NOS inhibitors were also effective in modulating dural blood flow [63].

This model mainly tests the action of systemic administration of potential anti-migraine compounds. Since CGRP receptor antagonists are clearly effective in acute migraine treatment [64, 65], the pharmacology of the mechanisms responsible for CGRP release from trigeminal fibres is of direct relevance to the development of newer migraine therapies, particularly for the newly described CGRP antibodies that currently undergo clinical trials [29, 30]. The model has the proven ability to predict the anti-migraine therapeutic potential of compounds, as triptans and dihydroergotamine were effective in inhibiting NDV, potentially by inhibiting the pre-synaptic release of CGRP from trigeminal fibres [47, 52, 53, 66–68]. The lack of an inhibitory NDV effect following neurokinin-1 receptor (NK1) antagonism, which will block the vasodilatory effects induced by substance P (SP) [66, 69], similarly to the poor results obtained with the use of NK1 antagonists in clinical trials [70, 71], further validates the good efficacy of this model. Thus, a series of compounds which include calcium channel blockers [72], cannabinoids [73], adenosine A1 receptor agonists [56], orexin 1 receptor agonists [74] and $5\text{-}HT_{1F}$ and $5\text{-}HT_7$

agonists [75] and nonsteroidal anti-inflammatory agents (NSAIDS) [76] that are able to inhibit NDV may indeed represent potential new migraine therapeutics.

The NDV model can be used in combination with the vascular model of CGRP infusion to compare a vascular over a neuronal side of action of potential therapeutics. Thus, the use of intravital microscopy can further facilitate dissecting the pharmacology of the trigeminovascular system and the potential site of action (vascular and/or neurogenic) of therapeutic compounds. Compounds that attenuate NDV, but not CGRP-induced dilation, are more likely to have a direct action on CGRP release from the pre-junctional site of the trigeminal fibres. Examples of such compounds include clinically active therapeutics, such as topiramate, rizatriptan and sumatriptan [44, 67], as well as potential anti-migraine treatments such as orexin 1 receptor agonists and calcium channel blockers [74, 77]. Additionally, from a biological perspective, female hormones were shown to enhance NDV through increased CGRP release from perivascular nerves and not through vascular changes, suggesting a trigeminal neuronal mechanism through which female hormones may exacerbate migraine in women [78].

Despite being proven as a highly predictive model of anti-migraine efficacy of treatments acting peripherally, the NDV model has important limitations. Clinically active compounds that have central neural system site of actions cannot be screened using this model. For example, the efficacy of many potential anti-migraine compounds that act centrally such as 5-HT_{1F} receptor agonists, dopamine receptor agonists and kainate receptor antagonists are not seen using this model [56, 68, 75, 79, 80]. Thus, caution must be used when testing compounds using the NDV model, which should be used along other models for a better understanding of the actual site of action of different therapeutics. Similarly, clinically active preventatives, such as propranolol, valproate and flunarizine, were unsuccessful at inhibiting NDV, which suggests a lack of action at the peripheral end of the trigeminal nerve [45, 56, 68]. On the other hand, as these preventatives have a clinically effective action in reducing the frequency of attacks over prolonged administration, the acute nature of the model cannot be used to investigate the long-term effects of treatments. As migraine is believed to be a disorder of the brain, the translational effectiveness of the model can be somewhat questioned.

2.2.2.2 Neurogenic Inflammation and Plasma Protein Extravasation Model

An earlier theory in migraine suggested that activation of trigeminal sensory fibres leads to sterile neurogenic inflammation characterised by plasma protein extravasation, vasodilation and mast cell degranulation within the meningeal environment. This is thought to be mediated by neuropeptide release from trigeminal sensory fibres and that it could induce pain [68, 81–83]. This theory has derived from indirect evidence mainly from preclinical studies, in which trigeminal ganglion stimulation or chemical activation of meningeal trigeminal fibres induces vascular and mast cell changes, with a concurrent vasodilation due to increased release of CGRP and SP. Release of tachykinins and endothelin-3 further promotes vascular permeability leading to protein leakage from post-capillary venules, also known as plasma protein extravasation (PPE), and activation of dural mast cells [84]. Activation of dural mast cells will result in the release of inflammatory mediators that, along with other inflammatory mediators of neurogenic origin, such as SP and CGRP, could produce long-lasting activation and sensitisation of trigeminal nociceptors [85]. These events are in line with the occurrence of nociceptive neurogenic inflammation of the dura mater in rodents [86]. The rodent model considers inflammation to play a key role in migraine pathophysiology; however, although it is generally accepted that the initiation of a sterile inflammatory response could induce nociceptive-like behaviour in animals [87], it is not clear whether this is sufficient to induce migraine. More importantly, how neurogenic inflammation may be induced during migraine attacks cannot be reliably answered. In patients, despite the efficacy of nonsteroidal anti-inflammatory drugs (NSAIDs) in providing, at least some, pain relief of migraine headache, no PPE has been detected with retinal angiography in patients during acute attacks of migraine or cluster headache [15], further questioning the validity of the model. In this model, SP seems to be the primary mediator responsible for PPE, as gene knockout studies confirm that neurogenic inflammation is via tachykinin receptor activation on endothelial cells [88], whereas CGRP alone does not induce PPE [86]. Interestingly, in contrast to CGRP, SP levels are only moderately different during migraine attacks [89].

In the neurogenic inflammatory model, plasma protein extravasation into the meninges is mainly induced by electrical stimulation of the trigeminal ganglion. PPE is detected by measuring the amount of extravasated albumin in the dura mater, using radiolabelled bovine serum albumin or of Evans blue which can bind directly to albumin. After unilateral stimulation of the trigeminal ganglion, the dura mater is removed and the ratio of the average labelled intensity of the stimulated side compared to the non-stimulated side is calculated. Intravenous administration of serotonin and of various neuropeptides, including SP, neurokinin A and bradykinin can additionally lead to meningeal PPE. Parasympathetic activation seems to also result in neurogenically mediated meningeal PPE [90].

The hypothesis of a sterile neurogenic inflammation in migraine was supported initially by the fact that PPE can be blocked by a number of clinically active treatments, such as ergot alkaloids [81]; cyclooxygenases1/2 (COX1/2) inhibitors, including indomethacin [91, 92]; and 5-$HT1_{B/D/F}$ agonists [93–95] including sumatriptan [96]. However, the subsequent failure of neurogenic inflammation blockade to demonstrate an efficacy in a clinical setting further contributed to the unreliability of the model and the characterisation of the theory of neurogenic inflammation in migraine pathophysiology as inadequate. Neurokinin-1 antagonists which block PPE in rodents [97, 98] were ineffective in patients [71, 99], while the specific PPE blockers, CP-122,288 and 4991w93, and a number of endothelin antagonists which prevent tachykinins and endothelin-3 from inducing PPE were not effective in aborting migraine attacks in the clinic [100], despite blocking neurogenic inflammation in this animal model. For these reasons, the neurogenic inflammation model following electrical stimulation of the trigeminal ganglion and assessment of PPE stimulation is now considered redundant.

2.3 Trigeminovascular Activation Model for Central Neural System Mechanisms

Migraine is now established as a brain disorder, given the lack of clinical evidence of peripheral changes that could trigger a migraine attack [101, 102]. A number of brain areas are consistently seen to be activated during all phases of a migraine attack [7, 20, 103], suggesting an important role of CNS structures in the development of migraine pathophysiology. The field is thus moving towards investigations that will allow the development of novel CNS-based therapeutic approaches. Studying the central components of the trigeminovascular pathway and modelling of central events that are likely to occur during a migraine attack offers good potentials for the development of new, efficient therapeutics.

The trigeminovascular activation model is thought to mainly reproduce the process of events that at least replicate components of the pain perception that occurs during the headache phase of the migraine attack. It is based on the fact that stimulation of the meninges, particularly around blood vessels, induces headache-like pain in humans [104–106]. The pain arises from the direct stimulation of trigeminal fibres that innervate the meninges [106] and is a result of the activation of the ascending trigeminothalamic pathway [13]. It is reasonable to assume that such stimulation is also nociceptive in other species and will activate the same sensory pathways involved in the perception of pain during migraine. In animals, activation of the trigeminovascular system can be induced by a variety of stimuli: trigeminal ganglion electrical stimulation through bipolar stimulating electrodes positioned stereotaxically in the ganglia, chemical stimulation of the meninges by direct application of an inflammatory soup or capsaicin on the dura mater, electrical stimulation of perivascular dural areas through a bipolar stimulating electrode or even mechanical stimulation evoked by repeated von Frey filaments testing on dural structures. Although the latter two approaches have been widely used, the choice of stimulation needs to be carefully selected. The lack of an inflammatory response in humans during migraine attacks and considerations for the integrity of the BBB function following chemical stimulation make mechanical and electrical stimulation of dural sites the only model for which objective, clinical evidence exists and confirms its painful nature, although it is also a non-physiological stimulus. As discussed earlier, although it is highly doubted that a significant sterile inflammatory response occurs on the meninges during a migraine attack, some kind of local release of peripheral inflammatory markers which will induce abnormal neuronal hyperexcitability in the TCC may explain sensitisation in migraine [87].

Central activation of the trigeminovascular system in this model can be assessed with multiple assays, analysed below in detail.

2.3.1 *Fos Immunoreactivity*

Fos is the protein product of the immediate early gene *c-fos,* which is rapidly and transiently activated in response to several forms of stimuli, including extracellular stimuli and intracellular second messenger systems. Extracellular stimuli activate membrane receptors and initiate a second messenger cascade that results in the upregulation of the immediate early genes, which produces transcription factors including the Fos protein. *c-fos* mRNA, which can be also detected in tissue, reaches its peak about 30–40 min poststimulus, while the levels of the nuclear protein Fos peak about 2 h poststimulation. Expression of the gene can be measured via Northern blot analysis or in situ hybridisation. The protein is normally visualised using immunocytochemical techniques. Quantification of Fos, although not accurate, is usually done through microscopic blinded counting of Fos-positive profiles and less frequently through Western blot analysis, both considered as semi-quantitative approaches.

Identification of Fos expression following trigeminovascular activation has been widely used in this model. Although Fos is not a specific marker of nociception, and other extracellular signal-regulated kinases have been implicated in mediating nociceptive signalling in the brain [107], this protein has been the choice of assessment cellular activation in a number of studies. In the field of migraine research, Fos immunoreactivity has been used to identify populations of cells activated in response to either electrical [108] or chemical [109] noxious stimulation of structures innervated by the trigeminal nerve. Fos has been shown to be expressed in the TCC [108, 110, 111] and other brain areas that could be activated with such nociceptive stimulations. For example, it was found that electrical stimulation of the superior sagittal sinus (SSS) resulted in increased expression of Fos in the periaqueductal grey [112] and in various hypothalamic nuclei, including the ventromedial nucleus, the supraoptic nucleus and the posterior hypothalamus [113]. Like other markers of cellular activation, Fos can be expressed in multiple locations as a result of polysynaptic activation. It can be thus used to map network pathways that become activated during trigeminovascular activation, and this has been one of the great advantages of using the Fos model. However, caution must be used when detecting this protein, since it is not expressed in all brain areas. For example, the basal thalamus where the ascending trigeminothalamic pathway projects does not appear to express Fos [114].

Beyond the anatomical studies, expression of Fos in the TCC, as a model of trigeminovascular activation, has been widely used for the investigation of various compounds. Triptans and ergotamine have been shown to reduce Fos expression in the TCC in response to electrical stimulation of the SSS or chemical stimulation of the dura mater [110, 115]. Interestingly, the less lipophilic sumatriptan did not have a significant effect in this model [116], thereby demonstrating a central site of action of second-generation triptans. Fos activation in the TCC seems to be dependent on serotonergic mechanisms, as serotonin-depleted rats expressed a greater quantity of Fos in response to application of inflammatory soup on the dura mater, as compared with controls [117]. Melatonin [118] and valproate [119] were also shown to reduce Fos expression in the TCC. Additionally, the CGRP antagonist olcegepant significantly inhibited the capsaicin-induced expression of Fos throughout the TCC [120]. More interestingly, neurokinin-1 receptor antagonists and PPE inhibitors had no effect over Fos immunoreactivity [121, 122], further demonstrating the potential of this model in screening possible therapeutics with a clinical benefit.

The main limitation of the Fos model is that it does not offer any further information on the type of activation, of the cell type or of the neurotransmitters involved. For example, glia cells will equally express Fos [123], as well as glutamatergic (excitatory) and GABAergic (inhibitory) neurons. Additionally, potent antinociceptive agents, such as the NMDA receptor antagonist MK-801, may also induce Fos expression in some brain regions of interest [124]. It is thus difficult to discriminate neurons activated in response to ascending or descending transmission of nociceptive or antinociceptive signals [125]. Additionally, induction of the Fos protein to quantifiable levels requires a strong consistent stimulation that may not be physiological [126] and that lack of Fos expression does not equate to lack of neuronal activity [111]. Nevertheless, its ability to respond to polysynaptic activation, enabling mapping of pathways of neuronal activation at many synapses down the line, is certainly a great advantage. Using combined immunohistochemical techniques, it can be further used for double labelling or for co-localisation studies. As the Fos staining occurs in the nucleus of the cell, it can be readily co-localised with many antigens that are expressed in the cytoplasm or on the cell surface.

2.3.2 *Metabolic and Blood Flow Measurements in Central Nuclei*

Trigeminal nerve activation in animals, through stimulation either of the SSS or of the trigeminal ganglion, has been shown to cause an increase in regional cerebral blood flow and metabolic activity in the TCC, brainstem regions, thalamus and frontal and parietal cortex [127, 128]. This has been considered to model changes in regional cerebral blood flow during migraine attacks, where in the headache phase cerebral blood flow may be abnormally high, often outlasting the headache phase [129]. Facial blood flow was also shown to increase during trigeminal ganglion stimulation [130], perhaps reflecting the facial and neck tenderness seen in some patients during the headache phase of migraine. Metabolic activity is measured using 2-deoxyglucose autoradiography and quantitative densitometry, while a laser Doppler probe is placed in the nucleus of interest for measuring blood flow changes. Autoradiography is not as widely used these days, given its risk. Microdialysis through an appropriate pore cannula in the region of the TCC has been also used to detect changes in neurotransmitter release [131]. In this model, sumatriptan, dihydroergotamine and *N*-Methyl-D-Aspartate (NMDA) receptor antagonism [132] have been shown to inhibit blood flow changes in response to trigeminal ganglion.

This model is now considered somewhat redundant, as in humans reductions of cerebral blood flow during migraine with acute anti-migraine treatments are not reported [133], suggesting that blood flow changes do not adequately reflect the clinical manifestation of head pain. This model served as a precursor to the electrophysiological measures of trigeminovascular activation in the central pathway [83, 128].

2.3.3 *Electrophysiological Recordings in the Ascending Trigeminothalamic Pathway*

Electrical, mechanical or chemical stimulation of dural vessels in animals, mainly of the SSS and the MMA, excites second-order neurons of the ascending trigeminothalamic pathway. State-of-the-art electrophysiological techniques can thus be used to record this neuronal activity [134]. More specifically, trigeminovascular stimulation has been shown to excite neurons in the TCC [135–138] by releasing glutamate along with CGRP from primary Aδ- and C-trigeminal fibres [139], in the thalamus, mainly the ventroposteromedial and posterior nuclei [140, 141], and more recently in different cortical regions, primarily the somatosensory S1 cortex [142]. Activated neurons in the TCC and thalamic nuclei are either nociceptive-specific or wide-dynamic range. Electrical stimulation is usually the preferred method of trigeminovascular stimulation, as poststimulus analysis of the activation latency can better determine the activation of neurons in response to stimulation of Aδ- and/or C-trigeminal fibres, while extracellular single neuron electrophysiological recordings have been used by most laboratories using this model.

The trigeminovascular model in combination with electrophysiological recordings has been proved a highly reliable model for testing a wide range of compounds, in particular of potential anti-migraine therapeutics that cross the BBB. The efficacy of compounds in this model is assessed in their ability to attenuate evoked trigeminovascular activation. Experimental pharmacological studies have shown that abortive anti-migraine drugs, such as dihydroergotamine [134], second-generation triptans [141, 143] and other 5-$HT1_{B/D}$ receptor agonists [144], act on second- and third-order neurons to inhibit neuronal activation. Systemic administrations of CGRP antagonists, which are effective in clinical trials of migraine treatment [64], were also shown to decrease the activity of neurons with meningeal input [145]. Further to predicting anti-migraine efficacy of acute treatments, topiramate, a preventive anti-migraine compound, was also effective in inhibiting trigeminovascular activity in the TCC and VPM [146].

A number of potential anti-migraine compounds and the pharmacology of their receptors have also been studied in this model. As glutamatergic transmission plays a key role in trigeminovascular nociception [147], ionotropic glutamate receptors (iGluR) [146–153] were shown to be involved in trigeminovascular nociceptive transmission and, among others, demonstrated a central mechanism of action of kainate receptor antagonists [80]. Inhibitors of the orexin hypothalamic peptides were recently shown to suppress trigeminovascular activation recorded in the TCC [154]. Pharmacological modulation by adenosine A, cannabinoid 1, TRPV1 and dopamine 2 receptors was also found to induce neuronal inhibition, without concomitant vasoconstriction, suggesting a novel avenue for the treatment of migraine [73, 155, 156].

Beyond systemic administration of compounds, the combined use of electrophysiological recordings with microiontophoresis allowed an even greater understanding of the involved pharmacology along the trigeminothalamic pathway. Microiontophoresis allows the direct application of charged compounds onto neurons while recording their electrophysiological properties. As recordings in the TCC or thalamus following systemic administration of compounds cannot provide proof of concept on the actual site of action of test compounds, which could be local or on multiple nuclei that could modulate the ascending pathway, the employment of microiontophoresis can confine the action of compounds to second- and/or third-order neurons. Furthermore, microiontophoresis allows the dissection of the inhibition of investigational compounds upon post- or presynaptic receptors. Microiontophoretic application of ergot alkaloids [157] and triptans in the TCC [158] and VPM [141] reversibly inhibited second-order trigeminal neurons demonstrating a central action of these compounds. These studies gave impetus to the development of more brain-penetrant 5-HT$1_{B/D}$ receptor agonists [159]. Furthermore, microiontophoresis of the CGRP receptor antagonist olcegepant in the TCC reversibly inhibited SSS electrical stimulation-induced trigeminocervical activation [160], and the same was shown for the thalamus, given the presence of CGRP receptors within the VPM nucleus [161]. In addition, the clinically active preventives propranolol [140] and valproate [162], but not gabapentin, were able to inhibit responses to L-glutamate and to trigeminovascular stimulation in the VPM, while topiramate was shown to act both in the TCC and VPM by blocking kainate receptors [153].

In addition to the use of this model in pharmacological screening of potential anti-migraine compounds, a small number of studies attempted to model some aspects of the physiological properties of second- or third-order neurons with regard to convergent inputs from the periphery or from brainstem and midbrain nuclei. This can be achieved by simultaneous electrophysiological recordings of neurons of the ascending trigeminothalamic pathway, while modulating electrically or pharmacologically the activity of distal nuclei or peripheral nerves. This approach led to the demonstration of convergent inputs from trigeminal sensory afferents that innervate both dural and facial structures [137], in particular those innervated by the ophthalmic division of the trigeminal nerve over second-order neurons. Additionally, these neurons receive afferents arising from the greater occipital nerve (GON) of the C2 dorsal root, which innervate cervical structures [163]. This property of second-order neurons is considered of great importance for the efficacy of occipital nerve blocks and stimulation in the treatment of chronic migraine [164]. The convergence of trigeminal and occipital fibres on second-order neurons might be also involved in the referral of pain from trigeminal to cervical structures and contribute to the clinical phenomena of cervical hypersensitivity in migraine [138].

Direct electrophysiological recordings from trigeminocervical neurons and modulation of higher modulatory nuclei have provided important insights into migraine pathophysiology. This approach has been employed as a great limitation of the available brain imaging techniques used in humans is the lack of good spatial resolution that will allow the identification of the exact nucleus or the nature of metabolic activation involved. In PET studies, brainstem nuclei have been shown to be activated during migraine headache [165] and following successful treatment [165]. A number of studies in the trigeminovascular model have suggested that the activity of the TCC is modulated through inputs from a variety of modulatory nuclei in the brainstem, such as the PAG [166], locus coeruleus [167] and dorsal raphe nucleus

[168]. Hypothalamic nuclei have been also shown to modulate the activity of the TCC, such as the A11 dopaminergic nucleus [169] and the paraventricular hypothalamic nucleus [170]. Although the VPM and Po thalamic nuclei are also known to receive a variety of inputs form midbrain and brainstem nuclei [171], their modulation by such pathways has not been investigated in the trigeminovascular model. Importantly though, a recent study demonstrated a convergence of third-order trigeminothalamic neurons in the posterior thalamic nucleus, with axons originating in retinal ganglion cells. This convergence has been suggested to be responsible for the modulation of thalamic neurons, activated in response to trigeminovascular activation, by light [172]. The interpretation of this outcome is that the convergence of photic signals onto dural nociceptive trigeminothalamic neurons that project to the somatosensory cortex exacerbates nociceptive processing, similar to the exacerbation of migraine headache by light [172].

Allodynia is an important phenomenon seen in migraineurs [173] and it is thought to be modelled in animals using chemical stimulation of the trigeminal fibres and electrophysiological recordings. Topical application of inflammatory agents on the rat dura has been shown to induce long-lasting activation of the trigeminovascular pathway and sensitisation of trigeminocervical neurons [109]. In rats, late sumatriptan intervention [174], but not dihydroergotamine [175], was not able to reverse trigeminocervical sensitisation, suggesting that, similarly to clinical observations, triptans might not be effective after the onset of central sensitisation. Intravenous or local meningeal application of COX1/2 inhibitors including indomethacin, naproxen and ketorolac was able to block sensitisation [176, 177]. In patients, it is believed that untreated migraine attacks could result in a spread of allodynia to the other side of the head or the forearm [173], indicating the potential spread of neuronal sensitisation from second-order neurons to third-order neurons in the thalamus [173]. Sensitisation of third-order neurons was shown by Burstein and colleagues using this model [178].

Currently, the trigeminovascular activation model assessed by electrophysiological recordings is considered as the most successful model of migraine headache. Perhaps, a disadvantage of the model is the use of anaesthetised animals, in which, depending on the choice of anaesthetic, nociceptive activation of the trigeminothalamic pathway may be, to some extent, suppressed. It also assumes that a decrease of the excitatory transmission of the trigeminothalamic pathway is antinociceptive; however, in a conscious model, such differences may not be observed.

2.4 Nitric Oxide Donors' Infusion Model

NO donors, such as NTG, have been shown to trigger an early-onset headache and migraine attack in sufferers after a delay of hours [5, 179]. This biphasic headache is not reproduced in healthy subjects; however, a mild early-onset headache is often reported [180], which is also associated with decreased thresholds to mechanical nociceptive stimuli [181]. NTG has been also reported to reproduce premonitory symptoms and nausea in some patients [6]. NO donors have never however been reported to induce a migraine aura, even in migraine with aura patients. Despite being characterised as the molecule of the year in 1992 [182] and having awarded the Nobel Prize to three American scientists for identifying its signalling pathway in the cardiovascular system, how NO triggers a migraine attack in migraineurs is yet unknown. In neurons, similar to the smooth muscle cells, NO was found to act through the stimulation of the enzyme-soluble guanylate cyclase, followed by the production of the second messenger cyclic guanosine monophosphate which activates protein kinase G. This results in the reuptake of Ca^{2+} and the opening of calcium-activated potassium channels. In the smooth muscle cells, this leads to vasodilation. In neurons, it is thought that protein kinase G may activate other transcription factors which can lead to changes in gene expression that alter the response of the cell to a variety of other stimuli.

A wide range of methods are employed in the NO donors' infusion animal model. In rodents, a NO donor is normally infused systemically at doses higher than those required to model headache in humans. The NO donor's dose, modality of administration and choice of the time of observations must be carefully controlled when adopting this model for the study of the trigeminovascular system. Over the years, this model has been developed to reflect a more representative disease phenotype and a number of different outcomes can be assessed, from immunohistochemistry to behavioural changes. Although in initial studies intraperitoneal injections were performed using enormous doses of NTG that may elicit blood pressure decrease [183, 184], a more realistic approach is now adopted where NO donors are intravenously or intracarotidly infused, using smaller doses that elicit minimum blood pressure effects [76, 184, 185].

In rodents, systemic NO donors, most commonly NTG and sodium nitroprusside (SNP), have been shown to induce Fos expression in different CNS areas, including the TCC, brainstem and hypothalamus [183, 186]. A sexual dimorphism in NTG-induced Fos expression in the TCC and in hypothalamic nuclei has been observed, with female rodents expressing a higher number of Fos-positive cells, a phenomenon modulated by estrogens [187]. Small changes in the expression of receptor and enzyme components, such as CGRP receptor subunits, the soluble guanylyl cyclase and the nitric oxide synthase, along the trigeminal system have been also reported following NO donors' infusion [188, 189]. However, similarly to the human model [190], the levels of CGRP itself do not appear to change following NTG infusion in rats [191]. Expression of the cellular activation marker Fos after any stimulus peaks at 2–4 h [107], and that is also the case in the NTG model. It is, thus, scientifically inappropriate to claim that the delayed expression of Fos in the trigeminocervical complex reflects the delayed onset of a migraine attack in humans. Nevertheless, a delayed occurrence of behavioural changes, which included the development of allodynia and hyperalgesia in freely moving animals, has been also observed following NO donors' administration [192, 193]. In a preliminary report, Akerman and colleagues have further demonstrated that NTG infusion induces a delayed increase of spontaneous and evoked firing of second-order neurons in the TCC [194]. It is thus likely that, either through local signalling pathways in the TCC or through interactions with modulatory pathways, NO alters the threshold of activation of the pain pathway involved in headaches. In which way NO donors may also induce premonitory symptoms [6] or nausea [7], it is not yet known. Andreou et al. suggested that NO interacts with dopaminergic hypothalamic pathways that modulate the activity of the TCC [195]. Indeed, in a brain imaging study in patients who developed premonitory symptom following NTG infusion, hypothalamic activation was a prominent outcome [20]. Interestingly, it has been shown that central dopaminergic neurotransmission is required for the NO-induced activation of c-*fos* in subcortical areas [196].

Sumatriptan has been shown to alleviate behavioural changes induced by NO donors' infusion [192], as well as to reduce Fos expression in the TCC [185]. NOS inhibitors and neurokinin-1 and CGRP receptor antagonists were demonstrated to reduce the number of Fos-expressing cells in the TCC [189]. A CGRP antagonist was also found to be effective in counteracting NTG-induced hyperalgesia [193]. A role for NSAIDs has been suggested for the NO-induced cellular changes [76].

Although the NO donors' infusion model is now widely used, mostly due to the relatively uncomplicated methods that can be used to assess its outcomes, several considerations need to be addressed. In healthy volunteers, NO donors induce only a short-lasting mild headache that does not respond to triptans or aspirin [180]. As the animals we use in the laboratory are otherwise healthy, it is rather difficult to interpret that NO infusion will result in a pure migraine model, and thus, the effectiveness of sumatriptan in this rodent model is further questioned. It is likely that the increased doses administered to animals compared to humans may contribute towards the expression of a more prominent phenotype; however, this is not established. In addition, the use of Fos as a system's activation tool needs careful consideration. As discussed earlier, its expression occurs following a variety of different stimuli, not just nociceptive-specific stimuli [107]. Thus, the expression of Fos in different CNS areas following infusion of NO donors just points to the areas that are susceptible to NO signalling. Nevertheless, from a practical point of view, the model offers the opportunity to study repeated behavioural changes in non-anaesthetised animals, which appear to have similar phenotype to that developed in humans [181]. As a pharmaceutical tool, the model needs to be further evaluated with treatments that have both a positive and negative benefit in the clinic.

2.5 Animal Models of the Aura Symptoms: Cortical Spreading Depression

Migraine aura involves transient focal neurological deficits, such as visual impairment and sensory or motor function impairment, and occurs in about 30 % of migraine patients just before or during the onset of the migraine headache [197]. Occasionally, these symptoms can even occur alone without the accompanying headache. The symptoms described as the migraine aura, with a cortical spreading rate of 2–6 mm^{-1}, are believed to be the result of cortical spreading depression (CSD), first identified by Leao in 1944 [198]. CSD itself is characterised as a slow wave of neuronal depolarisation and glial activation in the cortex, followed by a short-lasting depression and accompanied by blood flow changes. Olesen et al. [199] demonstrated in humans that aura is accompanied by a short phase of hyperaemia, followed by a slowly spreading oligaemia. It was thought that vascular changes are purely a response to metabolic changes due to neuronal discharge, but this view has been tackled by results from Brennan et al. [200], showing the possibility of a dissociation of the spread of regional cerebral blood flow (rCBF) changes and CSD. The occurrence of CSD causes profound temporary intra- and extracellular changes including pH changes, release of neurotransmitters and ionic shifts accompanied by cellular swelling [201]. CSD in the neocortex of a variety of species, including human, has been demonstrated to be dependent on activation of the NMDA receptor, and both glutamate and NMDA receptor agonists are capable of inducing CSD if applied cortically [202]. Interestingly, the NMDA receptor antagonist ketamine was tested clinically and found effective in treating the aura symptoms, but not the headache, of migraine patients with aura [203].

What triggers a CSD in patients has yet not been determined. It is thought that altered cortical excitability may be responsible for lowering the threshold of cortical activation in some patients. Such a hyperexcitable cortex may thus give rise to a CSD, but this theory needs to be further elaborated. In animals, CSD can be triggered by chemical (e.g. K^+, glutamate), mechanical or electrical stimulation of the cortex [83]. In animals, it was reported that CSD can be also triggered by sensory activation of the brainstem [204]; however, this needs to be further validated. Induction of CSD may be measured using electrophysiological techniques, commonly field potential recordings, and blood flow changes through laser Doppler flowmetry. Optical intrinsic signal imaging has been also used to monitor blood flow/metabolic changes that appear in connection with CSD [200]. In freely moving animals, induction of CSD was shown to induce motor freezing [205], without the development of cutaneous allodynia or any other nociceptive-like behaviour [206, 207].

The CSD model has been widely used to examine the efficacy of different treatments. Although CSD is a "yes or no" event, in the literature, treatments are considered as potentially effective not only when they block induction of CSD but also when the rate of propagation or amplitude of DC shift and blood flow changes are suppressed. A reduction of the sum of CSD waves in the case of K^+-induced CSD and an increase in the threshold of electrical activation of the cortex are also considered as important outcomes of therapeutic efficacy [208]. Drugs identified to have a prophylactic effect on CSD in rats after many weeks of daily treatment are valproate, topiramate, propranolol, amitriptyline and methysergide [209]. Lamotrigine was also shown to block K^+-induced CSD, possibly through interactions with the glutamatergic system [210]. CSD blockade with amiloride, via the acid-sensing ion channel 1a, was found to be attributed to its preventive role in a small open clinical trial [211, 212]. Apart from treatment with medications, cortical neuromodulation by transcranial magnetic stimulation has been found to be effective in both K^+- and mechanically induced CSD [213, 214]. Triptans are not expected to have an effect in CSD or the aura in humans [215], although a role for serotonin in the maintenance of balanced cortical activation has been suggested, given that animals depleted of serotonin demonstrate an enhanced cortical sensitivity to K^+ application [216].

Recent studies in the CSD model have focused on the potential mechanisms through which CSD may interact with the ascending trigeminothalamic pathway, either peripherally through activation of the trigeminovascular system or centrally through cortico-subcortical networks. Elevated Fos levels can be found due to CSD in various areas of the brain and in the TCC [217, 218], although contradictory preclinical data exist for the TCC [219]. Bolay et al. [220] demonstrated that CSD causes vasodilation of meningeal blood vessels, which is accompanied by activation of the TCC, manifested by the presence of Fos-positive cells. Recent electrophysiological studies, combining mechanical, chemical or electrically induced CSD and electrophysiological recordings from trigeminal ganglion or TCC neurons, demonstrated that meningeal nociceptors may be activated following CSD as spontaneous activity of neurons was facilitated 50 % of the time [221]. These outcomes suggest that CSD may induce activation of the ascending trigeminothalamic pathway through sensitisation of peripheral nociceptors, presumably following the CSD-induced release of substances from the cortex that may activate such channels through diffusion in the subarachnoid space, as initially suggested by Bolay et al. [220]. Karatas and colleagues [222] suggested recently that this occurs due to activation of the gap junction protein Pannexin 1, as inhibition of the signalling cascade activated by neuronal Pannexin 1 abolished CSD-induced trigeminovascular activation and dural mast cell degranulation [222]. The role of Pannexin 1 and gap junction molecules in migraine, however, remains to be determined. The gap junction channel modulator tonabersat (SB-220453) was shown to be decreasing the number of CSD waves induced by K^+ [223]; however, its efficacy in the clinic is questioned [224]. Other studies have shown that CSD can alter positively or negatively the activity of second-order neurons, without interactions with peripheral inputs, or the spread of CSD to subcortical areas in an otherwise healthy brain. This could be achieved through activation of corticospinal pathways, depending on the site of cortical stimulation [225], or indirectly through cortico-brainstem pathways [226, 227]. Andreou et al. have also shown that CSD may sensitise the ipsilateral sensory thalamic nuclei VPM and Po, through direct corticothalamic pathway activation [228, 229]. Whether these changes are sufficient to elicit the perception of migraine headache in patients is unclear.

Although to date the mechanisms and interactions of CSD with pain pathways have not been fully understood, recent advances shed some light on possible interactive mechanisms and provide important information about the phenomenon itself. The model serves well the scientific experimentation for the identification of new drugs specific for migraine aura; however, it remains to be established if agents that prevent the aura symptoms may also treat the migraine headache or even prevent a migraine attack from being triggered.

2.6 Genetic Models of Migraine

To date, genetic models of migraine include genetically modified mice in which known human mutations of familial hemiplegic migraine (FHM) and of familial advanced sleep phase (FASP) syndrome have been knocked in their genome. The genome-wide association studies (GWAS) performed to date have identified a number of genes that may be associated with more common forms of migraine. These studies might offer future opportunities for the development of further genetic models of migraine that could shed light on migraine pathophysiology and treatment development.

2.6.1 Familial Hemiplegic Migraine Models

Identification of autosomal-dominant gene mutations in familial hemiplegic migraine (FHM) patients, a rare subtype of migraine with prominent aura symptoms, allowed for the first time the development of genetic models of migraine [230]. FHM1 mutations affect the CACNA1A gene, FHM2 mutations affect the ATP1A2 gene and FHM3 mutations affect the SCN1A gene. So far, there have been no transgenic mice carrying the human FHM3 mutations. In contrast, knock-in models utilising two CACNA1A mutations of FHM1, the R192Q and S218L, have been developed, as well as one model of FHM2, carrying the human W887R mutation [85, 231]. These animal models have been used in conjunction with a number of assays described above, ranging from immunohistochemistry to behavioural tests, in a number of studies that aimed to gain insight into the pathophysiology of FHM and of more common types of migraine.

CSD experiments demonstrated a decreased threshold for CSD induction in the R192Q and S218L FHM1 mutant mice [232, 233]. Similarly to patients who carry the S218L mutation, S218L FHM1 mutant mice were also found to be predisposed to severe brain oedema [232]. One model of FHM2, carrying the human W887R mutation, has been also developed [231] and, likewise to the FHM1 models, is characterised by a decreased induction threshold of CSD and an increased velocity of propagation of the spreading wave [231]. Overall, both FHM1 and FHM2 genetic models appear to model significantly the clinical FHM phenotype as described in patients, making them excellent tools for further investigations that would aim to develop FHM-specific pharmacological treatment.

Although behavioural tests suggest that these animals demonstrate spontaneous nociceptive-like and photophobia-like behaviour [234, 235], their ability to model common types of migraine has been questioned. FHM shares many phenotypical similarities with common types of migraine, suggesting the existence of common neurobiological pathways. However, despite the well-established importance of CGRP in the pathophysiology of common types of migraine [89], immunohistological identification of CGRP in the TCC of FHM1 mice showed a reduced expression in trigeminal ganglia neurons and TCC [236]. In agreement with this, FHM patients with known mutations in the CACNA1A and ATP1A2 genes do not show hypersensitivity to either CGRP or NO donors' infusion, as characteristically seen in migraine patients [237, 238]. Furthermore, identification of Fos-positive cells in FHM mice following trigeminovascular stimulation demonstrated an unpredicted reduced number of positive profiles in the TCC compared to wild-type animals [239]. The opposite was also seen in the brainstem and hypothalamic nuclei where one would normally expect a reduced modulatory role [240]. These data indicate that the pathophysiological pathways underlying migraine headache in FHM may be different from the common types of migraine; however, further studies are needed to conclude to such a hypothesis. Nevertheless, given that all identified FHM mutations are highly susceptible to CSD propagation or to excitability of neuronal tissue, FHM genetic models can advance our understanding of migraine aura and its treatment.

2.6.2 *Casein Kinase 1δ (CK1δ) Model*

More recently, a mutation in the clock gene encoding casein kinase 1δ (T44A), which results in reduced enzymatic activity, has been described in patients with familial advanced sleep phase (FASP) syndrome. This syndrome is characterised by altered circadian rhythms, reflected in early morning waking and early sleep times [241]. In one family, FASP was found to be associated with migraine with and without aura [242], suggesting a functional relation of this mutation with migraine neurobiology. Genetically engineered mice that carry the T44A human gene mutation, similarly to FHM mice, also showed an increased susceptibility to CSD induction, and hypersensitive behavioural responses, associated with increased Fos expression in the TCC following infusion of NTG [242]. These data suggest a potential role of this gene in migraine neurobiology and the prospective use of this genetic model in studies looking into new treatments and the neurobiology involved in migraine. However, this model needs to be further validated in order to conclude if findings from this rare form of migraine may be extrapolated to more common forms of the disorder. The mutation itself needs to be further validated as a migraine mutation in a bigger population of patients.

2.7 Conscious Modėls of Episodic or Chronic Migraine Pain

A major limitation of the models described earlier is that they do not reflect the repeated episodic nature of migraine attacks and hence the neuronal plasticity may develop in migraine patients over the course of the disease. The need of a model that better represents the episodic nature of migraine attacks has been long stated [243]. Chronic models of migraine are generally developed using similar assays as those described above; however, they employ freely moving awake animals and usually behavioural tests in order to assess the model's phenotype and potential therapeutic outcomes.

Oshinsky and Gomonchareonsiri [243] developed probably the first model of episodic migraine, by inducing trigeminovascular activation through repeated applications of inflammatory soup through a cannula fixed on top of the dura mater of awake behaving rats. The authors demonstrated that over a period of 4 weeks, rats developed increased mechanical cutaneous sensitivity at the periorbital area following infusion of a NO donor. Using in vivo microdialysis, the authors further showed that these behavioural changes occur in parallel to increased glutamate levels in the TCC, suggesting a state of increased excitatory neurotransmission [243]. Using the same model, Stucky et al. [244] demonstrated pronounced sex differences, with female rats developing behavioural changes at a lower dose of inflammatory agents and for a longer duration. Additionally, using real-time polymerase chain reaction, female rats demonstrated lower mRNA levels of the CGRP receptor subunits. Beyond evoked pain, utilising video recordings and second-by-second analysis, this model was shown to present changes in spontaneous behaviour [245]. In these rats, behavioural observations indicated increased facial grooming ipsilateral to the cannula implantation, provoked following infusion of the inflammatory soup, compared to animals that were infused with saline. Additionally, these animals demonstrated an increased freezing and resting behaviour, which was significantly reduced by zolmitriptan and ketorolac, but not acetaminophen [245]. Similarly to the inflammatory soup repeated application, Dong and colleagues [246] employed electrical stimulation of dural vessels in awake rats that also induced trigeminovascular activation. They demonstrated that high-frequency stimulation can elicit increased facial grooming and head-flick behaviour that is blocked by morphine or rizatriptan. Pradhan and colleagues [247] characterised recently a model of chronic migraine, in which chronic intermitted administration of NO donors resulted in the development of chronic extracranial hyperalgesia, assessed with mechanical stimuli over the plantar surface. Female mice showed a stronger phenotype compared to male. Chronic hyperalgesia was found to be suppressed by sumatriptan, by the migraine preventative topiramate and by different δ-opioid receptor agonists [247, 248].

Finally, Oshinsky et al. [249] isolated a colony of Sprague-Dawley rats that appear to experience episodic trigeminal allodynia. Using von Frey mechanical stimulation of the trigeminal region, Oshinsky et al. characterised "spontaneous trigeminal allodynia" rats as those that have mechanical thresholds in the normal range (8–15 g) on some days and thresholds as low as 1.0 g on other days. Using this behavioural assay, a rat with spontaneous episodic trigeminal sensitivity was discovered and thought to represent a model of spontaneous trigeminal allodynia. Low withdrawal thresholds were also found in the masseter muscle region of the jaw but not in the hind paws. Subsequent mating of these rats showed that the trait is inherited, suggesting a similarity to the inherited nature of migraine in humans. These rats were also shown to have increased sensitivity to sound, similar to phonophobia in migraineurs [250]. Using von Frey filaments, the authors further found an increased sensitivity to the chemical headache triggers NTG and CGRP. Finally, the rats' periorbital mechanical threshold was normalised by clinically proven acute and preventive pharmacological migraine treatments. Sumatriptan, ketorolac and dihydroergotamine DHE transiently returned the periorbital nociceptive thresholds of the spontaneous trigeminal allodynia rats to normal levels. The migraine preventive treatment valproic acid restored the periorbital pain threshold to normal levels during treatment. Since the "spontaneous trigeminal allodynia" shares many of these phenotypes and sensitivities of migraineurs, the authors suggest that they can be a good model for studying the pathophysiology and drug discovery for migraine.

The main limitation of these models, despite their usefulness in exploring episodic trigeminovascular activation, lies on the interpretation of animal behaviour and its translation to human pain. Measuring hypersensitivity in rodents can be challenging and requires expert training and good laboratory practices, particularly for assessing hypersensitivity in the craniofacial region [251]. Indeed, animal studies of migraine have predominantly concentrated on modelling trigeminovascular nociception in the anaesthetised animal; as such, many of the important neurological features that accompany migraine are overlooked [83]. However, proof of concept for the effectiveness of these models needs to be provided by other laboratories, and this needs to include evaluation of the models using nonclinically active treatments. Adaptation of the models with assays that demonstrate activation of the trigeminothalamic pathway will be an advantage in the field of conscious migraine behavioural research.

2.8 Medication Overuse: Latent Sensitisation Model

Developed in the Porecca's lab, the latent sensitisation animal model aimed to represent the medication overuse headache, which is often developed in migraine patients following chronic administration of triptans and other acute painkillers [1]. The prevalence of cutaneous allodynia in MOH patients is higher than in episodic migraine sufferers [252]. Central sensitisation is thought to be the underlying mechanism for the development of a chronic migraineurs status.

In the latent sensitisation model, rats that were exposed to chronic administration of triptans [253, 254] or morphine [255] developed behavioural signs of cutaneous allodynia. Additionally, these animals were more sensitive to NO donors' infusion and demonstrated further signs of central sensitisation following exposure to stress stimuli [253–256]. These behavioural changes were accompanied by increased expression of CGRP and nNOS in the trigeminal ganglia and TCC [253–255]. Importantly, these changes persisted over a prolonged period of time following the cease of chronic drug administration. This long-lasting state of hypersensitivity to different stimuli was considered to reveal a state of "latent sensitisation" [256]. More recent data from the same laboratory also suggest that rats exposed to chronic sumatriptan administration have lower threshold of CSD induction compared to animals that received chronic administration of saline that could be reversed by topiramate administration [257].

Although the development of a medication overuse model in migraine is certainly desirable, particularly for testing the mechanisms of chronic migraine, the latent sensitisation model needs to be validated by other laboratories as well. More importantly, the efficacy of the model needs to be tested against treatments that are not known to induce MOH, such as migraine preventives that are taken on a daily basis, and even against chronic exposure to non-pain-related treatments. Additionally, given reports on the rather unlikely central action of sumatriptan due to its inability to cross the BBB, further clarifications are needed as to why this drug may interfere with the induction properties of CSD. Nevertheless, the model has yet a lot to offer, particularly in terms of neuronal plasticity that most likely underlies the development of medication overuse altered phenotype.

2.9 Conclusions

Preclinical investigations in migraine research involve animal models that have been developed over the years to better model our current understanding of disease mechanisms. Modelling of migraine in animal models has been based on clinical evidence coming from migraineurs. While the effectiveness of some of the newer described models is still pending, the field is blessed with well-established models that reliably replicate aspects of the migraine phenotype and screen satisfactorily potential therapeutics. In particular, animal models of migraine that involve activation of the trigeminovascular system, and CSD, the migraine aura model, are considered well-reliable pharmacological tools. One of the biggest challenges in developing a suitable animal model for the study of migraine is the extent of clinical symptoms required to be present in order to fulfil the diagnosis criteria [1]. Therefore, an ideal animal model for the study of migraine should resemble as close as possible this multi-symptom complexity in the form of quantifiable correlates [83]. The need for animal models which have aspects of migraine symptoms other than the pain is thus urgent. Future studies should aim to model these features along with the use of cutting edge techniques such as optogenetics, with the scope of facilitating our understanding of migraine, and to better replicate brain activation as seen in patients. Given the ongoing GWAS in migraine, the development of genetically modified animal models that will replicate the phenotype of common forms of migraine should be anticipated in the years to come.

References

1. Headache Classification Committee of the International Headache, S (2013) The international classification of headache disorders, 3rd edn (beta version). Cephalalgia 33:629–808
2. Blau JN (1986) Clinical characteristics of premonitory symptoms in migraine. In: Amery WK, Waquir A (eds) The prelude to the migraine attack. Balliere Tindall, London, pp 39–43
3. Coppola G, Di Lorenzo C, Schoenen J, Pierelli F (2013) Habituation and sensitization in primary headaches. J Headache Pain 14:65
4. Crawley J et al (2013) Protective effects of non-anticoagulant activated protein C variant (D36A/L38D/A39V) in a murine model of ischaemic stroke. J Thromb Haemost 11:64 (Wiley-Blackwell 111 River St, Hoboken 07030-5774, NJ USA, 2013)
5. Iversen HK, Olesen J, Tfelt-Hansen P (1989) Intravenous nitroglycerin as an experimental model of vascular headache. Basic characteristics. Pain 38:17–24
6. Afridi SK, Kaube H, Goadsby PJ (2004) Glyceryl trinitrate triggers premonitory symptoms in migraineurs. Pain 110:675–680
7. Maniyar FH, Sprenger T, Schankin C, Goadsby PJ (2014) The origin of nausea in migraine-A PET study. J Headache Pain 15:84

8. Pietrobon D (2007) Familial hemiplegic migraine. Neurotherapeutics 4:274–284
9. Kilkenny C, Browne WJ, Cuthill IC, Emerson M, Altman DG (2010) Improving bioscience research reporting: the ARRIVE guidelines for reporting animal research. PLoS Biol 8:e1000412
10. N.C.R.R.G.W. Group (2010) Animal research: reporting in vivo experiments: the ARRIVE guidelines. Exp Physiol 95:842–844
11. Danos O, Davies K, Lehn P, Mulligan R (2010) The ARRIVE guidelines, a welcome improvement to standards for reporting animal research. J Gene Med 12:559–560
12. Ray BS, Wolff HG (1940) Experimental studies on headache. Pain sensitive structures of the head and their significance in headache. Arch Surg 41:813–856
13. Goadsby P, Charbit A, Andreou A, Akerman S, Holland P (2009) Neurobiology of migraine. Neuroscience 161:327–341
14. Akerman S, Holland PR, Goadsby PJ (2011) Diencephalic and brainstem mechanisms in migraine. Nat Rev Neurosci 12:570–584
15. May A et al (1998) Retinal plasma extravasation in animals but not in humans: implications for the pathophysiology of migraine. Brain 121(Pt 7):1231–1237
16. Olesen J (1998) Regional cerebral blood flow and oxygen metabolism during migraine with and without aura. Cephalalgia 18:2–4
17. Leão AA (1944) Spreading depression of activity in cerebral cortex. J Neurophysiol 7:359–390
18. Alstadhaug KB (2009) Migraine and the hypothalamus. Cephalalgia 29:809–817
19. Denuelle M, Fabre N, Payoux P, Chollet F, Geraud G (2007) Hypothalamic activation in spontaneous migraine attacks. Headache 47:1418–1426
20. Maniyar FH, Sprenger T, Monteith T, Schankin C, Goadsby PJ (2014) Brain activations in the premonitory phase of nitroglycerin-triggered migraine attacks. Brain 137:232–241
21. De Vries P, Villalon CM, Saxena PR (1999) Pharmacological aspects of experimental headache models in relation to acute antimigraine therapy. Eur J Pharmacol 375:61–74
22. Goadsby PJ, Edvinsson L, Ekman R (1990) Vasoactive peptide release in the extracerebral circulation of humans during migraine headache. Ann Neurol 28:183–187
23. Schoonman GG et al (2008) Migraine headache is not associated with cerebral or meningeal vasodilatation–a 3T magnetic resonance angiography study. Brain 131:2192–2200
24. Rahmann A et al (2008) Vasoactive intestinal peptide causes marked cephalic vasodilation, but does not induce migraine. Cephalalgia 28:226–236
25. Schytz HW et al (2009) PACAP38 induces migraine-like attacks in patients with migraine without aura. Brain 132:16–25
26. Bigal ME, Lipton RB (2006) The preventive treatment of migraine. Neurologist 12:204–213
27. Nilsson T, Longmore J, Shaw D, Olesen IJ, Edvinsson L (1999) Contractile 5-HT1B receptors in human cerebral arteries: pharmacological characterization and localization with immunocytochemistry. Br J Pharmacol 128:1133–1140
28. Sams A, Jansen-Olesen I (1998) Expression of calcitonin receptor-like receptor and receptor-activity-modifying proteins in human cranial arteries. Neurosci Lett 258:41–44
29. Bigal ME et al (2013) Safety and tolerability of LBR-101, a humanized monoclonal antibody that blocks the binding of CGRP to its receptor: results of the phase 1 program. Cephalalgia 34:483–492
30. Dodick DW et al (2014) Safety and efficacy of ALD403, an antibody to calcitonin gene-related peptide, for the prevention of frequent episodic migraine: a randomised, double-blind, placebo-controlled, exploratory phase 2 trial. Lancet Neurol 13:1100–1107
31. Reuter U (2014) Anti-CGRP antibodies: a new approach to migraine prevention. Lancet Neurol 13:857–859
32. Heyck H (1969) Pathogenesis of migraine. Res Clin Stud Headache 2:1–28
33. Drummond PD, Lance JW (1984) Facial temperature in migraine, tension-vascular and tension headache. Cephalalgia 4:149–158
34. Guo S et al (2014) Prevalence of right-to-left shunts on transcranial Doppler in chronic migraine and medication-overuse headache. Cephalalgia 34:37–41

35. Den Boer MO et al (1993) On the preservation and regulation of vascular tone in arteriovenous anastomoses during anesthesia. J Appl Physiol 75:782–789
36. Villalon CM et al (1999) Canine external carotid vasoconstriction to methysergide, ergotamine and dihydroergotamine: role of 5-HT1B/1D receptors and alpha2-adrenoceptors. Br J Pharmacol 126:585–594
37. Kapoor K et al (2004) Assessment of anti-migraine potential of a novel alpha-adrenoceptor agonist S19014: effects on porcine carotid and regional haemodynamics and human coronary artery. Cephalalgia 24:425–438
38. Verheggen R, Hundeshagen AG, Brown AM, Schindler M, Kaumann AJ (1998) 5-HT1B receptor-mediated contractions in human temporal artery: evidence from selective antagonists and 5-HT receptor mRNA expression. Br J Pharmacol 124:1345–1354
39. Muller-Schweinitzer E, Weidmann H (1977) Regional differences in the responsiveness of isolated arteries from cattle, dog and man. Agents Actions 7:383–389
40. Franco-Cereceda A, Rudehill A, Lundberg JM (1987) Calcitonin gene-related peptide but not substance P mimics capsaicin-induced coronary vasodilation in the pig. Eur J Pharmacol 142:235–243
41. Petersen KA, Nilsson E, Olesen J, Edvinsson L (2005) Presence and function of the calcitonin gene-related peptide receptor on rat pial arteries investigated in vitro and in vivo. Cephalalgia 25:424–432
42. Faraci FM, Breese KR (1994) Dilatation of cerebral arterioles in response to N-methyl-D-aspartate: role of CGRP and acetylcholine. Brain Res 640:93–97
43. Busija DW, Chen J (1992) Effects of trigeminal neurotransmitters on piglet pial arterioles. J Dev Physiol 18:67–72
44. Gupta S et al (2006) Intravital microscopy on a closed cranial window in mice: a model to study trigeminovascular mechanisms involved in migraine. Cephalalgia 26:1294–1303
45. Akerman S, Williamson DJ, Kaube H, Goadsby PJ (2002) The effect of anti-migraine compounds on nitric oxide-induced dilation of dural meningeal vessels. Eur J Pharmacol 452:223–228
46. Williamson DJ, Hargreaves RJ, Hill RG, Shepheard SL (1997) Intravital microscope studies on the effects of neurokinin agonists and calcitonin gene-related peptide on dural vessel diameter in the anaesthetized rat. Cephalalgia 17:518–524
47. Petersen KA, Birk S, Doods H, Edvinsson L, Olesen J (2004) Inhibitory effect of BIBN4096BS on cephalic vasodilatation induced by CGRP or transcranial electrical stimulation in the rat. Br J Pharmacol 143:697–704
48. Gupta S, Villalon CM (2010) The relevance of preclinical research models for the development of antimigraine drugs: focus on 5-HT(1B/1D) and CGRP receptors. Pharmacol Ther 128:170–190
49. Gupta S, Bhatt DK, Boni LJ, Olesen J (2010) Improvement of the closed cranial window model in rats by intracarotid infusion of signalling molecules implicated in migraine. Cephalalgia 30:27–36
50. Tvedskov JF et al (2005) No increase of calcitonin gene-related peptide in jugular blood during migraine. Ann Neurol 58:561–568
51. Edvinsson L, Goadsby PJ (1994) Neuropeptides in migraine and cluster headache. Cephalalgia 14:320–327
52. Messlinger K, Hotta H, Pawlak M, Schmidt RF (1997) Effects of the 5-HT1 receptor agonists, sumatriptan and CP 93,129, on dural arterial flow in the rat. Eur J Pharmacol 332:173–181
53. Kurosawa M, Messlinger K, Pawlak M, Schmidt RF (1995) Increase of meningeal blood flow after electrical stimulation of rat dura mater encephali: mediation by calcitonin gene-related peptide. Br J Pharmacol 114:1397–1402
54. Geppetti P, Rossi E, Chiarugi A, Benemei S (2012) Antidromic vasodilatation and the migraine mechanism. J Headache Pain 13:103–111
55. Levy D, Burstein R, Strassman AM (2005) Calcitonin gene-related peptide does not excite or sensitize meningeal nociceptors: implications for the pathophysiology of migraine. Ann Neurol 58:698–705

56. Akerman S, Williamson DJ, Hill RG, Goadsby PJ (2001) The effect of adrenergic compounds on neurogenic dural vasodilatation. Eur J Pharmacol 424:53–58
57. Escott KJ, Connor HE, Brain SD, Beattie DT (1995) The involvement of calcitonin gene-related peptide (CGRP) and substance P in feline pial artery diameter responses evoked by capsaicin. Neuropeptides 29:129–135
58. Nagy I, Friston D, Valente JS, Torres Perez JV, Andreou AP (2014) Pharmacology of the capsaicin receptor, transient receptor potential vanilloid type-1 ion channel. Prog Drug Res Fortschritte der Arzneimittelforschung Progres des recherches pharmaceutiques 68:39–76
59. Summ O, Akerman S, Holland PR, Goadsby PJ (2009) The TRPV1 receptor antagonist, A-993610, shows no effect on neurogenic dural dilation but is able to block capsaicin induced dilation. Cephalalgia 29:136
60. Neeb L, Reuter U (2007) Nitric oxide in migraine. CNS Neurol Disord Drug Targets 6:258–264
61. Bellamy J, Bowen EJ, Russo AF, Durham PL (2006) Nitric oxide regulation of calcitonin gene-related peptide gene expression in rat trigeminal ganglia neurons. Eur J Neurosci 23:2057–2066
62. Li J, Vause CV, Durham PL (2008) Calcitonin gene-related peptide stimulation of nitric oxide synthesis and release from trigeminal ganglion glial cells. Brain Res 1196:22–32
63. Messlinger K, Suzuki A, Pawlak M, Zehnter A, Schmidt RF (2000) Involvement of nitric oxide in the modulation of dural arterial blood flow in the rat. Br J Pharmacol 129:1397–1404
64. Olesen J et al (2004) Calcitonin gene-related peptide receptor antagonist BIBN 4096 BS for the acute treatment of migraine. N Engl J Med 350:1104–1110
65. Ho TW et al (2009) Efficacy and tolerability of MK-0974 (telcagepant), a new oral antagonist of calcitonin gene-related peptide receptor, compared with zolmitriptan for acute migraine: a randomised, placebo-controlled, parallel-treatment trial. Lancet 372:2115–2123
66. Williamson DJ, Hargreaves RJ, Hill RG, Shepheard SL (1997) Sumatriptan inhibits neurogenic vasodilation of dural blood vessels in the anaesthetized rat–intravital microscope studies. Cephalalgia 17:525–531
67. Akerman S, Goadsby PJ (2005) Topiramate inhibits trigeminovascular activation: an intravital microscopy study. Br J Pharmacol 146:7–14
68. Williamson DJ, Hargreaves RJ (2001) Neurogenic inflammation in the context of migraine. Microsc Res Tech 53:167–178
69. Carmody J, Pawlak M, Messlinger K (1996) Lack of a role for substance P in the control of dural arterial flow. Exp Brain Res 111:424–428
70. Goldstein DJ et al (2001) Lanepitant, an NK-1 antagonist, in migraine prevention. Cephalalgia 21:102–106
71. Goldstein DJ et al (1997) Ineffectiveness of neurokinin-1 antagonist in acute migraine: a crossover study. Cephalalgia 17:785–790
72. Akerman S, Holland PR, Lasalandra MP, Goadsby PJ (2010) Inhibition of trigeminovascular dural nociceptive afferents by Ca(2+)-activated K(+) (MaxiK/BK(Ca)) channel opening. Pain 151:128–136
73. Akerman S, Holland PR, Goadsby PJ (2007) Cannabinoid (CB1) receptor activation inhibits trigeminovascular neurons. J Pharmacol Exp Ther 320:64–71
74. Holland PR, Akerman S, Goadsby PJ (2005) Orexin 1 receptor activation attenuates neurogenic dural vasodilation in an animal model of trigeminovascular nociception. J Pharmacol Exp Ther 315:1380–1385
75. Shepheard S et al (1999) Possible antimigraine mechanisms of action of the 5HT1F receptor agonist LY334370. Cephalalgia 19:851–858
76. Summ O, Andreou AP, Akerman S, Goadsby PJ (2010) A potential nitrergic mechanism of action for indomethacin, but not of other COX inhibitors: relevance to indomethacin-sensitive headaches. J Headache Pain 11:477–483
77. Akerman S, Williamson DJ, Goadsby PJ (2003) Voltage-dependent calcium channels are involved in neurogenic dural vasodilatation via a presynaptic transmitter release mechanism. Br J Pharmacol 140:558–566

78. Gupta S et al (2007) Female sex hormones and rat dural vasodilatation to CGRP, periarterial electrical stimulation and capsaicin. Headache 47:225–235
79. Akerman S, Goadsby PJ (2005) The role of dopamine in a model of trigeminovascular nociception. J Pharmacol Exp Ther 314:162–169
80. Andreou AP, Holland PR, Goadsby PJ (2009) Activation of iGluR5 kainate receptors inhibits neurogenic dural vasodilatation in an animal model of trigeminovascular activation. Br J Pharmacol 157:464–473
81. Markowitz S, Saito K, Moskowitz MA (1988) Neurogenically mediated plasma extravasation in dura mater: effect of ergot alkaloids. A possible mechanism of action in vascular headache. Cephalalgia 8:83–91
82. Moskowitz MA (1993) Neurogenic inflammation in the pathophysiology and treatment of migraine. Neurology 43:S16–S20
83. Andreou AP, Summ O, Charbit AR, Romero-Reyes M, Goadsby PJ (2010) Animal models of headache: from bedside to bench and back to bedside. Expert Rev Neurother 10:389–411
84. Dimitriadou V, Buzzi MG, Theoharides TC, Moskowitz MA (1992) Ultrastructural evidence for neurogenically mediated changes in blood vessels of the rat dura mater and tongue following antidromic trigeminal stimulation. Neuroscience 48:187–203
85. Pietrobon D, Moskowitz MA (2013) Pathophysiology of migraine. Annu Rev Physiol 75:365–391
86. Markowitz S, Saito K, Moskowitz MA (1987) Neurogenically mediated leakage of plasma protein occurs from blood vessels in dura mater but not brain. J Neurosci 7:4129–4136
87. Strassman AM, Raymond SA, Burstein R (1996) Sensitization of meningeal sensory neurons and the origin of headaches. Nature 384:560–564
88. Kandere-Grzybowska K et al (2003) Stress-induced dura vascular permeability does not develop in mast cell-deficient and neurokinin-1 receptor knockout mice. Brain Res 980:213–220
89. Goadsby PJ, Edvinsson L (1998) Neuropeptides in headache. Eur J Neurol 5:329–341
90. Delepine L, Aubineau P (1997) Plasma protein extravasation induced in the rat dura mater by stimulation of the parasympathetic sphenopalatine ganglion. Exp Neurol 147:389–400
91. Buzzi MG, Sakas DE, Moskowitz MA (1989) Indomethacin and acetylsalicylic acid block neurogenic plasma protein extravasation in rat dura mater. Eur J Pharmacol 165:251–258
92. Schuh-Hofer S, Tayefeh M, Reuter U, Dirnagl U, Arnold G (2006) Effects of parecoxib on plasma protein extravasation and c-fos expression in the rat. Headache 46:276–285
93. Phebus LA et al (1997) Characterization of LY344864 as a pharmacological tool to study 5-HT1F receptors: binding affinities, brain penetration and activity in the neurogenic dural inflammation model of migraine. Life Sci 61:2117–2126
94. Yu XJ, Cutrer FM, Moskowitz MA, Waeber C (1997) The 5-HT1D receptor antagonist GR-127,935 prevents inhibitory effects of sumatriptan but not CP-122,288 and 5-CT on neurogenic plasma extravasation within guinea pig dura mater. Neuropharmacology 36:83–91
95. Cutrer FM, Yu XJ, Ayata G, Moskowitz MA, Waeber C (1999) Effects of PNU-109,291, a selective 5-HT1D receptor agonist, on electrically induced dural plasma extravasation and capsaicin-evoked c-fos immunoreactivity within trigeminal nucleus caudalis. Neuropharmacology 38:1043–1053
96. Buzzi MG, Moskowitz MA (1990) The antimigraine drug, sumatriptan (GR43175), selectively blocks neurogenic plasma extravasation from blood vessels in dura mater. Br J Pharmacol 99:202–206
97. Polley JS et al (1997) The activity of GR205171, a potent non-peptide tachykinin NK1 receptor antagonist, in the trigeminovascular system. Regul Pept 68:23–29
98. Phebus LA et al (1997) The non-peptide NK-1 receptor antagonist LY303870 inhibits neurogenic dural inflammation in guinea pigs. Life Sci 60:1553–1561
99. Goldstein DJ, Offen WW, Klein EG, Phebus LA (1999) Lanepitant an NK-1 antagonist in migraine prophylaxis. Cephalalgia 19:377
100. Peroutka SJ (2005) Neurogenic inflammation and migraine: implications for the therapeutics. Mol Interv 5:304–311
101. Kruuse C, Thomsen LL, Birk S, Olesen J (2003) Migraine can be induced by sildenafil without changes in middle cerebral artery diameter. Brain 126:241–247

102. Goadsby PJ (2007) Recent advances in understanding migraine mechanisms, molecules and therapeutics. Trends Mol Med 13:39–44
103. Afridi SK, Goadsby PJ (2006) Neuroimaging of migraine. Curr Pain Headache Rep 10:221–224
104. Wolff HG (1948) Headache and other head pain. Oxford University Press, New York
105. Penfield W (1934) A contribution to the mechanism of intracranial pain. In: Proceedings of the association for research in nervous and mental disease, pp 399–415
106. Penfield W, McNaughton F (1940) Dural headache and innervation of the dura matter. Arch Neurol Psychiatry 44:43–75
107. White JP, Cibelli M, Fidalgo AR, Nagy I (2011) Extracellular signal-regulated kinases in pain of peripheral origin. Eur J Pharmacol 650:8–17
108. Kaube H, Keay KA, Hoskin KL, Bandler R, Goadsby PJ (1993) Expression of c-Fos-like immunoreactivity in the caudal medulla and upper cervical spinal cord following stimulation of the superior sagittal sinus in the cat. Brain Res 629:95–102
109. Burstein R, Yamamura H, Malick A, Strassman AM (1998) Chemical stimulation of the intracranial dura induces enhanced responses to facial stimulation in brain stem trigeminal neurons. J Neurophysiol 79:964–982
110. Hoskin KL, Bulmer DC, Goadsby PJ (1999) Fos expression in the trigeminocervical complex of the cat after stimulation of the superior sagittal sinus is reduced by L-NAME. Neurosci Lett 266:173–176
111. Hunt SP, Pini A, Evan G (1987) Induction of c-fos-like protein in spinal cord neurons following sensory stimulation. Nature 328:632–634
112. Keay KA, Bandler R (1998) Vascular head pain selectively activates ventrolateral periaqueductal gray in the cat. Neurosci Lett 245:58–60
113. Malick A, Jakubowski M, Elmquist JK, Saper CB, Burstein R (2001) A neurohistochemical blueprint for pain-induced loss of appetite. Proc Natl Acad Sci U S A 98:9930–9935
114. Pearse DD, Bushell G, Leah JD (2001) Jun, Fos and Krox in the thalamus after C-fiber stimulation: coincident-input-dependent expression, expression across somatotopic boundaries, and nucleolar translocation. Neuroscience 107:143–159
115. Knyihar-Csillik E et al (2007) Prevention of electrical stimulation-induced increase of c-fos immunoreaction in the caudal trigeminal nucleus by kynurenine combined with probenecid. Neurosci Lett 418:122–126
116. Hoskin KL, Goadsby PJ (1998) Comparison of more and less lipophilic serotonin (5HT1B/1D) agonists in a model of trigeminovascular nociception in cat. Exp Neurol 150:45–51
117. Maneepak M, le Grand S, Srikiatkhachorn A (2009) Serotonin depletion increases nociception-evoked trigeminal NMDA receptor phosphorylation. Headache 49: 375–382
118. Tanuri FC et al (2009) Melatonin treatment decreases c-fos expression in a headache model induced by capsaicin. J Headache Pain 10:105–110
119. Cutrer FM, Limmroth V, Ayata G, Moskowitz MA (1995) Attenuation by valproate of c-fos immunoreactivity in trigeminal nucleus caudalis induced by intracisternal capsaicin. Br J Pharmacol 116:3199–3204
120. Sixt ML, Messlinger K, Fischer MJ (2009) Calcitonin gene-related peptide receptor antagonist olcegepant acts in the spinal trigeminal nucleus. Brain 132:3134–3141
121. Goadsby PJ, Hoskin KL (1999) Differential effects of low dose CP122,288 and eletriptan on fos expression due to stimulation of the superior sagittal sinus in cat. Pain 82:15–22
122. Goadsby PJ, Hoskin KL, Knight YE (1998) Substance P blockade with the potent and centrally acting antagonist GR205171 does not effect central trigeminal activity with superior sagittal sinus stimulation. Neuroscience 86:337–343
123. Edling Y, Ingelman-Sundberg M, Simi A (2007) Glutamate activates c-fos in glial cells via a novel mechanism involving the glutamate receptor subtype mGlu5 and the transcriptional repressor DREAM. Glia 55:328–340
124. Dragunow M, Faull RL (1990) MK801 induces c-fos protein in thalamic and neocortical neurons of rat brain. Neurosci Lett 113:144–150

125. Flores C, Arvanitogiannis A, Shizgal P (1997) Fos-like immunoreactivity in forebrain regions following self-stimulation of the lateral hypothalamus and the ventral tegmental area. Behav Brain Res 87:239–251
126. Lima D, Avelino A (1994) Spinal c-fos expression is differentially induced by brief or persistent noxious stimulation. Neuroreport 5:1853–1856
127. Goadsby PJ, Zagami AS (1991) Stimulation of the superior sagittal sinus increases metabolic activity and blood flow in certain regions of the brainstem and upper cervical spinal cord of the cat. Brain 114(Pt 2):1001–1011
128. Lambert GA, Goadsby PJ, Zagami AS, Duckworth JW (1988) Comparative effects of stimulation of the trigeminal ganglion and the superior sagittal sinus on cerebral blood flow and evoked potentials in the cat. Brain Res 453:143–149
129. Olesen J et al (1990) Timing and topography of cerebral blood flow, aura, and headache during migraine attacks. Ann Neurol 28:791–798
130. Escott KJ, Beattie DT, Connor HE, Brain SD (1995) Trigeminal ganglion stimulation increases facial skin blood flow in the rat: a major role for calcitonin gene-related peptide. Brain Res 669:93–99
131. Oshinsky ML, Luo J (2006) Neurochemistry of trigeminal activation in an animal model of migraine. Headache 46(Suppl 1):S39–S44
132. Goadsby PJ, Edvinsson L (1993) The trigeminovascular system and migraine: studies characterizing cerebrovascular and neuropeptide changes seen in humans and cats. Ann Neurol 33:48–56
133. Diener HC et al (1991) Ergotamine, flunarizine and sumatriptan do not change cerebral blood flow velocity in normal subjects and migraneurs. J Neurol 238:245–250
134. Hoskin KL, Kaube H, Goadsby PJ (1996) Central activation of the trigeminovascular pathway in the cat is inhibited by dihydroergotamine. A c-Fos and electrophysiological study. Brain 119(Pt 1):249–256
135. Davis KD, Dostrovsky JO (1986) Activation of trigeminal brain-stem nociceptive neurons by dural artery stimulation. Pain 25:395–401
136. Strassman A, Mason P, Moskowitz M, Maciewicz R (1986) Response of brainstem trigeminal neurons to electrical stimulation of the dura. Brain Res 379:242–250
137. Bolton S, O'Shaughnessy CT, Goadsby PJ (2005) Properties of neurons in the trigeminal nucleus caudalis responding to noxious dural and facial stimulation. Brain Res 1046:122–129
138. Bartsch T, Goadsby PJ (2003) Increased responses in trigeminocervical nociceptive neurons to cervical input after stimulation of the dura mater. Brain 126:1801–1813
139. Xiao Y, Richter JA, Hurley JH (2008) Release of glutamate and CGRP from trigeminal ganglion neurons: role of calcium channels and 5-HT1 receptor signaling. Mol Pain 4:12
140. Shields KG, Goadsby PJ (2005) Propranolol modulates trigeminovascular responses in thalamic ventroposteromedial nucleus: a role in migraine? Brain 128:86–97
141. Shields KG, Goadsby PJ (2006) Serotonin receptors modulate trigeminovascular responses in ventroposteromedial nucleus of thalamus: a migraine target? Neurobiol Dis 23:491–501
142. Lambert GA et al (2014) Stimulation of dural vessels excites the SI somatosensory cortex of the cat via a relay in the thalamus. Cephalalgia 34:243–257
143. Cumberbatch MJ, Hill RG, Hargreaves RJ (1998) Differential effects of the 5HT1B/1D receptor agonist naratriptan on trigeminal versus spinal nociceptive responses. Cephalalgia 18:659–663
144. Cumberbatch MJ, Williamson DJ, Mason GS, Hill RG, Hargreaves RJ (1999) Dural vasodilation causes a sensitization of rat caudal trigeminal neurones in vivo that is blocked by a 5-HT1B/1D agonist. Br J Pharmacol 126:1478–1486
145. Fischer MJ, Koulchitsky S, Messlinger K (2005) The nonpeptide calcitonin gene-related peptide receptor antagonist BIBN4096BS lowers the activity of neurons with meningeal input in the rat spinal trigeminal nucleus. J Neurosci 25:5877–5883
146. Andreou A, Goadsby P (2010) Topiramate acts on kainate receptors within the trigeminothalamic pathway. Headache 50:S5 (Wiley-Blackwell Publishing, Inc Commerce Place, 350 Main St, Malden 02148, MA USA, 2010)

147. Andreou AP, Goadsby PJ (2009) Therapeutic potential of novel glutamate receptor antagonists in migraine. Expert Opin Investig Drugs 18:789–803
148. Storer RJ, Goadsby PJ (2009) N-Methyl-d-Aspartate receptor channel complex blockers including memantine and magnesium inhibit nociceptive traffic in the trigeminocervical complex of the rat. Cephalalgia 29:135
149. Andreou A, Goadsby P (2009) LY466195, a clinically active compound in the acute treatment of migraine, inhibits activation in the trigeminocervical complex and the ventroposteromedial thalamus after nociceptive trigeminovascular activation. Cephalalgia 29:132 (Wiley-Blackwell Publishing, Inc Commerce Place, 350 Main St, Malden 02148, MA USA, 2009)
150. Andreou A, Goadsby P, Holland P (2007) Pre- and post-synaptic involvement of GluR5 kainate receptors in trigeminovascular nociceptive processing. Cephalalgia 27:605
151. Andreou A, Holland P, Goadsby P (2008) iGluR5 kainate receptors modulate trigeminovascular nociceptive transmission in thalamic ventroposteromedial nucleus. Headache 48:S5–S6 (Blackwell Publishing 9600 GARSINGTON RD, Oxford OX4 2DQ, Oxon, England, 2008)
152. Andreou A, Storer R, Holland P, Goadsby P (2006) CNQX inhibits trigeminovascular neurons in the rat: a microiontophoresis study. Cephalalgia 26:1383 (Blackwell Publishing 9600 GARSINGTON RD, Oxford OX4 2DQ, Oxon, England, 2006)
153. Andreou AP, Goadsby PJ (2011) Topiramate in the treatment of migraine: a kainate (glutamate) receptor antagonist within the trigeminothalamic pathway. Cephalalgia 31:1343–1358
154. Hoffmann J et al (2014) Evidence for orexinergic mechanisms in migraine. Neurobiol Dis 74C:137–143
155. Charbit A, Akerman S, Goadsby P (2009) Comparison of the effects of central and peripheral dopamine receptor activation on evoked firing in the trigeminocervical complex. J Pharmacol Exp Ther 331(2):752–763
156. Lambert GA et al (2009) The effects of the TRPV1 receptor antagonist SB-705498 on trigeminovascular sensitisation and neurotransmission. Naunyn Schmiedebergs Arch Pharmacol 380:311–325
157. Lambert GA, Lowy AJ, Boers PM, Angus-Leppan H, Zagami AS (1992) The spinal cord processing of input from the superior sagittal sinus: pathway and modulation by ergot alkaloids. Brain Res 597:321–330
158. Storer RJ, Goadsby PJ (1997) Microiontophoretic application of serotonin (5HT)1B/1D agonists inhibits trigeminal cell firing in the cat. Brain 120(Pt 12):2171–2177
159. Goadsby PJ, Lipton RB, Ferrari MD (2002) Migraine–current understanding and treatment. N Engl J Med 346:257–270
160. Storer RJ, Akerman S, Goadsby PJ (2004) Calcitonin gene-related peptide (CGRP) modulates nociceptive trigeminovascular transmission in the cat. Br J Pharmacol 142:1171–1181
161. Summ O, Charbit AR, Andreou AP, Goadsby PJ (2010) Modulation of nocioceptive transmission with calcitonin gene-related peptide receptor antagonists in the thalamus. Brain 133:2540–2548
162. Andreou AP, Shields KG, Goadsby PJ (2010) GABA and valproate modulate trigeminovascular nociceptive transmission in the thalamus. Neurobiol Dis 37:314–323
163. Bartsch T, Goadsby PJ (2002) Stimulation of the greater occipital nerve induces increased central excitability of dural afferent input. Brain 125:1496–1509
164. Bartsch T, Paemeleire K, Goadsby PJ (2009) Neurostimulation approaches to primary headache disorders. Curr Opin Neurol 22:262–268
165. Weiller C et al (1995) Brain stem activation in spontaneous human migraine attacks. Nat Med 1:658–660
166. Knight YE, Bartsch T, Kaube H, Goadsby PJ (2002) P/Q-type calcium-channel blockade in the periaqueductal gray facilitates trigeminal nociception: a functional genetic link for migraine? J Neurosci 22:RC213
167. Goadsby PJ, Lambert GA, Lance JW (1982) Differential effects on the internal and external carotid circulation of the monkey evoked by locus coeruleus stimulation. Brain Res 249: 247–254

168. Supronsinchai W et al (2013) GABAA receptors in the nucleus raphe magnus modulate firing of neurons in the trigeminocervical complex. J Headache Pain 14:P67
169. Charbit AR, Akerman S, Holland PR, Goadsby PJ (2009) Neurons of the dopaminergic/calcitonin gene-related peptide A11 cell group modulate neuronal firing in the trigeminocervical complex: an electrophysiological and immunohistochemical study. J Neurosci 29:12532–12541
170. Robert C et al (2013) Paraventricular hypothalamic regulation of trigeminovascular mechanisms involved in headaches. J Neurosci 33:8827–8840
171. Noseda R, Kainz V, Borsook D, Burstein R (2014) Neurochemical pathways that converge on thalamic trigeminovascular neurons: potential substrate for modulation of migraine by sleep, food intake, stress and anxiety. PLoS One 9:e103929
172. Noseda R et al (2010) A neural mechanism for exacerbation of headache by light. Nat Neurosci 13:239–245
173. Burstein R, Cutrer MF, Yarnitsky D (2000) The development of cutaneous allodynia during a migraine attack clinical evidence for the sequential recruitment of spinal and supraspinal nociceptive neurons in migraine. Brain 123(Pt 8):1703–1709
174. Burstein R, Jakubowski M (2004) Analgesic triptan action in an animal model of intracranial pain: a race against the development of central sensitization. Ann Neurol 55:27–36
175. Pozo-Rosich P, Oshinsky M (2005) Dihydroergotamine (DHE) reverses central sensitization in the trigeminal nucleus caudalis. Headache 45:767
176. Jakubowski M et al (2005) Terminating migraine with allodynia and ongoing central sensitization using parenteral administration of COX1/COX2 inhibitors. Headache 45:850–861
177. Levy D, Zhang XC, Jakubowski M, Burstein R (2008) Sensitization of meningeal nociceptors: inhibition by naproxen. Eur J Neurosci 27:917–922
178. Burstein R et al (2010) Thalamic sensitization transforms localized pain into widespread allodynia. Ann Neurol 68:81–91
179. Thomsen LL, Kruuse C, Iversen HK, Olesen J (1994) A nitric oxide donor (nitroglycerin) triggers genuine migraine attacks. Eur J Neurol 1:73–80
180. Tvedskov JF, Iversen HK, Olesen J, Tfelt-Hansen P (2010) Nitroglycerin provocation in normal subjects is not a useful human migraine model? Cephalalgia 30:928–932
181. Thomsen LL, Brennum J, Iversen HK, Olesen J (1996) Effect of a nitric oxide donor (glyceryl trinitrate) on nociceptive thresholds in man. Cephalalgia 16:169–174
182. Culotta E, Koshland DE (1992) No news is good-news. Science 258:1862–1865
183. Tassorelli C, Greco R, Cappelletti D, Sandrini G, Nappi G (2005) Comparative analysis of the neuronal activation and cardiovascular effects of nitroglycerin, sodium nitroprusside and L-arginine. Brain Res 1051:17–24
184. Olesen J, Jansen-Olesen I (2012) Towards a reliable animal model of migraine. Cephalalgia 32:578–580
185. Ramachandran R et al (2012) A naturalistic glyceryl trinitrate infusion migraine model in the rat. Cephalalgia 32:73–84
186. Tassorelli C, Joseph SA (1996) Systemic nitroglycerin activates peptidergic and catecholaminergic pathways in rat brain. Peptides 17:443–449
187. Greco R et al (2013) Effect of sex and estrogens on neuronal activation in an animal model of migraine. Headache 53:288–296
188. Dieterle A, Fischer MJ, Link AS, Neuhuber WL, Messlinger K (2011) Increase in CGRP- and nNOS-immunoreactive neurons in the rat trigeminal ganglion after infusion of an NO donor. Cephalalgia 31:31–42
189. Ramachandran R et al (2014) Nitric oxide synthase, calcitonin gene-related peptide and NK-1 receptor mechanisms are involved in GTN-induced neuronal activation. Cephalalgia 34:136–147
190. Kruuse C, Iversen HK, Jansen-Olesen I, Edvinsson L, Olesen J (2010) Calcitonin gene-related peptide (CGRP) levels during glyceryl trinitrate (GTN)-induced headache in healthy volunteers. Cephalalgia 30:467–474
191. Offenhauser N et al (2005) CGRP release and c-fos expression within trigeminal nucleus caudalis of the rat following glyceryltrinitrate infusion. Cephalalgia 25:225–236

192. Bates E et al (2009) Sumatriptan alleviates nitroglycerin-induced mechanical and thermal allodynia in mice. Cephalalgia 30(2):170–178
193. Greco R et al (2013) Effects of CGRP receptor antagonism in nitroglycerin-induced hyperalgesia. Cephalalgia 34:594–604
194. Akerman S, Goadsby PJ (2014) Acute anti-migraine treatments abort established central sensitization of trigeminovascular neurons: validation of a novel translational approach. Headache 54:2–3
195. Andreou AP, Chamberlain J (2014) Nitric oxide alters the neuronal firing of the dopaminergic hypothalamic nucleus A11. Headache 54:6–7
196. Greco R et al (2008) Role of central dopaminergic circuitry in pain processing and nitroglycerin-induced hyperalgesia. Brain Res 1238:215–223
197. Rasmussen BK, Olesen J (1992) Migraine with aura and migraine without aura: an epidemiological study. Cephalalgia 12:221–228
198. Leao AAP (1944) Spreading depression of activity in cerebral cortex. J Neurophysiol 7:359–390
199. Olesen J (1991) Cerebral and extracranial circulatory disturbances in migraine: pathophysiological implications. Cerebrovasc Brain Metab Rev 3:1–28
200. Brennan KC et al (2007) Distinct vascular conduction with cortical spreading depression. J Neurophysiol 97:4143–4151
201. Haghir H, Kovac S, Speckmann EJ, Zilles K, Gorji A (2009) Patterns of neurotransmitter receptor distributions following cortical spreading depression. Neuroscience 163(4): 1340–1352
202. Gorji A et al (2001) Spreading depression in human neocortical slices. Brain Res 906:74–83
203. Kaube H, Herzog J, Kaufer T, Dichgans M, Diener HC (2000) Aura in some patients with familial hemiplegic migraine can be stopped by intranasal ketamine. Neurology 55:139–141
204. Vinogradova LV, Kuznetsova GD, Coenen AM (2009) Unilateral cortical spreading depression induced by sound in rats. Brain Res 1286:201–207
205. Bolay H, Akcali D, Yalcinkaya D, Sara Y (2009) Behavioral changes associated with cortical spreading depression in awake rats. Cephalalgia 29:142
206. Akcali D, Sayin A, Sara Y, Bolay H (2010) Does single cortical spreading depression elicit pain behaviour in freely moving rats? Cephalalgia 30:1195–1206
207. Fioravanti B et al (2011) Evaluation of cutaneous allodynia following induction of cortical spreading depression in freely moving rats. Cephalalgia 31:1090–1100
208. Ayata C (2009) Spreading depression: from serendipity to targeted therapy in migraine prophylaxis. Cephalalgia 29:1095–1114
209. Ayata C, Jin H, Kudo C, Dalkara T, Moskowitz MA (2006) Suppression of cortical spreading depression in migraine prophylaxis. Ann Neurol 59:652–661
210. Bogdanov VB et al (2011) Migraine preventive drugs differentially affect cortical spreading depression in rat. Neurobiol Dis 41:430–435
211. Andreou A et al (2010) Acid-sensing ion channel 1-A potential site of action of amiloride in migraine with aura. J Headache Pain 11:S125 (Springer-Verlag ITALIA SRL VIA DECEMBRIO, 28, Milan, 20137, Italy, 2010)
212. Holland PR et al (2012) Acid-sensing ion channel 1: a novel therapeutic target for migraine with aura. Ann Neurol 72:559–563
213. Holland PR, Schembri C, Fredrick J, Goadsby PJ (2009) Transcranial magnetic stimulation for the treatment of migraine aura? Cephalalgia 29:22
214. Andreou AP, Summ O, Schembri CT, Fredrick JP, Goadsby PJ (2010) Transcranial magnetic stimulation inhibits cortical spreading depression but not trigeminocervical activation in animal models of migraine. Headache 50:S58
215. Ingvardsen BK, Laursen H, Olsen UB, Hansen AJ (1997) Possible mechanism of c-fos expression in trigeminal nucleus caudalis following cortical spreading depression. Pain 72:407–415
216. Supornsilpchai W, Sanguanrangsirikul S, Maneesri S, Srikiatkhachorn A (2006) Serotonin depletion, cortical spreading depression, and trigeminal nociception. Headache 46:34–39

217. Moskowitz MA, Nozaki K, Kraig RP (1993) Neocortical spreading depression provokes the expression of c-fos protein-like immunoreactivity within trigeminal nucleus caudalis via trigeminovascular mechanisms. J Neurosci 13:1167–1177
218. Herrera DG, Robertson HA (1996) Activation of c-fos in the brain. Prog Neurobiol 50:83–107
219. Ebersberger A, Schaible HG, Averbeck B, Richter F (2001) Is there a correlation between spreading depression, neurogenic inflammation, and nociception that might cause migraine headache? Ann Neurol 49:7–13
220. Bolay H et al (2002) Intrinsic brain activity triggers trigeminal meningeal afferents in a migraine model. Nat Med 8:136–142
221. Zhang X et al (2011) Activation of central trigeminovascular neurons by cortical spreading depression. Ann Neurol 69:855–865
222. Karatas H et al (2013) Spreading depression triggers headache by activating neuronal Panx1 channels. Science 339:1092–1095
223. Read SJ, Hirst WD, Upton N, Parsons AA (2001) Cortical spreading depression produces increased cGMP levels in cortex and brain stem that is inhibited by tonabersat (SB-220453) but not sumatriptan. Brain Res 891:69–77
224. Goadsby PJ, Ferrari MD, Csanyi A, Olesen J, Mills JG (2009) Randomized, double-blind, placebo-controlled, proof-of-concept study of the cortical spreading depression inhibiting agent tonabersat in migraine prophylaxis. Cephalalgia 29(7):742–750
225. Noseda R, Constandil L, Bourgeais L, Chalus M, Villanueva L (2010) Changes of meningeal excitability mediated by corticotrigeminal networks: a link for the endogenous modulation of migraine pain. J Neurosci 30:14420–14429
226. Lambert GA, Hoskin KL, Zagami AS (2008) Cortico-NRM influences on trigeminal neuronal sensation. Cephalalgia 28:640–652
227. Lambert GA, Michalicek J, Storer RJ, Zagami AS (1999) Effect of cortical spreading depression on activity of trigeminovascular sensory neurons. Cephalalgia 19:631–638
228. Andreou AP, Sprenger T, Goadsby PJ (2012) Cortical spreading depression-evoked discharges on trigeminothalamic neurons. Headache 52:900
229. Andreou AP, Sprenger T, Goadsby PJ (2013) Cortical modulation of thalamic function during cortical spreading depression- unraveling a new central mechanism involved in migraine aura. J Headache Pain 14:16
230. van den Maagdenberg AM, Haan J, Terwindt GM, Ferrari MD (2007) Migraine: gene mutations and functional consequences. Curr Opin Neurol 20:299–305
231. Leo L et al (2011) Increased susceptibility to cortical spreading depression in the mouse model of familial hemiplegic migraine type 2. PLoS Genet 7:e1002129
232. Eikermann-Haerter K et al (2009) Genetic and hormonal factors modulate spreading depression and transient hemiparesis in mouse models of familial hemiplegic migraine type 1. J Clin Invest 119:99–109
233. Tottene A et al (2009) Enhanced excitatory transmission at cortical synapses as the basis for facilitated spreading depression in Ca(v)2.1 knockin migraine mice. Neuron 61:762–773
234. Langford DJ et al (2010) Coding of facial expressions of pain in the laboratory mouse. Nat Methods 7:447–449
235. Chanda ML et al (2013) Behavioral evidence for photophobia and stress-related ipsilateral head pain in transgenic Cacna1a mutant mice. Pain 154:1254–1262
236. Mathew R et al (2011) Immunohistochemical characterization of calcitonin gene-related peptide in the trigeminal system of the familial hemiplegic migraine 1 knock-in mouse. Cephalalgia 31:1368–1380
237. Hansen JM et al (2008) Familial hemiplegic migraine type 2 does not share hypersensitivity to nitric oxide with common types of migraine. Cephalalgia 28:367–375
238. Hansen JM, Thomsen LL, Olesen J, Ashina M (2008) Calcitonin gene-related peptide does not cause the familial hemiplegic migraine phenotype. Neurology 71:841–847
239. Park J et al (2014) Differential trigeminovascular nociceptive responses in the thalamus in the familial hemiplegic migraine 1 knock-in mouse: a Fos protein study. Neurobiol Dis 64:1–7

240. Moon H-S et al (2010) Altered responses in the descending modulatory system of transgenic mice with the CACNA1A mutation. Headache 50:67
241. Xu Y et al (2005) Functional consequences of a CKIdelta mutation causing familial advanced sleep phase syndrome. Nature 434:640–644
242. Brennan KC et al (2013) Casein kinase idelta mutations in familial migraine and advanced sleep phase. Sci Transl Med 5:183ra156, 181-111
243. Oshinsky ML, Gomonchareonsiri S (2007) Episodic dural stimulation in awake rats: a model for recurrent headache. Headache 47:1026–1036
244. Stucky NL et al (2011) Sex differences in behavior and expression of CGRP-related genes in a rodent model of chronic migraine. Headache 51:674–692
245. Melo-Carrillo A, Lopez-Avila A (2013) A chronic animal model of migraine, induced by repeated meningeal nociception, characterized by a behavioral and pharmacological approach. Cephalalgia 33:1096–1105
246. Dong Z, Jiang L, Wang X, Wang X, Yu S (2011) Nociceptive behaviors were induced by electrical stimulation of the dura mater surrounding the superior sagittal sinus in conscious adult rats and reduced by morphine and rizatriptan benzoate. Brain Res 1368:151–158
247. Pradhan AA et al (2014) Characterization of a novel model of chronic migraine. Pain 155:269–274
248. Pradhan AA, Smith ML, Zyuzin J, Charles A (2014) delta-Opioid receptor agonists inhibit migraine-related hyperalgesia, aversive state and cortical spreading depression in mice. Br J Pharmacol 171:2375–2384
249. Oshinsky ML et al (2012) Spontaneous trigeminal allodynia in rats: a model of primary headache. Headache 52:1336–1349
250. Ashkenazi A, Mushtaq A, Yang I, Oshinsky ML (2009) Ictal and interictal phonophobia in migraine-a quantitative controlled study. Cephalalgia 29:1042–1048
251. Romero-Reyes M et al (2013) Spontaneous behavioral responses in the orofacial region: a model of trigeminal pain in mouse. Headache 53:137–151
252. Lipton RB et al (2008) Cutaneous allodynia in the migraine population. Ann Neurol 63:148–158
253. De Felice M et al (2010) Triptan-induced enhancement of neuronal nitric oxide synthase in trigeminal ganglion dural afferents underlies increased responsiveness to potential migraine triggers. Brain 133:2475–2488
254. De Felice M et al (2010) Triptan-induced latent sensitization: a possible basis for medication overuse headache. Ann Neurol 67:325–337
255. De Felice M, Porreca F (2009) Opiate-induced persistent pronociceptive trigeminal neural adaptations: potential relevance to opiate-induced medication overuse headache. Cephalalgia 29:1277–1284
256. De Felice M, Ossipov MH, Porreca F (2011) Persistent medication-induced neural adaptations, descending facilitation, and medication overuse headache. Curr Opin Neurol 24:193–196
257. Green AL et al (2013) Increased susceptibility to cortical spreading depression in an animal model of medication-overuse headache. Cephalalgia 34:594–604

Chapter 3
Animal Models of Tension-Type Headache and Trigeminal Autonomic Cephalalgias

Cristina Tassorelli, Rosaria Greco, and Simon Akerman

3.1 Introduction

Primary headache disorders, according to the International Classification of Headache Disorders 3rd edition (ICHD-III-beta), include migraine, tension-type headache (TTH), trigeminal autonomic cephalalgias (TACs), and other primary headaches [1]. Primary headaches represent a common and major health problem worldwide and significantly impair patients' quality of life [2, 3]. In the more common forms of primary headaches, phenotype is caused by an interaction of various genetic variants, each of them having a small effect with different environmental factors. Genetic association studies reported that genes involved in vascular, neuronal, and endocrine functions may have a significant function in primary headaches [4, 5]. Great advances have been made over the past 25 years in understanding the pathophysiology of headaches, and several animal models have provided a translational knowledge on migraine pathophysiology. However, currently available animal models for investigating the pathophysiology of other primary headaches, such TTH and trigeminal

C. Tassorelli, MD, PhD (✉)
Laboratory of Neurophysiology of Integrative Autonomic Systems, Headache Science Centre, National Neurological Institute C. Mondino, Via Mondino 2, Pavia 27100, Italy

Department of Brain and Behavioural Sciences, University of Pavia, Pavia, Italy
e-mail: cristina.tassorelli@mondino.it

R. Greco, PhD
Laboratory of Neurophysiology of Integrative Autonomic Systems, Headache Science Centre, National Neurological Institute C. Mondino, Via Mondino 2, Pavia 27100, Italy
e-mail: rosaria.greco@mondino.it

S. Akerman, PhD (✉)
Department of Oral and Maxillofacial Pathology, Radiology and Medicine, New York University College of Dentistry, 345 East 24th Street, New York, NY 10010, USA
e-mail: simon.akerman@nyu.edu

M. Ashina, P. Geppetti (eds.), *Pathophysiology of Headaches: From Molecule to Man*, Headache, DOI 10.1007/978-3-319-15621-7_3

autonomic cephalalgias (TACs), are very limited, but what is known will be discussed in this chapter. More rare headache types, including cough and exercise headache, headache related to sexual activity, thunderclap headache, and cold-stimulus headache, are just a few others also bracketed under "other primary headaches," and even less is understood about their pathophysiology, and as a consequence they are difficult to model in animals and therefore will not be discussed here.

TTH is the most common type of primary headache: its lifetime prevalence in the general population ranges in different studies from 30 to 78 %. At the same time, it is the least studied of the primary headache disorders, despite the fact it has the highest socioeconomic impact. TACs are a group of primary headaches characterized by lateralized symptoms: prominent headache and ipsilateral cranial autonomic features, such as conjunctiva injection, lacrimation, and rhinorrhea. Pathophysiological mechanisms of primary headaches remain obscure, although numerous hypotheses have been postulated over the years. Apparently, migraine and cluster headache patients do not have any peculiar anatomical or biochemical features that makes them different from normal subjects [6]. Pain activation and autonomic pathways, and release of peptides mediating vascular, as well as sympathetic and parasympathetic responses, occur in such patients during migraine or cluster headache attacks, but the primary cause remains unclear [7]. Despite the fact that TTH is the most common type of headache, the information about key pathophysiological issues, such as the nature and the site of the noxious stimulus, is surprisingly limited. However, since TTH is a disease in humans and not known in animals, experimental animal models are of limited value when evaluating the underlying pathophysiology. Fortunately, quantitative analyses of mechanical and thermal pain thresholds in humans can be used for this purpose. As the extracranial tissues are readily available in humans, previously evaluated psychophysical examinations were applied to these regions to gain information about the pain processing in TTH. However, the reproducibility and validity of these diagnostic tests represented a problem, as objective measures of pain intensity and quality were not accessible.

3.2 Tension-Type Headache (TTH): Putative Mechanisms

TTH is the most common type of primary headache [8]; its lifetime prevalence in the general population ranges in different studies from 30 to 78 %. While this type of headache was previously considered to be primarily psychogenic, several studies strongly suggest a neurobiological basis, at least for the more severe subtypes of TTH. TTH is also a graded phenomenon in which pain severity increases with headache frequency. At one extreme are rare episodes of slight pain and discomfort in the head; at the other are daily, disabling headaches with considerable social and personal impact [9]. Due to this, and also to the very high prevalence, TTH may be regarded as the most important type of headache. For decades it has been a matter of debate whether the pain in TTH originates from myofascial tissues or from central mechanisms in the brain [10, 11]. Experimental findings to substantiate any of these hypotheses are scarce, and the pain mechanisms in TTH are practically unknown. Although the general information of nociception and pain has enhanced, the understanding of deep pain processing in visceral and myofascial tissues is still fairly limited. In a previous study, it was demonstrated that in patients with chronic TTH, nitroglycerin (NTG), a "nitric oxide donor," produces an immediate headache and, after several hours, a typical TTH [12]. The immediate headache is not accompanied by an increase in pericranial tenderness [13], but it might be associated with endogenous production of nitric oxide and sensitization of perivascular sensory afferent nerves [14]. The cooccurrence suggested that, similar to migraine, chronic TTH might be associated with central supersensitivity to nitric oxide. Previous researchers have suggested that the pain in TTH is similar to myofascial pain elicited from other parts of the body, but whether it is strictly localized to muscle tissues or to other deep tissues is still uncertain [15, 16]. In addition, although the pain clinically resembles pain from the myofascial tissues, components of both peripheral and central origin may contribute. However, on the basis of reduced duration of the late exteroceptive silent period (Es) of temporalis muscle activity in patients with chronic TTH, it was suggested that the limbic pathways to the brainstem were disturbed and that studies of Es may offer key information about the central mechanisms [17, 18].

There are several evidences from animal experiments that sensitization, windup, or expansion of receptive fields of central nervous system (CNS) neurons plays an important role in pain induction and maintenance [19–23]. It is known from animal experiments that input from deep myofascial tissue is much more effective in inducing central sensitization than cutaneous input [24–26]. This sensitization observed at the central level, however, not only reflects peripheral changes but rather an additional change in CNS activity. La Motte et al. reported that articular nociceptive fibers and their corresponding spinal neurons can be sensitized within 2–3 h after a chemical inflammation of the joint [27]. Subsequently, some studies reported that the activity in high-threshold mechanoreceptor fibers was shifted to include low-threshold mechanoreceptor fiber activity from sensitized muscles in experimental animal models [27–29]. Repetitive headache attacks associated with peripheral sensitization in myofascial tissues may trigger the process of central sensitization and the corresponding shift to chronic TTH. Besides its putative involvement in TTH, it was suggested that neck muscle pain may be an associated feature of migraine headache [30, 31]. Increased tenderness of pericranial muscles in patients is reported in numerous studies and positively associated with both the intensity and frequency of TTH [15, 16, 31].

3.3 Animal Models of Tension-Type Headache

Intramuscular infusion of α,β-methylene adenosine 5′-triphosphate (α,β-meATP) has been proposed as a translational mouse model for investigating the putative pathophysiological mechanisms of TTH that addresses the impact of nociceptive afferent input from neck muscles on CNS nociceptive processing monitored by the jaw-opening reflex [32–37]. This reflex is an accepted model for the investigation of altered excitability in sensory brainstem neurons with convergent afferent input from different craniofacial tissues such as neck muscles. This indicates heterosynaptic facilitation due to access of nociceptive afferents from neck muscle to the reflex neuronal network in the brainstem. Facilitation of neck muscle nociceptive processing is induced via bilateral infusion of α,β-meATP (100 nM, 20 μl each) into semispinalis neck muscles in mouse during a time period of 1 min. Brainstem nociception is monitored by the jaw-opening reflex elicited via electrical tongue stimulation.

Another animal model of TTH has been developed [34, 35], and it allows investigation of the interactions between peripheral myofascial factors and central sensitization [34, 36]. Local administration of nerve growth factor (NGF) into neck muscles induces strong and sustained facilitation of brainstem nociception as monitored by the sensorimotor jaw-opening reflex (JOR) in anesthetized mice [35]. NGF indeed interacts with tyrosine kinase A (TrkA) and p75 receptors in muscle [38] and excites nociceptive input to the spinal cord and the brainstem via group IV fiber afferents [33]. The peripheral afferent fibers convey inputs originating in muscles and joints, and within a very short time, these messages reach upper levels of the CNS such as the thalamus and the somatosensory cortex after a relay in the dorsal horn of the spinal cord or in the trigeminal ganglion.

3.4 Trigeminal Autonomic Cephalalgias (TACs): Putative Mechanisms

TACs [39] are highly disabling primary headache disorders characterized by severe unilateral head pain that is sometimes described as the worst pain experienced by humans, which occurs in association with ipsilateral cranial autonomic features [1]. Their pathophysiology is characterized by three major clinical features: unilateral trigeminal distribution of pain; lateralized associated symptoms, including cranial autonomic features [40, 41]; and an episodic pattern of attacks [42, 43]. The different TACs are differentiated from each other by their highly individual characteristic attack patterns and also to some extent by their response to treatments, which is summarized in Table 3.1 and reviewed in detail elsewhere [43–45].

In order to develop and describe animal models of TACs, it is necessary to have a clear understanding of their clinical features and a grasp of the anatomy and physiology of the potential pathways involved. The advantage of studying headache disorders is that there is a very clear classification of symptoms relevant to each headache type. With respect to TACs, this includes the combination of lateralization: of pain, associated features and cranial autonomic features, and some degree of episodicity [42, 43]. Other aspects of our understanding of their pathophysiology come from clinical research. It is known there is activation in the posterior hypothalamic gray matter during the pain in cluster headache [46], paroxysmal hemicrania [47], short-lasting unilateral neuralgiform headache attacks with conjunctival injec-

Table 3.1 Clinical features and treatments of trigeminal autonomic cephalalgias

	Cluster headache	Paroxysmal hemicrania	SUNCT/SUNA
Sex F:M	1:3	1:1	1:1.2
Pain type	Stabbing, boring	Throbbing, boring, stabbing	Burning, stabbing, sharp
Severity	Excruciating	Excruciating	Severe to excruciating
Site	Orbit, temple	Orbit, temple	Periorbital
Attack frequency	1/alternate days–8/day	1–40/day (>5/day most of the time)	3–200/day
Duration of attack	15–180 min	2–30 min	5–240 s
Autonomic features	Yes	Yes	Yes (mainly conjunctival injection and lacrimation – SUNCT)
Abortive treatments	Sumatriptan, oxygen	None	None
Preventive treatments	Verapamil, methysergide, lithium	Indomethacin (absolute response)	Lamotrigine, topiramate, gabapentin

Adapted from Cohen et al. [44], with permission

SUNCT short-lasting unilateral neuralgiform headache attacks with conjunctival injection and tearing, *SUNA* short-lasting unilateral neuralgiform headache

tion and tearing [48, 49], and hemicrania continua [50], and deep brain stimulation of this region can relieve cluster headache symptoms [51, 52]. Furthermore, there is release of calcitonin gene-related peptide (CGRP) and vasoactive intestinal peptide (VIP) into the extracranial vasculature during cluster headache [53] and chronic paroxysmal hemicrania [54]. In experimental clinical studies that try to replicate the pain in TACs in the trigeminal ophthalmic distribution, capsaicin injection in the forehead produces many of the vascular changes that are present during TACs, but with no hypothalamic activation [55]. The implication being that the trigeminally mediated vascular changes are not the cause of the hypothalamic activation, but rather hypothalamic activation causes the subsequent activation of trigeminovascular and cranial autonomic pathways, which results in TACs symptoms. What is also clear is that the different TACs have a very clear response to treatments, which differ from each other, but also differ from other headache disorders and can be used as tools to develop a specific animal model of TACs, compared to other headache disorders such as migraine and TTH.

Preclinical studies into general primary headache disorders have helped us have a very clear understanding of the anatomy and physiology of the trigeminovascular and cranial autonomic systems that are likely shared across many headache disorders, which helps explain TACs symptoms. The excruciating trigeminal distribution of pain is likely to be a consequence of activation of the trigeminovascular system. From here there are ascending projections up to the higher brainstem and diencephalic nuclei [56–58], as well as a reflex connection with the superior salivatory nucleus within the pons, which is the origin of cells of the parasympathetic autonomic vasodilator pathway [59]. In TACs lateralized cranial autonomic features are a significant and defining feature of these disorders, and they are believed to result from activation of this trigeminal autonomic reflex arc to the superior salivatory nucleus and its projection to the cranial vessels and lacrimal glands. Imaging studies during TACs suggest that the hypothalamus is also important in their pathophysiology, and it is likely that the episodic and circadian nature of attacks is in some way related to the internal control of biological rhythms, through hypothalamic nuclei. Anatomically there are reciprocal functional connections between the trigeminal nucleus caudalis and various hypothalamic nuclei that receive dural nociceptive information and provide descending control of trigeminovascular nociceptive traffic. The superior salivatory nucleus also receives descending projections from various hypothalamic nuclei, including the paraventricular hypothalamic nuclei (PVN) [60], as well as limbic and cortical areas [59, 61, 62]. It therefore seems that the superior salivatory nucleus is ideally placed to integrate and relay nociceptive and autonomic information to and from the trigeminovascular system as well as being under descending control of the hypothalamus, in the pathophysiology of TACs. This anatomy and physiology is summarized in Fig. 3.1.

3.5 Animal Models of the Trigeminal Autonomic Cephalalgias

3.5.1 Trigeminal Distribution of Pain: Dural Nociceptive Activation

One approach to studying the symptoms in TACs in animal models is to investigate them individually. Trigeminal distribution of pain is perhaps the most studied, although most of the assays were designed predominantly to study migraine pathophysiology and screen potential migraine therapeutics. Dural nociceptive trigeminovascular activation, driven either electrically or with chemical mediators, is thought to activate trigeminovascular nociceptive afferents that are activated during headache. This produces dural vasodilation [69] and neuronal activation in the trigeminocervical complex [70–72], as well as neuronal activation of the SuS [65] and higher pain processing structures, such as midbrain periaqueductal gray [65, 73] and hypothalamic [63, 64] and thalamic nuclei [74, 75]. Furthermore, dural electrical stimulation produces release of CGRP and VIP from the extracranial vasculature, similar to TACs [76]. The limitations of this assay is that it does not completely replicate the pathophysiology of TACs, in that imaging studies during TACs do not demonstrate midbrain activation; this is more commonly associated with migraine pathophysiology. Also, its response to treatments does not fully match that of TACs, with 100 % oxygen treatment in particular, known to specifically relieve symptoms of cluster headache, whereas it is unable to inhibit neurogenic dural vasodilation or neuronal activation in the trigeminocervical complex [77]. In fact only drugs that are effective in treating both migraine and TACs, such as triptans [78–82],

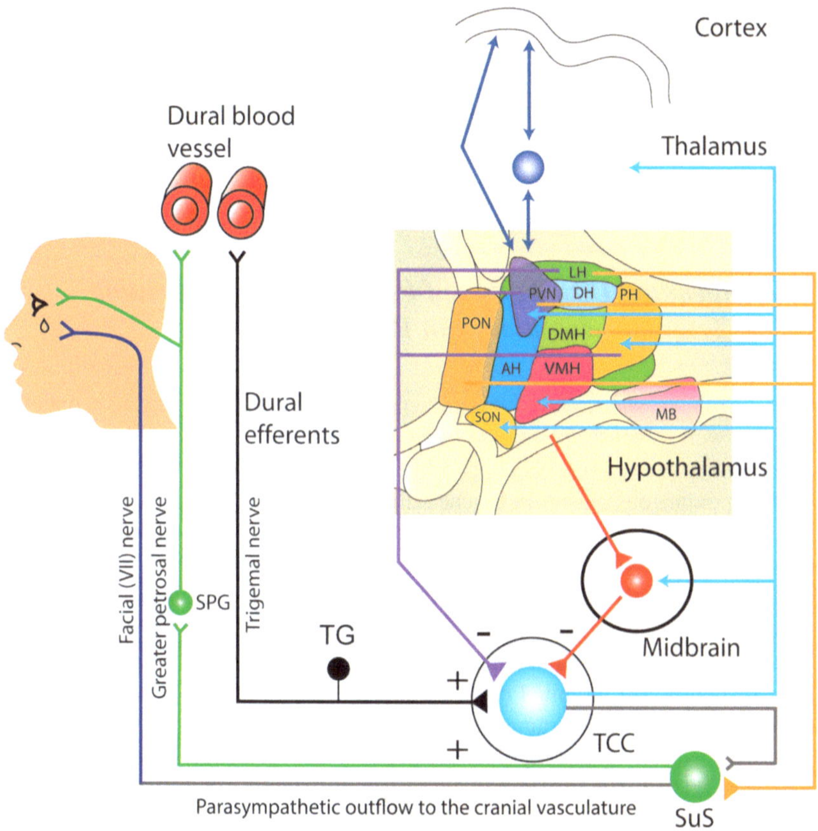

Fig. 3.1 Anatomy and pathophysiology of trigeminal autonomic cephalalgias. The trigeminal distribution of pain in trigeminal autonomic cephalalgias (TACs) is likely mediated by activation of the trigeminovascular system. This includes the rich plexus of nociceptive nerve fibers that originate in the trigeminal ganglion and innervate the peripheral cranial vasculature, including the pain-producing cranial vessels of the dura mater, and the central projection to the trigeminal nucleus caudalis and its extension to the cervical spinal cord, the trigeminocervical complex (TCC). Nociceptive incoming signals to the TCC ascend to higher brain structures including the midbrain and specific hypothalamic nuclei; the posterior (PH), supraoptic (SON) [63], ventromedial (VMH), and paraventricular hypothalamic nuclei (PVN) [64]; and thalamocortical neurons. Dural nociceptive activation also causes neuronal activation in the superior salivatory nucleus (SuS) within the pons [65], through a trigeminal autonomic reflex arc, which is the origin of cells of the autonomic parasympathetic projection to the cranial vasculature [59]. This efferent projection is predominantly through the greater petrosal nerve, a branch of the facial (VIIth) cranial nerve, and its projection through the sphenopalatine (sometimes called pterygopalatine) ganglion [66]. Cranial autonomic symptoms in TACs are believed to result from activation of this trigeminal autonomic reflex arc to the superior salivatory nucleus and its projection to the cranial vessels and lacrimal glands. Descending projections from PH [67, 68], PVN, and lateral hypothalamic nuclei [60] to the TCC are involved in modulating trigeminovascular nociceptive traffic. Furthermore, descending projections to the SuS from LH, PVN, dorsomedial (DMH), and preoptic hypothalamic nuclei (*PON*) [59–62] may control and modulate parasympathetic autonomic projections to the cranial vasculature that result in autonomic symptoms in TACs. Hypothalamic activation is likely important in triggering TACs and their symptoms, through its descending projections to the TCC and SuS, and the episodic and circadian nature of attacks is likely related to the internal control of biological rhythms, through hypothalamic nuclei. AH anterior hypothalamic area, DH dorsal hypothalamus, MB mammillary body

nonsteroidal anti-inflammatory drugs (NSAIDs) [83–86], and topiramate [87, 88], are effective in this assay. While nociceptive trigeminovascular activation is very relevant to the pathophysiology of TACs, this model perhaps lends itself more to understanding migraine, and screening for migraine therapeutics, rather than TACs. A further limitation is that without a measure of autonomic symptoms, it is difficult to translate this animal model to wider aspects of TACs pathophysiology.

3.5.2 Trigeminal Distribution of Pain: Nitrergic Activation of Trigeminovascular Pain Pathways

Nitric oxide donors, such as NTG, are known to provoke cluster headache in patients [89, 90], as well as cause the release of CGRP from the extracerebral vasculature during an attack phase [90]. NTG-provoked cluster headache physiologically also resembles spontaneous cluster headache with craniovascular vasodilation and neuronal activation in the brainstem, thalamus, and cortical structures [91]. In preclinical studies, Nitric oxide donors cause craniovascular vasodilation [92, 93] and activation and sensitization of central trigeminovascular neurons [94–97], with increased immunoreactivity for CGRP in the trigeminal ganglion [98] and depletion of CGRP stores in the trigeminal nucleus caudalis [99]. Furthermore, some of these nitrergic responses are inhibited by triptans [92, 100]; NSAIDs, specifically indomethacin [92, 101, 102]; topiramate [87]; and also CGRP receptor antagonists [103, 104]. Perhaps a disadvantage to this assay is that when cluster headache patients are in remission, NTG does not trigger a cluster attack or cause the release of CGRP or produce hypothalamic activation [46, 90, 91], and we would assume animals are in a "naïve" state and thus not suffering from cluster headache or any other TAC. Furthermore, NTG is also known to trigger migraine [105] and TTH [12] in sufferers of these primary headache types, and the craniovascular and trigeminovascular neuronal changes that NTG produces in animals are not specific to TACs. It seems NTG in animal models is useful in helping to understand the craniovascular and trigeminovascular neuronal changes that take place in primary headaches, but it is more difficult to generalize these changes to one specific primary headache. Further validation with treatments specific to TACs, such as oxygen ventilation, may help dissect the specificity of nitrergic activation in animal models to TACs.

3.5.3 Trigeminal Distribution of Pain with Cranial Autonomic Features: Oral or Nasal Capsaicin Injection

Perhaps what is more relevant to a specific animal model of TACs is to demonstrate more than one symptom, such as trigeminal distribution of pain and cranial autonomic symptoms. Two animal models have been developed, which use differing approaches. Similar to the capsaicin injections in the forehead of patients, demonstrating trigeminally mediated pain, an animal approach has used oral or nasal capsaicin injection in order to activate both trigeminal and autonomic systems [106]. In this model, blood flow changes in dural arteries were measured as a response to trigeminovascular activation and lacrimation measured by placement of filter paper to the medial angle of the eye and the change in weight used as a measure of lacrimation and activation of the autonomic pathway. Oral and intranasal capsaicin injection caused increases in dural and cortical blood flow as well as lacrimation [106]. These responses were reversed by systemic injection of hexamethonium bromide, an autonomic ganglion blocker. The implication is that oral or intranasal capsaicin causes activation of the trigeminal autonomic reflex, probably the parasympathetic projection, to produce cranial vascular changes and lacrimation. While this model does demonstrate several symptoms of TACs, there are reservations to its clinical relevance. Firstly, clinical studies with capsaicin produced craniovascular changes but did not produce hypothalamic activation known to characterize TACs, implying the response merely demonstrates trigeminally mediated pain. Secondly, the vascular changes are not specific to TACs, but generically trigeminally mediated neurovascular activation and pain [91], and these vascular changes are likely an epiphenomenon to a neurally mediated trigeminovascular activation. Thirdly, the model has not been validated with TACs specific treatments such as oxygen or indomethacin. Perhaps this model is a good example of trigeminal autonomic activation, but TACs probably require a central component, given the hypothalamic activation present during attacks of these headaches.

3.5.4 *Trigeminal Distribution of Pain with Cranial Autonomic Features: Superior Salivatory Nucleus Stimulation*

A second animal model that measures both trigeminal distribution of pain and cranial autonomic symptoms uses superior salivatory nucleus (SuS) stimulation [77, 83]. In this model, dural meningeal artery vasodilation and neuronal firing in the trigeminocervical complex were used to measure trigeminal distribution of pain and changes in blood flow in the lacrimal gland/duct as a measure of cranial autonomic activation. This model represents an approach where a primary activation in the brain is the cause of activation of nociceptive and parasympathetic pathways, which result in TAC symptoms. Using intravital microscopy of the dural vasculature SuS stimulation caused a modest (3.3 %) but significant increase in meningeal diameter [77]. Dural stimulation causes significant vasodilation mediated by CGRP release [107]. However, activation of the cranial parasympathetic projection is thought to cause the release of VIP [108], and exogenous VIP is known to be a much less potent vasodilator of the meninges [109] than exogenous CGRP, which may explain this moderate response.

Electrophysiological methods were used to measure single-unit central trigeminovascular neuronal activity in the TCC in response to SuS stimulation. Neurons were first characterized as having cutaneous facial receptive fields in the ophthalmic division of the trigeminal nerve. Two distinct populations of neurons were determined after SuS stimulation: those with short latency of action (between 3 and 20 ms, average 12.1 ms) and those with a much longer latency of action (7–40 ms, average 20.4 ms). When 100 % inhaled oxygen was used to characterize the response as being specific to TACs, only the longer latency response was inhibited; the shorter latency response was unaffected [77, 83]. Further characterization with the autonomic ganglion blocker, hexamethonium bromide, determined that again only the longer latency responses were inhibited. These data imply that the longer latency neuronal response is mediated by activation of the parasympathetic outflow to the cranial vasculature and that the locus of action of oxygen is likely via this pathway. The shorter latency response is most likely via antidromic activation of the trigeminal autonomic reflex. To validate the longer latency response as a specific model of TACs, several therapeutics were used that are beneficial in TACs compared to those also beneficial in migraine. Firstly a triptan was significantly more efficacious compared to a CGRP receptor antagonist, and secondly the cyclooxygenase (COX) inhibitor, indomethacin, was also significantly more efficacious compared to another NSAID, naproxen [83]. These data validate the specificity of the model to TAC treatments and highlight their mechanism of action which may be via the parasympathetic projection.

The autonomic response was determined by measurement of blood flow changes around the lacrimal gland/duct. SuS stimulation caused characteristic changes in flow that were reproducible over 30 min, indicative of an autonomic response [77]. Both 100 % inhaled oxygen and hexamethonium bromide significantly inhibited the responses [77, 83], indicating this autonomic response is likely mediated by activation of the parasympathetic outflow to the cranial vasculature. Furthermore, both a triptan and indomethacin also inhibited these responses, whereas the CGRP receptor antagonist and naproxen had no effect. These data validate the autonomic changes in this model as similar to those during TACs. Overall this model of TACs seems to represent most closely the known pathophysiology of TACs with trigeminally mediated pain and autonomic symptoms as well as responsiveness to treatments. It also uses a central site of origin for the initiation of the symptoms. At present its one failure might be that there is currently no evidence of hypothalamic activation; however, given that it is known the SuS receives direct projections from the PVN, it offers the possibility that a PVN-mediated effect may drive these changes or at the least SuS stimulation would activate PVN nuclei. The episodic and seasonal nature of attacks is difficult to demonstrate in an animal model, but perhaps this model comes closest to matching the generalized pathophysiology of TACs.

3.6 Conclusion

Primary headaches share many similarities, primarily trigeminovascular activation, which means that developing animal models for different primary headaches will always have some overlap, and share an element of common anatomy and pathophysiology. While migraine is the most studied of all primary headaches, from both a clinical and preclinical perspective, there have still been advances in our understanding of the pathophysiology of tension-type headache and the trigeminal autonomic cephalalgias, through a combination of clinical studies and preclinical animal models. Animals models of other primary headaches such as TTH and TACs have tended to focus on their defining symptoms. In the case of TTH, this is the pain which seems to emanate from the visceral or myofascial tissue of the neck, and so animal models have relied upon injection of noxious substances into the neck muscles. Likewise, TACs are defined by a combination of lateralized headache and autonomic symptoms, with a likely central site of origin, and so animal models have concentrated on the trigeminal autonomic reflex as a mediator of symptoms. While the current animal models are far from ideal, particularly for TTH, they do represent what we understand of their pathophysiology and symptomatology. As we understand more about the pathophysiology of these complex headache disorders, it is hoped that the development of new animal models and adaptations to the existing models will become more subtle and will therefore represent single primary headache disorders and generalize less to simply headache *per se*.

References

1. Headache Classification Committee of the International Headache Society (2013) The International Classification of Headache Disorders, 3rd edition (beta version). Cephalalgia 33(9):629–808
2. Lanteri-Minet M, Duru G, Mudge M, Cottrell S (2011) Quality of life impairment, disability and economic burden associated with chronic daily headache, focusing on chronic migraine with or without medication overuse: a systematic review. Cephalalgia 31(7):837–850
3. Raggi A, Giovannetti AM, Quintas R, D'Amico D, Cieza A, Sabariego C et al (2012) A systematic review of the psychosocial difficulties relevant to patients with migraine. J Headache Pain 13(8):595–606
4. Colson NJ, Lea RA, Quinlan S, Griffiths LR (2006) No role for estrogen receptor 1 gene intron 1 Pvu II and exon 4 C325G polymorphisms in migraine susceptibility. BMC Med Genet 7:12
5. Maher BH, Griffiths LR (2011) Identification of molecular genetic factors that influence migraine. Mol Genet Genomics 285(6):433–446
6. Buzzi MG, Bonamini M, Cerbo R (1993) The anatomy and biochemistry of headache. Funct Neurol 8(6):395–402
7. Buzzi MG, Bonamini M, Moskowitz MA (1995) Neurogenic model of migraine. Cephalalgia 15(4):277–280
8. Rasmussen BK, Jensen R, Olesen J (1991) Questionnaire versus clinical interview in the diagnosis of headache. Headache 31(5):290–295
9. Rasmussen BK, Jensen R, Olesen J (1992) Impact of headache on sickness absence and utilisation of medical services: a Danish population study. J Epidemiol Community Health 46(4):443–446
10. Pikoff H (1984) Is the muscular model of headache still viable? A review of conflicting data. Headache 24(2):186–198
11. Langemark M, Bach FW, Jensen TS, Olesen J (1993) Decreased nociceptive flexion reflex threshold in chronic tension-type headache. Arch Neurol 50(10):1061–1064

12. Ashina M, Bendtsen L, Jensen R, Olesen J (2000) Nitric oxide-induced headache in patients with chronic tension-type headache. Brain 123(Pt 9):1830–1837
13. Ashina M, Bendtsen L, Jensen R, Sakai F, Olesen J (2000) Possible mechanisms of glyceryl-trinitrate-induced immediate headache in patients with chronic tension-type headache. Cephalalgia 20(10):919–924
14. Ashina M, Simonsen H, Bendtsen L, Jensen R, Olesen J (2004) Glyceryl trinitrate may trigger endogenous nitric oxide production in patients with chronic tension-type headache. Cephalalgia 24(11):967–972
15. Jensen R (1999) Pathophysiological mechanisms of tension-type headache: a review of epidemiological and experimental studies. Cephalalgia 19(6):602–621
16. Fumal A, Schoenen J (2008) Tension-type headache: current research and clinical management. Lancet Neurol 7(1):70–83
17. Wang W, Schoenen J (1994) Reduction of temporalis exteroceptive suppression by peripheral electrical stimulation in migraine and tension-type headaches. Pain 59(3):327–334
18. Schoenen J, Gerard P, De Pasqua V, Sianard-Gainko J (1991) Multiple clinical and paraclinical analyses of chronic tension-type headache associated or unassociated with disorder of pericranial muscles. Cephalalgia 11(3):135–139
19. Torebjork HE, Lundberg LE, LaMotte RH (1992) Central changes in processing of mechanoreceptive input in capsaicin-induced secondary hyperalgesia in humans. J Physiol 448: 765–780
20. Torebjork HE, LaMotte RH, Robinson CJ (1984) Peripheral neural correlates of magnitude of cutaneous pain and hyperalgesia: simultaneous recordings in humans of sensory judgments of pain and evoked responses in nociceptors with C-fibers. J Neurophysiol 51(2): 325–339
21. Owens CM, Zhang D, Willis WD (1992) Changes in the response states of primate spinothalamic tract cells caused by mechanical damage of the skin or activation of descending controls. J Neurophysiol 67(6):1509–1527
22. Hoheisel U, Mense S (1989) Long-term changes in discharge behaviour of cat dorsal horn neurones following noxious stimulation of deep tissues. Pain 36(2):239–247
23. Yu XM, Mense S (1990) Response properties and descending control of rat dorsal horn neurons with deep receptive fields. Neuroscience 39(3):823–831
24. Wall PD, Woolf CJ (1984) Muscle but not cutaneous C-afferent input produces prolonged increases in the excitability of the flexion reflex in the rat. J Physiol 356:443–458
25. Woolf CJ (1996) Windup and central sensitization are not equivalent. Pain 66(2–3):105–108
26. Treede RD, Meyer RA, Raja SN, Campbell JN (1992) Peripheral and central mechanisms of cutaneous hyperalgesia. Prog Neurobiol 38(4):397–421
27. Mense S (1993) Spinal mechanisms of muscle pain and hyperalgesia. In: Vecchiet L, Albe-Fessard D, Lindblom U, Giamberadino MA (eds) New trends in referred pain and hyperalgesia. Elsevier, Amsterdam, pp 25–34
28. Hoheisel U, Koch K, Mense S (1994) Functional reorganization in the rat dorsal horn during an experimental myositis. Pain 59(1):111–118
29. McHaffie JG, Larson MA, Stein BE (1994) Response properties of nociceptive and low-threshold neurons in rat trigeminal pars caudalis. J Comp Neurol 347(3):409–425
30. Rasmussen BK (2001) Epidemiology of headache. Cephalalgia 21(7):774–777
31. Bendtsen L (2000) Central sensitization in tension-type headache – possible pathophysiological mechanisms. Cephalalgia 20(5):486–508
32. Ellrich J, Fischer A, Gilsbach JM, Makowska A, Spangenberg P (2010) Inhibition of nitric oxide synthases prevents and reverses alpha, beta-meATP-induced neck muscle nociception in mice. Cephalalgia 30(10):1225–1232
33. Ellrich J, Makowska A (2007) Nerve growth factor and ATP excite different neck muscle nociceptors in anaesthetized mice. Cephalalgia 27(11):1226–1235
34. Makowska A, Panfil C, Ellrich J (2005) Long-term potentiation of orofacial sensorimotor processing by noxious input from the semispinal neck muscle in mice. Cephalalgia 25(2): 109–116

35. Makowska A, Panfil C, Ellrich J (2005) Nerve growth factor injection into semispinal neck muscle evokes sustained facilitation of the jaw-opening reflex in anesthetized mice – possible implications for tension-type headache. Exp Neurol 191(2):301–309
36. Makowska A, Panfil C, Ellrich J (2006) ATP induces sustained facilitation of craniofacial nociception through P2X receptors on neck muscle nociceptors in mice. Cephalalgia 26(6):697–706
37. Ristic D, Spangenberg P, Ellrich J (2011) Acetylsalicylic acid inhibits alpha, beta-meATP-induced facilitation of neck muscle nociception in mice – implications for acute treatment of tension-type headache. Eur J Pharmacol 673(1–3):13–19
38. Chao MV (2003) Neurotrophins and their receptors: a convergence point for many signalling pathways. Nat Rev Neurosci 4(4):299–309
39. Goadsby PJ, Lipton RB (1997) A review of paroxysmal hemicranias, SUNCT syndrome and other short-lasting headaches with autonomic feature, including new cases. Brain 120 (Pt 1):193–209
40. Irimia P, Cittadini E, Paemeleire K, Cohen AS, Goadsby PJ (2008) Unilateral photophobia or phonophobia in migraine compared with trigeminal autonomic cephalalgias. Cephalalgia 28(6):626–630
41. Lai T-H, Fuh J-L, Wang S-J (2009) Cranial autonomic symptoms in migraine: characteristics and comparison with cluster headache. J Neurol Neurosurg Psychiatry 80:1116–1119
42. Goadsby PJ (2002) Pathophysiology of cluster headache: a trigeminal autonomic cephalalgia. Lancet Neurol 1(4):251–257
43. Leone M, Bussone G (2009) Pathophysiology of trigeminal autonomic cephalalgias. Lancet Neurol 8(8):755–764
44. Cohen AS, Matharu MS, Goadsby PJ (2007) Trigeminal autonomic cephalalgias: current and future treatments. Headache 47(6):969–980
45. May A (2006) Update on the diagnosis and management of trigemino-autonomic headaches. J Neurol 253(12):1525–1532
46. May A, Bahra A, Buchel C, Frackowiak RS, Goadsby PJ (1998) Hypothalamic activation in cluster headache attacks. Lancet 352(9124):275–278
47. Matharu MS, Cohen AS, Frackowiak RS, Goadsby PJ (2006) Posterior hypothalamic activation in paroxysmal hemicrania. Ann Neurol 59(3):535–545
48. May A, Bahra A, Buchel C, Turner R, Goadsby PJ (1999) Functional magnetic resonance imaging in spontaneous attacks of SUNCT: short-lasting neuralgiform headache with conjunctival injection and tearing. Ann Neurol 46(5):791–794
49. Sprenger T, Valet M, Platzer S, Pfaffenrath V, Steude U, Tolle TR (2005) SUNCT: bilateral hypothalamic activation during headache attacks and resolving of symptoms after trigeminal decompression. Pain 113(3):422–426
50. Matharu MS, Cohen AS, McGonigle DJ, Ward N, Frackowiak RSJ, Goadsby PJ (2004) Posterior hypothalamic and brainstem activation in hemicrania continua. Headache 44:747–761
51. Leone M, Proietti Cecchini A, Franzini A, Broggi G, Cortelli P, Montagna P et al (2008) Lessons from 8 years' experience of hypothalamic stimulation in cluster headache. Cephalalgia 28(7):787–797; discussion 98
52. Leone M, Franzini A, Bussone G (2001) Stereotactic stimulation of posterior hypothalamic gray matter in a patient with intractable cluster headache. N Engl J Med 345(19): 1428–1429
53. Goadsby PJ, Edvinsson L (1994) Human in vivo evidence for trigeminovascular activation in cluster headache. Neuropeptide changes and effects of acute attacks therapies. Brain 117(Pt 3): 427–434
54. Goadsby PJ, Edvinsson L (1996) Neuropeptide changes in a case of chronic paroxysmal hemicrania – evidence for trigemino-parasympathetic activation. Cephalalgia 16(6): 448–450
55. May A, Kaube H, Buchel C, Eichten C, Rijntjes M, Juptner M et al (1998) Experimental cranial pain elicited by capsaicin: a PET study. Pain 74(1):61–66

56. Bernstein C, Burstein R (2012) Sensitization of the trigeminovascular pathway: perspective and implications to migraine pathophysiology. J Clin Neurol 8(2):89–99
57. Goadsby PJ, Lipton RB, Ferrari MD (2002) Migraine – current understanding and treatment. N Engl J Med 346(4):257–270
58. Akerman S, Holland PR, Goadsby PJ (2011) Diencephalic and brainstem mechanisms in migraine. Nat Rev Neurosci 12(10):570–584
59. Spencer SE, Sawyer WB, Wada H, Platt KB, Loewy AD (1990) CNS projections to the pterygopalatine parasympathetic preganglionic neurons in the rat: a retrograde transneuronal viral cell body labeling study. Brain Res 534(1–2):149–169
60. Robert C, Bourgeais L, Arreto CD, Condes-Lara M, Noseda R, Jay T et al (2013) Paraventricular hypothalamic regulation of trigeminovascular mechanisms involved in headaches. J Neurosci 33(20):8827–8840
61. Hosoya Y, Matsushita M, Sugiura Y (1983) A direct hypothalamic projection to the superior salivatory nucleus neurons in the rat. A study using anterograde autoradiographic and retrograde HRP methods. Brain Res 266(2):329–333
62. Hosoya Y, Sugiura Y, Ito R, Kohno K (1990) Descending projections from the hypothalamic paraventricular nucleus to the A5 area, including the superior salivatory nucleus, in the rat. Exp Brain Res 82(3):513–518
63. Benjamin L, Levy MJ, Lasalandra MP, Knight YE, Akerman S, Classey JD et al (2004) Hypothalamic activation after stimulation of the superior sagittal sinus in the cat: a Fos study. Neurobiol Dis 16(3):500–505
64. Malick A, Jakubowski M, Elmquist JK, Saper CB, Burstein R (2001) A neurohistochemical blueprint for pain-induced loss of appetite. Proc Natl Acad Sci U S A 98(17):9930–9935
65. Knight YE, Classey JD, Lasalandra MP, Akerman S, Kowacs F, Hoskin KL et al (2005) Patterns of fos expression in the rostral medulla and caudal pons evoked by noxious craniovascular stimulation and periaqueductal gray stimulation in the cat. Brain Res 1045(1–2):1–11
66. Gray H (1918) Anatomy of the human body. Lea and Febiger, Philadelphia
67. Bartsch T, Levy MJ, Knight YE, Goadsby PJ (2004) Differential modulation of nociceptive dural input to [hypocretin] orexin A and B receptor activation in the posterior hypothalamic area. Pain 109(3):367–378
68. Bartsch T, Levy MJ, Knight YE, Goadsby PJ (2005) Inhibition of nociceptive dural input in the trigeminal nucleus caudalis by somatostatin receptor blockade in the posterior hypothalamus. Pain 117(1–2):30–39
69. Akerman S, Williamson DJ, Kaube H, Goadsby PJ (2002) Nitric oxide synthase inhibitors can antagonize neurogenic and calcitonin gene-related peptide induced dilation of dural meningeal vessels. Br J Pharmacol 137(1):62–68
70. Goadsby PJ, Zagami AS (1991) Stimulation of the superior sagittal sinus increases metabolic activity and blood flow in certain regions of the brainstem and upper cervical spinal cord of the cat. Brain 114(Pt 2):1001–1011
71. Kaube H, Keay KA, Hoskin KL, Bandler R, Goadsby PJ (1993) Expression of c-Fos-like immunoreactivity in the caudal medulla and upper cervical spinal cord following stimulation of the superior sagittal sinus in the cat. Brain Res 629(1):95–102
72. Burstein R, Yamamura H, Malick A, Strassman AM (1998) Chemical stimulation of the intracranial dura induces enhanced responses to facial stimulation in brain stem trigeminal neurons. J Neurophysiol 79(2):964–982
73. Hoskin KL, Bulmer DCE, Lasalandra M, Jonkman A, Goadsby PJ (2001) Fos expression in the midbrain periaqueductal grey after trigeminovascular stimulation. J Anat 198:29–35
74. Burstein R, Jakubowski M, Garcia-Nicas E, Kainz V, Bajwa Z, Hargreaves R et al (2010) Thalamic sensitization transforms localized pain into widespread allodynia. Ann Neurol 68(1):81–91
75. Zagami AS, Lambert GA (1990) Stimulation of cranial vessels excites nociceptive neurones in several thalamic nuclei of the cat. Exp Brain Res 81(3):552–566
76. Zagami AS, Goadsby PJ, Edvinsson L (1990) Stimulation of the superior sagittal sinus in the cat causes release of vasoactive peptides. Neuropeptides 16(2):69–75

77. Akerman S, Holland PR, Lasalandra MP, Goadsby PJ (2009) Oxygen inhibits neuronal activation in the trigeminocervical complex after stimulation of trigeminal autonomic reflex, but not during direct dural activation of trigeminal afferents. Headache 49(8):1131–1143
78. Burstein R, Jakubowski M (2004) Analgesic triptan action in an animal model of intracranial pain: a race against the development of central sensitization. Ann Neurol 55(1):27–36
79. Goadsby PJ, Hoskin KL (1996) Inhibition of trigeminal neurons by intravenous administration of the serotonin (5HT)1B/D receptor agonist zolmitriptan (311C90): are brain stem sites therapeutic target in migraine? Pain 67(2–3):355–359
80. Goadsby PJ, Knight YE (1997) Inhibition of trigeminal neurones after intravenous administration of naratriptan through an action at 5-hydroxy-tryptamine (5-HT(1B/1D)) receptors. Br J Pharmacol 122(5):918–922
81. Hoskin KL, Kaube H, Goadsby PJ (1996) Sumatriptan can inhibit trigeminal afferents by an exclusively neural mechanism. Brain 119(Pt 5):1419–1428
82. Levy D, Jakubowski M, Burstein R (2004) Disruption of communication between peripheral and central trigeminovascular neurons mediates the antimigraine action of 5HT 1B/1D receptor agonists. Proc Natl Acad Sci U S A 101(12):4274–4279
83. Akerman S, Holland PR, Summ O, Lasalandra MP, Goadsby PJ (2012) A translational in vivo model of trigeminal autonomic cephalalgias: therapeutic characterization. Brain 135(Pt 12):3664–3675
84. Kaube H, Hoskin KL, Goadsby PJ (1993) Intravenous acetylsalicylic acid inhibits central trigeminal neurons in the dorsal horn of the upper cervical spinal cord in the cat. Headache 33(10):541–544
85. Jakubowski M, Levy D, Goor-Aryeh I, Collins B, Bajwa Z, Burstein R (2005) Terminating migraine with allodynia and ongoing central sensitization using parenteral administration of COX1/COX2 inhibitors. Headache 45(7):850–861
86. Jakubowski M, Levy D, Kainz V, Zhang XC, Kosaras B, Burstein R (2007) Sensitization of central trigeminovascular neurons: blockade by intravenous naproxen infusion. Neuroscience 148(2):573–583
87. Akerman S, Goadsby PJ (2005) Topiramate inhibits trigeminovascular activation: an intravital microscopy study. Br J Pharmacol 146(1):7–14
88. Storer RJ, Goadsby PJ (2004) Topiramate inhibits trigeminovascular neurons in the cat. Cephalalgia 24(12):1049–1056
89. Ekbom K (1968) Nitrolglycerin as a provocative agent in cluster headache. Arch Neurol 19(5):487–493
90. Fanciullacci M, Alessandri M, Figini M, Geppetti P, Michelacci S (1995) Increase in plasma calcitonin gene-related peptide from the extracerebral circulation during nitroglycerin-induced cluster headache attack. Pain 60(2):119–123
91. May A, Bahra A, Buchel C, Frackowiak RS, Goadsby PJ (2000) PET and MRA findings in cluster headache and MRA in experimental pain. Neurology 55(9):1328–1335
92. Akerman S, Williamson DJ, Kaube H, Goadsby PJ (2002) The effect of anti-migraine compounds on nitric oxide-induced dilation of dural meningeal vessels. Eur J Pharmacol 452(2):223–228
93. Strecker T, Dux M, Messlinger K (2002) Increase in meningeal blood flow by nitric oxide – interaction with calcitonin gene-related peptide receptor and prostaglandin synthesis inhibition. Cephalalgia 22(3):233–241
94. Tassorelli C, Joseph SA (1995) Systemic nitroglycerin induces Fos immunoreactivity in brain-stem and forebrain structures of the rat. Brain Res 682(1–2):167–181
95. Koulchitsky S, Fischer MJ, De Col R, Schlechtweg PM, Messlinger K (2004) Biphasic response to nitric oxide of spinal trigeminal neurons with meningeal input in rat – possible implications for the pathophysiology of headaches. J Neurophysiol 92(3):1320–1328
96. Lambert GA, Donaldson C, Boers PM, Zagami AS (2000) Activation of trigeminovascular neurons by glyceryl trinitrate. Brain Res 887(1):203–210
97. Akerman S, Hoffmann J, Goadsby PJ (2013) A translational approach to studying triptan-induced reversal of established central sensitization of trigeminovascular neurons. Cephalalgia 33(8(S1)):211

98. Dieterle A, Fischer MJ, Link AS, Neuhuber WL, Messlinger K (2011) Increase in CGRP- and nNOS-immunoreactive neurons in the rat trigeminal ganglion after infusion of an NO donor. Cephalalgia 31(1):31–42
99. Pardutz A, Multon S, Malgrange B, Parducz A, Vecsei L, Schoenen J (2002) Effect of systemic nitroglycerin on CGRP and 5-HT afferents to rat caudal spinal trigeminal nucleus and its modulation by estrogen. Eur J Neurosci 15(11):1803–1809
100. Akerman S, Goadsby PJ (2014) Acute anti-migraine treatments abort established central sensitization of trigeminovascular neurons: validation of a novel translational approach. Headache 54(S1):2–3
101. Summ O, Andreou AP, Akerman S, Goadsby PJ (2010) A potential nitrergic mechanism of action for indomethacin, but not other COX inhibitors – relevance to indomethacin-sensitive headaches. J Headache Pain 11(6):477–483
102. Summ O, Andreou AP, Akerman S, Hoffmann J, Goadsby PJ (2011) Effects of indomethacin, naproxen and ibuprofen on no-induced trigeminal firing recorded in the trigeminocervical complex. Cephalalgia 31(S1):10
103. Akerman S, Kaube H, Goadsby PJ (2004) Anandamide is able to inhibit trigeminal neurons using an in vivo model of trigeminovascular-mediated nociception. J Pharmacol Exp Ther 309:56–63
104. Koulchitsky S, Fischer MJ, Messlinger K (2009) Calcitonin gene-related peptide receptor inhibition reduces neuronal activity induced by prolonged increase in nitric oxide in the rat spinal trigeminal nucleus. Cephalalgia 29(4):408–417
105. Iversen HK, Olesen J, Tfelt-hansen P (1989) Intravenous nitroglycerin as an experimental-model of vascular headache – basic characteristics. Pain 38(1):17–24
106. Gottselig R, Messlinger K (2004) Noxious chemical stimulation of rat facial mucosa increases intracranial blood flow through a trigemino-parasympathetic reflex – an experimental model for vascular dysfunctions in cluster headache. Cephalalgia 24(3):206–214
107. Williamson DJ, Hargreaves RJ, Hill RG, Shepheard SL (1997) Sumatriptan inhibits neurogenic vasodilation of dural blood vessels in the anaesthetized rat – intravital microscope studies. Cephalalgia 17(4):525–531
108. Goadsby PJ, MacDonald GJ (1985) Extracranial vasodilation mediated by vasoactive intestinal polypeptide (VIP). Brain Res 329(1–2):285–288
109. Boni LJ, Ploug KB, Olesen J, Jansen-Olesen I, Gupta S (2009) The in vivo effect of VIP, PACAP-38 and PACAP-27 and mRNA expression of their receptors in rat middle meningeal artery. Cephalalgia 29(8):837–847

Chapter 4
Genetics of Headache

Cherubino Di Lorenzo, Filippo M. Santorelli, and Arn M.J.M. van den Maagdenberg

4.1 Introduction

4.1.1 Migraine as a Disease

Migraine is an episodic brain disorder with disabling attacks of headache that are associated with nausea, vomiting, and hypersensitivity to light, sound, and smell. Clinically, migraine is divided into two main subtypes that are based on the absence (migraine without aura, MO) or presence (migraine with aura, MA) of an aura [1]. Aura symptoms typically have a duration between 5 and 60 min and almost always include visual symptoms but may also include sensory and aphasic symptoms. A diagnosis of migraine is made according to criteria of the International Classification of Headache Disorders (ICHD-II) from the International Headache Society (IHS) [1]. Overall, migraine has a variable prevalence worldwide with slight, but not significant, differences according to the race [2, 3]. There is a peak in the prevalence of migraine between age 20 and 50 years [4]. Women are affected three times more often than men [5]. Migraine has a profound effect on well-being and general functioning, not only during attacks, but also in terms of work performance, family and social relationships, and school achievement [6]. Consequently,

C. Di Lorenzo
Department of Molecular Medicine, Don Gnocchi Foundation, Milan, Italy

F.M. Santorelli (✉)
Departments of Child Neurology and Molecular Medicine, IRCCS Stella Maris, via dei Giacinti 2, 56128 Pisa, Italy
e-mail: f.santorelli@fsm.unipi.it

Arn M.J.M.van den Maagdenberg (✉)
Departments of Human Genetics and Neurology, Leiden University Medical Centre, PO Box 9600, Leiden 2300 RC, The Netherlands
e-mail: maagdenberg@lumc.nl

M. Ashina, P. Geppetti (eds.), *Pathophysiology of Headaches: From Molecule to Man*, Headache, DOI 10.1007/978-3-319-15621-7_4

the WHO expert panel rates migraine among the most disabling and costly chronic disorders [7]. The burden of migraine in many patients is even larger as they also suffer from comorbid disorders such as epilepsy, stroke, and depression [8].

4.1.2 Migraine Pathophysiology

The neurobiological mechanisms underlying migraine have been unraveled only to certain extent. It is commonly accepted that the migraine aura is caused by cortical spreading depression (CSD), a wave of neuronal and glial depolarization that moves slowly over the cortex [9]. Although CSD can be easily investigated in experimental animals, evidence that it occurs in humans is still scarce. Using functional magnetic resonance imaging (fMRI), Hadjikhani and colleagues [10] were able to detect local increases in blood-oxygen-level-dependent (BOLD) signals that spread through the visual cortex of a patient with MA at a rate (3.5 mm/min) similar to what is seen in experimentally induced CSD in animals. The headache itself is caused by activation of the trigeminovascular system that consists of the neurons innervating the cerebral vessels whose cell bodies are located in the trigeminal ganglion (reviewed in [11]). The ganglion contains bipolar cells with peripheral fibers making synaptic connections particularly with pain-producing large cranial vessels and dura mater and centrally projecting fibers synapsing on neurons in the caudal brain stem and high cervical cord. Trigeminal innervation predominantly is to forebrain but extends to posterior areas to the rostral basilar artery [11].

4.1.3 Migraine Is a Genetic Disease

Migraine has a strong genetic component as evidenced by observations that migraine runs in families and that about half of migraineurs have a first-degree relative also affected by a similar condition [12]. In addition to genetic determinants, migraine risk is also conferred by environmental factors. It is believed that their interplay has a major causal role. Notably, epidemiological evidences suggest a close gene-environment interaction (endogenous or exogenous), among which several predisposing or triggering factors of which only some can be avoided, such as gender [13]. In the case of menstrually related migraine, it is certainly plausible that the environment (the hormonal milieu) affects gene regulation, leading to a dysregulation of the nervous system and a subsequent migraine attack [14].

For genetic study designs, it is important whether MO and MA should be seen as different disease entities or if they represent different expressions of the same disease. Although there is epidemiological support for the first [15–17], most recent studies seem to suggest that pure MO to pure MA are at both ends of a clinical spectrum [18–20]. Clinical observations support this view as both subtypes share identical headache symptoms and frequently co-occur in an individual. Genetic studies are expected to shed light on this debate in the near future by showing if migraine susceptibility genes are shared by both migraine types.

The aim of this Chapter is to discuss current molecular genetic findings primarily in migraine, since most studies have addressed this headache type. Genetic research in other headache types lags behind. We will therefore mainly describe genetic findings in rare monogenic forms of migraine, such as familial hemiplegic migraine (FHM) and findings in common forms of migraine, particularly information gathered in recent years from genome-wide association studies. In addition, we will discuss the advent of pharmacogenomics studies in migraine. Finally, we will provide a forward look on how the innovative technical methodology of next-generation sequencing that has recently become feasible will shape future research in the field of migraine.

4.2 Genes and Pathways in Monogenic Forms of Migraine

4.2.1 Gene Discovery in Familial Hemiplegic Migraine (FHM)

FHM is a rare monogenic subtype of migraine with aura, characterized by at least some degree of weakness (hemiparesis) during the aura [1]. The hemiparesis may last from minutes to several hours or days. Apart from the hemiparesis, the headache and aura features of the FHM attack are identical to those of attacks of common forms of migraine [21]. In addition to hemiplegic attacks, the majority of FHM patients also experience attacks of "regular" migraine with or without aura [22]. Thus, from a clinical point of view, FHM seems part of the migraine spectrum and a valid model to study the common forms of migraine.

The first FHM gene (FHM1), *CACNA1A*, is located on chromosome 19p13 and codes for a subunit of neuronal voltage-gated $Ca_V2.1$ Ca^{2+} channels [23]. These channels modulate the release of neurotransmitters at most central synapses throughout the brain [24]. Clinical variability associated with FHM1 mutations can range from pure FHM to FHM with cerebellar ataxia, epilepsy, or even fatal coma due to excessive cerebral edema (for review, see [25]). Hemiplegia can also be associated with fever, drowsiness, or confusion that usually resolves within hours, days, or sometimes weeks [26].

The second FHM gene (FHM2), *ATP1A2*, is located on chromosome 1q23 and encodes the α2 subunit of a Na^+/K^+ pump ATPase [27]. This catalytic subunit binds Na^+, K^+, and ATP and utilizes ATP hydrolysis to exchange Na^+ ions out of the cell for K^+ ions into the cell. Thus, the ATPase is able to modulate the reuptake of potassium and glutamate from the synaptic cleft into the glial cell. FHM2 mutations have been associated with a pure disease [27–29] or a combination of FHM with cerebellar ataxia [30], recurrent comas, aphasia, behavioral changes [31], or impaired hearing, and vertigo [32].

The third FHM gene (FHM3), *SCN1A*, is located on chromosome 2q24 and encodes a subunit of voltage-gated $Na_V1.1$ Na^+ channels [33] that play an important role in the generation and propagation of action potentials. FHM3 mutations sometimes present with additional clinical features such as childhood epilepsy and transient blindness [33–35].

PRRT2, a gene located on chromosome 16p11, was recently put forward as the fourth hemiplegic migraine gene (FHM4) because truncating deletions were identified in few patients with symptoms of hemiplegic migraine [36]. A role of *PRRT2* in migraine is supported by the notion that the gene encodes a proline-rich transmembrane protein that interacts with SNAP25, a component of the SNARE protein complex that is involved in controlling the release of neurotransmitters. However, similar *PRRT2* deletions are also detected in hundreds of patients that do not report migraine, including cases with a clinical diagnosis of paroxysmal kinesigenic dyskinesia, benign familial infantile convulsions, and infantile convulsion choreoathetosis. This suggests that the relation of *PRRT2* and migraine remains complicated and require further investigations [37].

4.2.2 Functional Consequences of Familial Hemiplegic Migraine (FHM) Mutations

Functional consequences of FHM mutations have been intensively investigating in cellular models. Most studies of FHM1 mutations revealed a gain of neuronal $Ca_V2.1$ channel function caused by a shift in the voltage dependence toward more negative membrane potentials, while the channel open probability was increased [38]. Loss of glial Na^+/K^+ ATPase function was typically reported for FHM2 mutations [39]. Most FHM3 mutations seem to exert loss-of-function effects on $Na_V1.1$ sodium channels, likely on inhibitory neurons, but gain-of-function effects have also been proposed [25]. Taken together, the cellular studies predict increased neurotransmitter and potassium ion levels at the synaptic cleft, especially after high-intensity neuronal firing, which would facilitate CSD and thus could explain the occurrence of severe auras in FHM patients.

Knockin (KI) mouse models have been instrumental in unraveling in vivo consequences of FHM mutations in the intact animal. Two FHM1 KI models have been generated (harboring the R192Q or S218L missense mutations in the α1 $Ca_V2.1$ channel protein, respectively) [40, 41], but only FHM1 S218L mice exhibit susceptibility to seizures, cerebellar ataxia, and delayed cerebral edema after minor head trauma, which fits the clinical phenotype seen in human carriers of the *CACNA1A* S218L mutation [41, 42]. FHM1 R192Q mice show no overt phenotype [40] although signs of spontaneous pain that seem stress induced have been reported [43]. It is believed that an increased neuronal Ca^{2+} influx and neurotransmitter release are at the basis of these phenotypes [40, 41], a condition that also explains the increased susceptibility to experimentally induced CSD [40, 41, 44, 45]. An increased susceptibility to experimental CSD was also reported in heterozygous FHM2 KI that harbors the W887R mutation [46]. Homozygous mice die soon after birth, as seen also in homozygous *Atp1a2* knockout mice [47]. No FHM3 KI mice

Table 4.1 A short list of syndromic monogenic clinical conditions presenting with migraine headache

Syndromic migraine	Gene (chromosome) involved	Migraine features
Familial hemiplegic migraine (FHM) [25]	*CACNA1A* (19p13); *ATP1A2* (1q23); *SCN1A* (2q24)	Hemiplegic aura, increased prevalence of MA/MO
Mitochondrial encephalomyopathy, lactic acidosis, stroke-like episodes (MELAS) [49]	*MTTL1* (mtDNA)	Recurrent MA, focal neurological deficits, vomiting, convulsions
Cerebral autosomal dominant arteriopathy with subcortical infarcts and leukoencephalopathy (CADASIL) [50]	*NOTCH3* (19q13)	MA in about one third of mutation carriers
Retinal vasculopathy with cerebral leukodystrophy (RVCL) [51]	*TREX1* (3p21)	Increased prevalence of MO in mutation carriers

MA migraine with aura, *MO* migraine without aura

have been generated, yet, but a mouse carrying a mutation in *SCN1A* leading to haploinsufficiency has been generated as model of childhood epileptic encephalopathy [48].

4.2.3 Genetic Findings in Other Syndromic Forms of Migraine

Additional "syndromic" forms of migraine (see Table 4.1) are all characterized by headache attacks in addition to other neurological features that have either an autosomal or mitochondrial inheritance pattern. Three will be discussed here.

Cerebral autosomal dominant arteriopathy with subcortical infarcts and leukoencephalopathy (CADASIL) is a brain microangiopathy characterized by stroke-like episodes, cognitive decline, aura features (in about 35 % of cases), psychiatric disorders, and epilepsy [50]. Other symptoms are reversible acute encephalopathy [52], subclinical peripheral neuropathy [53], and subclinical retinal vascular abnormalities [54]. The CADASIL gene, *NOTCH3*, is located on chromosome 19p13 [55] and plays an important role in vascular smooth muscle cells of small blood vessels of the brain [56]. Notably, transgenic mice in which a CADASIL mutation was overexpressed showed a reduced threshold for CSD [57].

Retinal vasculopathy with cerebral leukodystrophy (RVCL) is a neurovascular syndrome caused by carboxyl terminal mutations in *TREX1* that lead to a subcellular mislocalization of Trex1 protein [51]. Trex1 is the major 3′–5′ mammalian exonuclease, an enzyme thought to be involved in DNA repair and apoptosis after DNA damage, but likely has additional functions such as clearance of cytosolic DNA. RVCL is characterized by a pronounced retinopathy but often also cognitive disturbances, focal neurological symptoms, kidney and liver dysfunction from Raynaud's phenomenon, and migraine [58–60]. Particularly, in a large Dutch RVCL family, a high prevalence of migraine was found, with 14 out of the 20 *TREX1* mutation carriers suffering from MO [60].

Mitochondrial encephalomyopathy, lactic acidosis, and stroke-like episodes (MELAS) syndrome is characterized by seizures, hemiparesis, hemianopia, cortical blindness, migraine, and episodic vomiting [49, 61]. Frequent systemic manifestations include cardiac, renal, endocrine, and gastrointestinal disorders [62]. Several mtDNA mutations have been detected in patients with MELAS of which the A-to-G transition at nucleotide position 3243 in the mitochondrial DNA (mtDNA) is the most common one [62]. Less commonly mtDNA mutations, such as the one at nt position 8344, are associated with myoclonic epilepsy associated with ragged-red fibers (MERRF) and often features such as MA [63, 64]. Vascular system involvement has been implicated for both syndromic conditions and decreased oxidative brain metabolism with ensuing excessive accumulation of reactive oxygen species (ROS) that likely plays a role in CSD initiation and appears to be relevant in the pathogenesis of migraine [65].

4.2.4 Genetic Finding in Familial Migraine

Direct sequencing of a large number of ion transporter genes in over 100 migraine patients revealed a single truncating nonfunctional F139WfsX24 mutation in the *KCNK18* gene. This gene was put forward as a potential etiology in familial migraine as it segregated as monogenic migraine trait [66]. The gene *KCNK18* encodes the TRESK protein, a member of the two-pore domain family of potassium channels, involved in neuronal excitability control. TRESK is now regarded as genetic modifier of a migraine phenotype, not the direct cause, because several rare variants in TRESK, including a change that showed complete loss of function, were identified in controls [67]. Apparently, nonfunctional TRESK copies are tolerated and do not lead to disease. Regardless, TRESK remains an interesting possible migraine target, already because of its role in neuronal excitability.

4.3 Genetic Discoveries in Common Migraine: The Era of Genome-Wide Association Studies

Since a few years, genome-wide association studies (GWAS) are considered the preferred methodology to detect genetic susceptibility loci for common disorders, including migraine. GWAS entail the simultaneous investigation of several hundred thousand single nucleotide polymorphisms (SNP), which are variants in the DNA sequence with a known genomic position, using microarray-based technology and large cohorts of migraine cases and controls. “Associated variants” show a difference in allele frequency between cases and controls for a specific SNP with a level of significance that survives correction for the many performed statistical tests; genome-wide significance in GWAS means that the *p*-value for association should be below 5×10^{-8}. GWA studies produce statistically robust findings, but the associated variants almost without exception have a small effect size (e.g., relative risk <1.2).

The current status of GWAS in migraine is best illustrated by the meta-analysis of large datasets from earlier migraine GWAS [68–70] that combined data of over 23,000 cases and 100,000 controls from 29 cohorts that was performed by the International Headache Genetics Consortium [71]. In total, 13 migraine susceptibility loci were identified for various migraine types (Table 4.2). The genes that were assigned to these loci seem to affect neuronal pathways (including glutamatergic neurotransmission and axon guidance pathways), metalloproteinases, and vascular pathways. Among the more interesting variants are SNP *rs1835740* that is located between the genes *MTDH/AEG1* and *PGCP* on chromosome 8q22 that are both involved in glutamate homeostasis and SNP *rs11172113* in the low-density lipoprotein receptor-related protein 1 (*LRP1*) gene that plays a role in neurotransmission in the brain. A third gene linked to neurotransmission, and identified through top SNP *rs3790455*, is *MEF2D. MEF2D* encodes a transcription factor that promotes survival of newly formed neurons in the brain. Finally, SNP *rs10166942*, near the *TRPM8* gene, seems to represent a clear link between genes explicitly involved in pain-related pathways and migraine.

However, given the small effect sizes, none of the associated variants can be considered conclusive genetic biomarkers of migraine as each of them, and even the 13 genetic factors together, explains only very little of the genetic variance and therefore hardly have predictive value. Instead, they may serve as clues to pinpoint specific biological pathways that are involved in the pathogenesis of migraine. Now, powerful whole-genome assay technologies are becoming rapidly less expensive, and the use of "hypothesis-driven" candidate gene association studies as an approach to identify migraine susceptibility genes seems superseded. Still, for specific clinical conditions or endophenotypes, it may be worthwhile to design array-based platforms to assay multiple "attractive" or "hypothesis-driven" candidates, especially when rarer alleles are investigated that are not present on commercially available GWAS chip arrays. In combination with large and phenotypically homogeneous cohort of patients, such custom-made array may prove useful in linking genetic information with disease characteristics.

4.4 Pharmacogenomics in Migraine

Pharmacogenomics aims to investigate the genetic background linked to variation in a patient's response to a specific drug. Studying pharmacogenomics in migraine seems to make perfect sense given the high number of preventive and symptomatic treatments available to patients, the use of several off-label medications adopted in general practice, and the fact that many migraine patients experience side effects or adverse events from treatments [72]. Despite the high heterogeneity in efficacy of available treatments in migraine patients, little effort has gone into investigating the genetics behind this variable response to drug treatment. This is particularly remarkable since such studies have been carried out for other common and socially relevant disease conditions, such as cardiovascular disorders and type 2 diabetes mellitus, or cognitive impairment [73].

Table 4.2 Selection of SNPs associated with MA or MO in GWAS studies

SNP	Chr	Position	Location	Gene	Minor allele	MAF	OR (95 % CI)	Additional significance	Ref.
rs2651899	1	3,073,572	Genic	*PRDM16*	C	0.41	1.09 (1.07–1.12)	All migraine	[63, 65]
rs10915437	1	4,082,866	Intergenic	Close to *AJAP1*	G	0.36	0.86 (0.82–0.91)	All migraine	[65]
rs12134493	1	115,479,469	Intergenic	Close to *TSPAN2*	A	0.46	1.14 (1.10–1.18)	All migraine	[65]
rs2274316	1	154,712,866	Genic	*MEF2D*	C	0.37	1.07 (1.04–1.09)	All migraine; MO	[64, 65]
rs7577262	2	234,483,608	Genic	*TRPM8*	A	0.10	0.87 (0.84–0.90)	MO	[63–65]
rs6790925	3	30,455,089	Intergenic	Close to *TGFBR2*	T	0.38	1.15 (1.10–1.21)	All migraine	[64, 65]
rs9349379	6	13,011,943	Genic	*PHACTR1*	G	0.40	0.86 (0.82–0.90)	All migraine	[64, 65]
rs13208321	6	96,967,075	Genic	*FHL5*	A	0.22	1.18 (1.13–1.24)	MO	[65]
rs4379368	7	40,432,725	Genic	*C7 or f10*	T	0.12	1.11 (1.08–1.15)	MO	[65]
rs1835740	8	98,236,089	Intergenic	*MTDH/AEG-1*	A	0.35	1.18 (1.13–1.24)	MA	[62]

rs10504861	8	89,617,048	Intergenic	Close to *MMP16*	T	0.16	0.86 (0.81–0.90)	MO	[65]
rs6478241	9	118,292,450	Genic	*ASTN2*	A	0.38	1.16 (1.11–1.22)	All migraine	[64, 65]
rs11172113	12	55,813,550	Genic	*LRP1*	C	0.43	0.90 (0.88–0.92)	MO	[63–65]

Reproduced with permission from Malkov et al. [65] (modified)
When multiple subgroups show significant association, *p*-values and odds ratios (OR) are shown for the analysis with the lowest *p*-value. ORs reported for the minor allele. Chromosomal positions are based on NCBI build 36. In the Location and Gene columns, SNPs located within the gene transcript ("Genic") list that gene, while for intergenic SNPs ("Intergenic"), the gene reported for that locus is listed
MAF minor allele frequency, *All migraine* no specification for migraine type, *MA* migraine with aura, *MO* migraine without aura

A recent review revealed that pharmacogenomics in migraine is still understudied: only seven studies investigated symptomatic (acute) medication, and only two studies investigated preventive medication; all investigated episodic migraine [73]. Unfortunately, the studies had methodological limitations, including extreme heterogeneity in study design, use of a self-assessed questionnaire rather than a clinical examination to define the type and severity of migraine, lack of appropriate statistical tests, and a limited number of participants (less than 100 in most cases), with no appropriate effort to replicate results. Various associations, though with design limitations, were found between genotype and acute symptomatic response including an inconsistent response to triptans linked to variants in the *SLC6A4* gene [74–76]. Nonetheless, if properly designed, pharmacogenomics studies should be encouraged because they may contribute to help optimizing health-care resources, improve treatment options in individual patients, and thereby reduce the burden of migraine.

4.5 Genetics in Other Primary Headaches

Genetic research in other primary headaches than migraine is advancing at a much slower rate. Although there is epidemiological evidence that in both tension-type headache (TTH) [77, 78] and cluster headache (CH) [79, 80] genetic factors seem to be involved, only for CH, a few molecular genetic studies have been performed. The previously mentioned mutations associated with MELAS were reported in a single CH patient with no family history of CH [81]. In addition, multiple deletions of the mitochondrial DNA were reported in a patient with CH and familial chronic progressive external ophthalmoplegia [82]. Both reports suggested an association of mtDNA abnormalities with CH, but subsequent studies could not confirm this in other patients with CH [83, 84]. Finally, a significant association was found in Italian patients between hypocretin receptor 2 (*HCRTR2*) gene polymorphism p.Ile308Val (G1246A) and CH [85], which was replicated in a large German cohort of CH patients [86] but not in another study [87]. A very recent meta-analysis that included a large Dutch cohort and had in total over 1100 cluster headache cases and more than 1600 controls from the six study populations showed association of p.Ile308Val with CH [88].

4.6 Future Perspective: New Horizon in Migraine Genetics

What will the future in migraine genetics bring? A recent revolution in DNA sequencing technology, commonly referred to as "next-generation sequencing (NGS)," will soon enter general medical practice. NGS allows the interrogation of the "exome," i.e., all exons which include the protein-coding parts of genes, or the "whole genome" in a single experiment. The ability to have comprehensive information on DNA variations in all our ~25,000 genes in a single test certainly will turn out a powerful and effective diagnostic tool for doctors. In the last 5 years, NGS has been very instrumental in establishing genetic diagnoses for rare, clinically, sometimes unrecognizable, monogenic disorders or puzzling disorders that are suspected to be genetic in origin [89]. The *pros* and *cons* of the clinical use of exome sequencing are carefully described in a recent review [90]. Numerous challenges remain, including the proper ethical handling of incidental findings and disease variants unsearched for, which needs active debate within the scientific and public community, and the difficulty to sort out DNA variants with pathogenic significance from the vast majority of variants that do not cause pathology. Much research effort goes into developing well-thought priority scores and bioinformatics or functional tools (as yet underdeveloped) to spot disease-related variants. Although no studies have been published thus far, it is likely that using NGS technology will impact significantly the near future research in genetic migraine, not only as a tool to identify novel FHM genes but likely also to identify variants with a medium-high effect size in more common forms of migraine headache.

A potentially promising endeavor might also be to design specific-targeted gene panels based on GWAS information combined with information from analyzing protein-to-protein networks. If successful for the common forms of migraine, NGS could resolve the current debate where the so-called missing heritability resides. Missing heritability refers to the puzzling observation that for most common disorders including migraine, less than 10 % of the genetic variance currently is explained by genetic factors that are captured by GWAS technology. It is perhaps a sobering thought that despite use research and financial investment, at this moment, we only have identified a tiny fraction of the genetic variants that confer migraine risk. As a consequence, the clinical relevance of the identified variants until now does not have meaningful clinical relevance. Although many patients have become aware of the possibility that one can obtain genetic information from their buccal swabs or saliva by sending biological samples to specialized companies that advertise on the web (see among others https://www.23andme.com), the scientific community should make clear that this sort of "self-investigation" will not be of much help in assessing a patient's disease risk and therefore discourage them to participate as the interpretation of the genetic information is far from being straightforward [91].

4.7 Conclusions and Perspectives

The last decade or so has greatly advanced our insight in genetic factors that contribute to migraine. Genetic studies of monogenic forms of migraine, foremost FHM, revealed neurotransmission as a key disease mechanism. Other monogenetic diseases in which migraine is prominent pointed toward abnormal vascular function being involved in migraine pathogenesis. The era of GWAS has already led to the discovery of over a dozen genes and gene variants and pinpointed multiple players involved in neurotransmission and vascular function as possible disease mechanisms in the common forms of migraine. The advent of novel methodologies such as NGS for the identification of migraine susceptibility gene variants and pharmacogenetics to help explaining altered drug responses is expected to shed important new light on the involved molecular mechanisms. As more and more genetic factors (and the molecular pathways they are involved in) become known, there might be exciting opportunities for personalized approaches to therapies and development of new drugs relieving pain and discomfort.

References

1. Headache Classification Committee of the International Headache Society (IHS) (2013) The International Classification of Headache Disorders, 3rd edition (beta version). Cephalalgia 33:629–808
2. Peng KP, Wang SJ (2014) Epidemiology of headache disorders in the Asia-pacific region. Headache 54:610–618
3. Stewart WF, Roy J, Lipton RB (2013) Migraine prevalence, socioeconomic status, and social causation. Neurology 81:948–955
4. Bigal ME, Lipton RB (2009) The epidemiology, burden, and comorbidities of migraine. Neurol Clin 27:321–334
5. Buse DC, Loder EW, Gorman JA, Stewart WF, Reed ML, Fanning KM, Serrano D, Lipton RB (2013) Sex differences in the prevalence, symptoms, and associated features of migraine, probable migraine and other severe headache: results of the American Migraine Prevalence and Prevention (AMPP) Study. Headache 53:1278–1299
6. D'Amico D, Tepper SJ (2008) Prophylaxis of migraine: general principles and patient acceptance. Neuropsychiatr Dis Treat 4:1155–1167
7. World Health Organization. World Health Report 2001: mental health: new understanding, new hope. Accessed: http://www.who.int/whr/2001/en/index.html. Accessed: 20/05/2012
8. Scher AI, Bigal ME, Lipton RB (2005) Comorbidity of migraine. Curr Opin Neurol 18:305–310
9. Lauritzen M (1994) Pathophysiology of the migraine aura. The spreading depression theory. Brain 117:199–210
10. Hadjikhani N, Sanchez Del Rio M, Wu O, Schwartz D, Bakker D, Fischl B, Kwong KK, Cutrer FM, Rosen BR, Tootell RB, Sorensen AG, Moskowitz MA (2001) Mechanisms of migraine aura revealed by functional MRI in human visual cortex. Proc Natl Acad Sci U S A 98:4687–4692
11. Noseda R, Burstein R (2013) Migraine pathophysiology: anatomy of the trigeminovascular pathway and associated neurological symptoms, CSD, sensitization and modulation of pain. Pain 54(Suppl 1):1–21
12. Schürks M (2012) Genetics of migraine in the age of genome-wide association studies. J Headache Pain 13:1–9
13. Magis D, Schoenen J (2012) Migraine: from genetics to environment. Rev Med Liege 67:349–358
14. Colson N, Fernandez F, Griffiths L (2010) Genetics of menstrual migraine: the molecular evidence. Curr Pain Headache Rep 14:389–395
15. Russell MB, Olesen J (1995) Increased familial risk and evidence of genetic factor in migraine. Br Med J 311:541–544
16. Russell MB, Iselius L, Olesen J (1996) Migraine without aura and migraine with aura are inherited disorders. Cephalalgia 16:305–309
17. Russell MB, Ulrich V, Gervil M, Olesen J (2002) Migraine without aura and migraine with aura are distinct disorders. A population-based twin survey. Headache 42:332–336

18. Kallela M, Wessman M, Havanka H, Palotie A, Farkkila M (2001) Familial migraine with and without aura: clinical characteristics and co-occurrence. Eur J Neurol 8:441–449
19. Nyholt DR, Gillespie NG, Heath AC, Merikangas KR, Duffy DL, Martin NG (2004) Latent class and genetic analysis does not support migraine with aura and migraine without aura as separate entities. Genet Epidemiol 26:231–244
20. Ligthart L, Boomsma DI, Martin NG, Stubbe JH, Nyholt DR (2006) Migraine with aura and migraine without aura are not distinct entities: further evidence from a large Dutch population study. Twin Res Hum Genet 9:54–63
21. Thomsen LL, Eriksen MK, Roemer SF, Andersen I, Olesen J, Russell MB (2002) A population-based study of familial hemiplegic migraine suggests revised diagnostic criteria. Brain 125:1379–1391
22. Thomsen LL, Ostergaard E, Olesen J, Russell MB (2003) Evidence for a separate type of migraine with aura: sporadic hemiplegic migraine. Neurology 60:595–601
23. Ophoff RA, Terwindt GM, Vergouwe MN, van Eijk R, Oefner PJ, Hoffman SM, Lamerdin JE, Mohrenweiser HW, Bulman DE, Ferrari M, Haan J, Lindhout D, van Ommen GJ, Hofker MH, Ferrari MD, Frants RR (1996) Familial hemiplegic migraine and episodic ataxia type-2 are caused by mutations in the Ca^{2+} channel gene CACNL1A4. Cell 87:543–552
24. Catterall WA (1998) Structure and function of neuronal Ca^{2+} channels and their role in neurotransmitter release. Cell Calcium 24:307–323
25. de Vries B, Frants RR, Ferrari MD, van den Maagdenberg AM (2009) Molecular genetics of migraine. Hum Genet 126:115–132
26. Joutel A, Bousser MG, Biousse V, Labauge P, Chabriat H, Nibbio A, Maciazek J, Meyer B, Bach MA, Weissenbach J (1993) A gene for familial hemiplegic migraine maps to chromosome 19. Nat Genet 5:40–45
27. De Fusco M, Marconi R, Silvestri L, Atorino L, Rampoldi L, Morgante L, Ballabio A, Aridon P, Casari G (2003) Haploinsufficiency of ATP1A2 encoding the Na^+/K^+ pump alpha2 subunit associated with familial hemiplegic migraine type 2. Nat Genet 33:192–196
28. Riant F, De Fusco M, Aridon P, Ducros A, Ploton C, Marchelli F, Maciazek J, Bousser MG, Casari G, Tournier-Lasserve E (2005) ATP1A2 mutations in 11 families with familial hemiplegic migraine. Hum Mutat 26:281
29. Vanmolkot KR, Kors EE, Turk U, Turkdogan D, Keyser A, Broos LA, Kia SK, van den Heuvel JJ, Black DF, Haan J, Frants RR, Barone V, Ferrari MD, Casari G, Koenderink JB, van den Maagdenberg AM (2006) Two de novo mutations in the Na, K-ATPase gene ATP1A2 associated with pure familial hemiplegic migraine. Eur J Hum Genet 14:555–560
30. Spadaro M, Ursu S, Lehmann-Horn F, Veneziano L, Antonini G, Giunti P, Frontali M, Jurkat-Rott K (2004) A G301R Na^+/K^+-Atpase mutation causes familial hemiplegic migraine type 2 with cerebellar signs. Neurogenetics 5:177–185
31. Echenne B, Ducros A, Rivier F, Joutel A, Humbertclaude V, Roubertie A, Azaïs M, Bousser MG, Tournier-Lasserve E (1999) Recurrent episodes of coma: an unusual phenotype of familial hemiplegic migraine with linkage to chromosome 1. Neuropediatrics 30:214–217
32. Jurkat-Rott K, Freilinger T, Dreier JP, Herzog J, Göbel H, Petzold GC, Montagna P, Gasser T, Lehmann-Horn F, Dichgans M (2004) Variability of familial hemiplegic migraine with novel A1A2 Na^+/K^+–ATPase variants. Neurology 62:1857–1861
33. Dichgans M, Freilinger T, Eckstein G, Babini E, Lorenz-Depiereux B, Biskup S, Ferrari MD, Herzog J, van den Maagdenberg AM, Pusch M, Strom TM (2005) Mutation in the neuronal voltage-gated sodium channel SCN1A in familial hemiplegic migraine. Lancet 366:371–377
34. Parihar R, Ganesh S (2013) The SCN1A gene variants and epileptic encephalopathies. J Hum Genet 58:573–580
35. Vahedi K, Depienne C, Le Fort D, Riant F, Chaine P, Trouillard O, Gaudric A, Morris MA, Leguern E, Tournier-Lasserve E, Bousser MG (2009) Elicited repetitive daily blindness: a new phenotype associated with hemiplegic migraine and SCN1A mutations. Neurology 72:1178–1183

36. Riant F, Roze E, Barbance C, Méneret A, Guyant-Maréchal L, Lucas C, Sabouraud P, Trébuchon A, Depienne C, Tournier-Lasserve E (2012) PRRT2 mutations cause hemiplegic migraine. Neurology 79:2122–2124
37. Pelzer N, de Vries B, Kamphorst JT, Vijfhuizen LS, Ferrari MD, Haan J, van den Maagdenberg AM, Terwindt GM (2014) PRRT2 and hemiplegic migraine: a complex association. Neurology 83:288–290
38. Pietrobon D (2010) Insights into migraine mechanisms and $Ca_V2.1$ calcium channel function from mouse models of familial hemiplegic migraine. J Physiol 588:1871–1878
39. Tavraz NN, Friedrich T, Durr KL, Koenderink JB, Bamberg E, Freilinger T, Dichgans M (2008) Diverse functional consequences of mutations in the Na^+/K^+ –ATPase alpha2-subunit causing familial hemiplegic migraine type 2. J Biol Chem 283:31097–31106
40. van den Maagdenberg AM, Pietrobon D, Pizzorusso T, Kaja S, Broos LAM, Cesetti T, van de Ven RCG, Tottene A, van der Kaa J, Plomp JJ, Frants RR, Ferrari MD (2004) A *Cacna1a* knockin migraine mouse model with increased susceptibility to cortical spreading depression. Neuron 41:701–710
41. van den Maagdenberg AM, Pizzorusso T, Kaja S, Terpolilli N, Shapovalova M, Hoebeek FE, Barrett CF, Gherardini L, van de Ven RCG, Todorov B, Broos LAM, Tottene A, Gao Z, Fodor M, de Zeeuw CI, Frants RR, Plesnila N, Plomp JJ, Pietrobon D, Ferrari MD (2010) High cortical spreading depression susceptibility and migraine-associated symptoms in $Ca_V2.1$ S218L mice. Ann Neurol 67:85–98
42. Stam AH, Luijckx GJ, Poll-The BT, Ginjaar IB, Frants RR, Haan J, Ferrari MD, Terwindt GM, van den Maagdenberg AM (2009) Early seizures and cerebral oedema after trivial head trauma associated with the *CACNA1A* S218L mutation. J Neurol Neurosurg Psychiatry 80: 1125–1129
43. Chanda ML, Tuttle AH, Baran I, Atlin C, Guindi D, Hathaway G, Israelian N, Levenstadt J, Low D, Macrae L, O'Shea L, Silver A, Zendegui E, Mariette Lenselink A, Spijker S, Ferrari MD, van den Maagdenberg AM, Mogil JS (2013) Behavioral evidence for photophobia and stress-related ipsilateral head pain in transgenic *Cacna1a* mutant mice. Pain 154:1254–1262
44. Eikermann-Haerter K, Dileköz E, Kudo C, Savitz SI, Waeber C, Baum MJ, Ferrari MD, van den Maagdenberg AM, Moskowitz MA, Ayata C (2009) Genetic and hormonal factors modulate spreading depression and transient hemiparesis in mouse models of familial hemiplegic migraine type 1. J Clin Invest 119:99–109
45. Tottene A, Conti R, Fabbro A, Vecchia D, Shapovalova M, Santello M, van den Maagdenberg AM, Ferrari MD, Pietrobon D (2009) Enhanced excitatory transmission at cortical synapses as the basis for facilitated spreading depression in $Ca_V2.1$ knockin migraine mice. Neuron 61:762–773
46. Leo L, Gherardini L, Barone V, De Fusco M, Pietrobon D, Pizzorusso T, Casari G (2011) Increased susceptibility to cortical spreading depression in the mouse model of familial hemiplegic migraine type 2. PLoS Genet 7(6):e1002129
47. James PF, Grupp IL, Grupp G, Woo AL, Askew GR, Croyle ML, Walsh RA, Lingrel JB (1999) Identification of a specific role for the Na, K-ATPase alpha 2 isoform as a regulator of calcium in the heart. Mol Cell 3:555–563
48. Ogiwara I, Miyamoto H, Morita N, Atapour N, Mazaki E, Inoue I, Takeuchi T, Itohara S, Yanegawa Y, Obata K, Furuichi T, Hensch TK, Yamakawa K (2007) $Na_V1.1$ localizes to axons of parvalbumin-positive inhibitory interneurons: a circuit basis for epileptic seizures in mice carrying an *Scn1a* gene mutation. J Neurosci 27:5903–5914
49. Montagna P, Gallassi R, Medori R, Govoni E, Zeviani M, Di Mauro S, Lugaresi E, Andermann F (1988) MELAS syndrome: characteristic migrainous and epileptic features and maternal transmission. Neurology 38:751–754
50. Dichgans M, Mayer M, Uttner I, Bruning R, Muller-Hocker J, Rungger G, Ebke M, Klockgether T, Gasser T (1998) The phenotypic spectrum of CADASIL: clinical findings in 102 cases. Ann Neurol 44:731–739
51. Richards A, van den Maagdenberg AM, Jen JC, Kavanagh D, Bertram P, Spitzer D, Liszewski MK, Barilla-Labarca ML, Terwindt GM, Kasai Y, McLellan M, Grand MG, Vanmolkot KR,

de Vries B, Wan J, Kane MJ, Mamsa H, Schäfer R, Stam AH, Haan J, de Jong PT, Storimans CW, van Schooneveld MJ, Oosterhuis JA, Gschwendter A, Dichgans M, Kotschet KE, Hodgkinson S, Hardy TA, Delatycki MB, Hajj-Ali RA, Kothari PH, Nelson SF, Frants RR, Baloh RW, Ferrari MD, Atkinson JP (2007) C-terminal truncations in human 3′–5′ DNA exonuclease TREX1 cause autosomal dominant retinal vasculopathy with cerebral leukodystrophy. Nat Genet 39:1068–1070

52. Feuerhake F, Volk B, Ostertag CB, Jungling FD, Kassubek J, Orszagh M, Dichgans M (2002) Reversible coma with raised intracranial pressure: an unusual clinical manifestation of CADASIL. Acta Neuropathol 103:188–192
53. Sicurelli F, Dotti MT, De Stefano N, Malandrini A, Mondelli M, Bianchi S, Federico A (2005) Peripheral neuropathy in CADASIL. J Neurol 252:1206–1209
54. Haritoglou C, Rudolph G, Hoops JP, Opherk C, Kampik A, Dichgans M (2004) Retinal vascular abnormalities in CADASIL. Neurology 62:1202–1205
55. Joutel A, Corpechot C, Ducros A, Vahedi K, Chabriat H, Mouton P, Alamowitch S, Domenga V, Cécillion M, Marechal E, Maciazek J, Vayssiere C, Cruaud C, Cabanis EA, Ruchoux MM, Weissenbach J, Bach JF, Bousser M-G, Tournier-Lasserve E (1996) Notch3 mutations in CADASIL, a hereditary adult-onset condition causing stroke and dementia. Nature 383:707–710
56. Yamamoto Y, Craggs L, Baumann M, Kalimo H, Kalaria RN (2011) Review: molecular genetics and pathology of hereditary small vessel diseases of the brain. Neuropathol Appl Neurobiol 37:94–113
57. Eikermann-Haerter K, Yuzawa I, Dileköz E, Joutel A, Moskowitz MA, Ayata C (2011) Cerebral autosomal dominant arteriopathy with subcortical infarcts and leukoencephalopathy syndrome mutations increase susceptibility to spreading depression. Ann Neurol 69:413–418
58. Grand MG, Kaine J, Fulling K, Atkinson J, Dowton SB, Farber M, Craver J, Rice K (1988) Cerebroretinal vasculopathy. A new hereditary syndrome. Ophthalmology 95:649–659
59. Jen J, Cohen AH, Yue Q, Stout JT, Vinters HV, Nelson S, Baloh RW (1997) Hereditary endotheliopathy with retinopathy, nephropathy, and stroke (HERNS). Neurology 49:1322–1330
60. Terwindt GM, Haan J, Ophoff RA, Groenen SM, Storimans CW, Lanser JB, Roos RA, Bleeker-Wagemakers EM, Frants RR, Ferrari MD (1998) Clinical and genetic analysis of a large Dutch family with autosomal dominant vascular retinopathy, migraine and Raynaud's phenomenon. Brain 121:303–316
61. Pavlakis SG, Phillips PC, DiMauro S, De Vivo DC, Rowland LP (1984) Mitochondrial myopathy, encephalopathy, lactic acidosis, and strokelike episodes: a distinctive clinical syndrome. Ann Neurol 16:481–488
62. Sproule DM, Kaufmann P (2008) Mitochondrial encephalopathy, lactic acidosis, and strokelike episodes: basic concepts, clinical phenotype, and therapeutic management of MELAS syndrome. Ann N Y Acad Sci 1142:133–158
63. Mancuso M, Filosto M, Mootha VK, Rocchi A, Pistolesi S, Murri L, DiMauro S, Siciliano G (2004) A novel mitochondrial tRNAPhe mutation causes MERRF syndrome. Neurology 62:2119–2121
64. Melone MA, Tessa A, Petrini S, Lus G, Sampaolo S, di Fede G, Santorelli FM, Cotrufo R (2004) Revelation of a new mitochondrial DNA mutation (G12147A) in a MELAS/MERFF phenotype. Arch Neurol 61:269–272
65. Malkov A, Ivanov AI, Popova I, Mukhtarov M, Gubkina O, Waseem T, Bregestovski P, Zilberter Y (2014) Reactive oxygen species initiate a metabolic collapse in hippocampal slices: potential trigger of cortical spreading depression. J Cereb Blood Flow Metab 34: 1540–1549
66. Lafrenière RG, Cader MZ, Poulin J-F, Andres-Enguix I, Simoneau M, Gupta N, Boisvert K, Lafrenière F, Mclaughlan S, Dubé M-P, Marcinkiewicz MM, Ramagopalan S, Ansorge O, Brais B, Sequeiros J, Pereira-Monteiro JM, Griffiths LR, Tucker SJ, Ebers G, Rouleau GA (2010) A dominant-negative mutation in the TRESK potassium channel is linked to familial migraine with aura. Nat Med 16:1157–1160

67. Andres-Enguix I, Shang L, Stansfeld PJ, Morahan JM, Sansom MSP, Lafrenière RG, Roy B, Griffiths LR, Rouleau GA, Ebers GC, Cader ZM, Tucker SJ (2012) Functional analysis of missense variants in the TRESK (KCNK18) K channel. Sci Rep 2:237
68. Anttila V, Stefansson H, Kallela M, Todt U, Terwindt GM, Calafato MS, Nyholt DR, Dimas AS, Freilinger T, Müller-Myhsok B, Artto V, Inouye M, Alakurtti K, Kaunisto MA, Hämäläinen E, de Vries B, Stam AH, Weller CM, Heinze A, Heinze-Kuhn K, Goebel I, Borck G, Göbel H, Steinberg S, Wolf C, Björnsson A, Gudmundsson G, Kirchmann M, Hauge A, Werge T, Schoenen J, Eriksson JG, Hagen K, Stovner L, Wichmann H-E, Meitinger T, Alexander M, Moebus S, Schreiber S, Aulchenko YS, Breteler MMB, Uitterlinden AG, Hofman A, van Duijn CM, Tikka-Kleemola P, Vepsäläinen S, Lucae S, Tozzi F, Muglia P, Barrett J, Kaprio J, Färkkilä M, Peltonen L, Stefansson K, Zwart J-A, Ferrari MD, Olesen J, Daly M, Wessman M, van den Maagdenberg AM, Dichgans M, Kubisch C, Dermitzakis ET, Frants RR, Palotie A, International Headache Genetics Consortium (2010) Genome-wide association study of migraine implicates a common susceptibility variant on 8q22.1. Nat Genet 42:869–873
69. Chasman DI, Schürks M, Anttila V, de Vries B, Schminke U, Launer LJ, Terwindt GM, van den Maagdenberg AMJM, Fendrich K, Völzke H, Ernst F, Griffiths LR, Buring JE, Kallela M, Freilinger T, Kubisch C, Ridker PM, Palotie A, Ferrari MD, Hoffmann W, Zee RYL, Kurth T (2011) Genome-wide association study reveals three susceptibility loci for common migraine in the general population. Nat Genet 43:695–698
70. Freilinger T, Anttila V, de Vries B, Malik R, Kallela M, Terwindt GM, Pozo-Rosich P, Winsvold B, Nyholt DR, van Oosterhout WPJ, Artto V, Todt U, Hämäläinen E, Fernández-Morales J, Louter MA, Kaunisto MA, Schoenen J, Raitakari O, Lehtimäki T, Vila-Pueyo M, Göbel H, Wichmann E, Sintas C, Uitterlinden AG, Hofman A, Rivadeneira F, Heinze A, Tronvik E, van Duijn CM, Kaprio J, Cormand B, Wessman M, Frants RR, Meitinger T, Müller-Myhsok B, Zwart J-A, Färkkilä M, Macaya A, Ferrari MD, Kubisch C, Palotie A, Dichgans M, van den Maagdenberg AM, International Headache Genetics Consortium (2012) Genome-wide association analysis identifies susceptibility loci for migraine without aura. Nat Genet 44:777–782
71. Anttila V, Winsvold BS, Gormley P, Kurth T, Bettella F, McMahon G, Kallela M, Malik R, de Vries B, Terwindt G, Medland SE, Todt U, McArdle WL, Quaye L, Koiranen M, Ikram MA, Lehtimäki T, Stam AH, Ligthart L, Wedenoja J, Dunham I, Neale BM, Palta P, Hämäläinen E, Schürks M, Rose LM, Buring JE, Ridker PM, Steinberg S, Stefansson H, Jakobsson F, Lawlor DA, Evans DM, Ring SM, Färkkilä M, Artto V, Kaunisto MA, Freilinger T, Schoenen J, Frants RR, Pelzer N, Weller CM, Zielman R, Heath AC, Madden PAF, Montgomery GW, Martin NG, Borck G, Göbel H, Heinze A, Heinze-Kuhn K, Williams FMK, Hartikainen A-L, Pouta A, van den Ende J, Uitterlinden AG, Hofman A, Amin N, Hottenga J-J, Vink JM, Heikkilä K, Alexander M, Müller-Myhsok B, Schreiber S, Meitinger T, Wichmann HE, Aromaa A, Eriksson JG, Traynor BJ, Trabzuni D, Rossin E, Lage K, Jacobs SBR, Gibbs JR, Birney E, Kaprio J, Penninx BW, Boomsma DI, van Duijn C, Raitakari O, Järvelin M-R, Zwart J-A, Cherkas L, Strachan DP, Kubisch C, Ferrari MD, van den Maagdenberg AM, Dichgans M, Wessman M, Smith GD, Stefansson K, Daly MJ, Nyholt DR, Chasman DI, Palotie A, North American Brain Expression Consortium, UK Brain Expression Consortium, North American Brain Expression Consortium, UK Brain Expression Consortium, International Headache Genetics Consortium (2013) Genome-wide meta-analysis identifies new susceptibility loci for migraine. Nat Genet 45:912–917
72. Luykx J, Mason M, Ferrari MD, Carpay J (2009) Are migraineurs at increased risk of adverse drug responses? A meta-analytic comparison of topiramate-related adverse drug reactions in epilepsy and migraine. Clin Pharmacol Ther 85:283–288
73. Viana M, Terrazzino S, Genazzani AA, Grieco GS, Cargnin S, Santorelli FM, Pierelli F, Tassorelli C, Nappi G, Di Lorenzo C (2014) Pharmacogenomics of episodic migraine: time has come for a step forward. Pharmacogenomics 15:541–549
74. Ishii M, Sakairi Y, Hara H, Imagawa A, Shimizu S, Takahashi J, Nagamine A, Naito Y, Masuda Y, Usami S, Kiuchi Y (2012) Negative predictors of clinical response to triptans in patients with migraine. Neurol Sci 33:453–461

75. Terrazzino S, Viana M, Floriddia E, Monaco F, Mittino D, Sances G, Tassorelli C, Nappi G, Rinaldi M, Canonico PL, Genazzani AA (2010) The serotonin transporter gene polymorphism STin2 VNTR confers an increased risk of inconsistent response to triptans in migraine patients. Eur J Pharmacol 641:82–87
76. Asuni C, Cherchi A, Congiu D, Piccardi MP, Del Zompo M, Stochino ME (2007) Association study between clinical response to rizatriptan and some candidate genes. J Headache Pain 8:185–189
77. Ulrich V, Gervil M, Olesen J (2004) The relative influence of environment and genes in episodic tension-type headache. Neurology 62:2065–2069
78. Russell MB, Saltyte-Benth J, Levi N (2006) Are infrequent episodic, frequent episodic and chronic tension-type headache inherited? A population-based study of 11 199 twin pairs. J Head Pain 7:119–126
79. Russell MB, Andersson PG, Thomsen LL (1995) Familial occurrence of cluster headache. J Neurol Neurosurg Psychiatry 58:341–343
80. Russell MB, Andersson PG, Thomsen LL, Iselius L (1995) Cluster headache is an autosomal dominantly inherited disorder in some families: a complex segregation analysis. J Med Genet 32:954–956
81. Shimomura T, Kitano A, Marukawa H, Mishima K, Isoe K, Adachi Y, Takahashi K (1994) Point mutation in platelet mitochondrial tRNA Leu(UUR) in patient with cluster headache. Lancet 344:625
82. Odawara M, Tamaoka A, Mizusawa H, Yamashita K (1997) A case of cluster headache associated with mitochondrial DNA deletions. Muscle Nerve 20:394–395
83. Cortelli P, Zacchini A, Barboni P, Malpassi P, Carelli V, Montagna P (1995) Lack of association between mitochondrial tRNA (Leu(UUR)) point mutation and cluster headache. Lancet 345:1120–1121
84. Seibel P, Grünewald T, Gundolla A, Diener HC, Reichmann H (1996) Investigation on the mitochondrial transfer RNA(Leu)(UUR) in blood cells from patients with cluster headache. J Neurol 243:305–307
85. Rainero I, Gallone S, Valfrè W, Ferrero M, Angilella G, Rivoiro C, Rubino E, De Martino P, Savi L, Ferrone M, Pinessi L (2004) A polymorphism of the hypocretin receptor 2 gene is associated with cluster headache. Neurology 63:1286–1288
86. Schürks M, Kurth T, Geissler I, Tessmann G, Diener HC, Rosskopf D (2006) Cluster headache is associated with the G1246A polymorphism in the hypocretin receptor 2 gene. Neurology 66:1917–1919
87. Baumber L, Sjöstrand C, Leone M, Harty H, Bussone G, Hillert J, Trembath RC, Russell MB (2006) A genome-wide scan and HCRTR2 candidate gene analysis in a European cluster headache cohort. Neurology 66:1888–1893
88. Weller CM, Wilbrink LA, Houwing-Duistermaat JJ, Koelewijn SC, Vijfhuizen LS, Haan J, Ferrari MD, Terwindt GM, van den Maagdenberg AM, de Vries B (2014). Cluster headache and the hypocretin receptor 2 reconsidered: a genetic association study and meta-analysis. Cephalalgia. pii: 0333102414557839. [Epub ahead of print]. PubMed PMID: 25398231
89. Xue Y, Ankala A, Wilcox WR, Hegde MR (2014) Solving the molecular diagnostic testing conundrum for Mendelian disorders in the era of next-generation sequencing: single-gene, gene panel, or exome/genome sequencing. Genet Med. doi:10.1038/gim.2014.122, [Epub ahead of print]. PubMed PMID: 25232854
90. Biesecker LG, Green RC (2014) Diagnostic clinical genome and exome sequencing. N Engl J Med 370:2418–2425
91. Annas GJ, Elias S (2014) 23 and me and the FDA. N Engl J Med 370:2248–2249

Chapter 5
Human Models of Primary Headaches

Henrik Winther Schytz and Guus G. Schoonman

5.1 Introduction

Why do humans suffer from primary headaches? What happens during the development of a migraine or tension-type headache? These are important questions given the high prevalence and need for better treatment among various primary headache disorders. However, the episodic and unpredictable nature of most primary headaches results in various logistic challenges when attempting to study the neurobiology of primary headaches. A powerful method to study primary headaches is using pharmacological triggers, which mimic endogenous triggers, to induce headache in humans. This chapter will describe and discuss established knowledge on human models of primary headache.

5.2 Methodology in Human Headache Models

In the most frequently used human headache model, patients are randomly allocated to receive intravenous infusion of the target substance or placebo (isotonic saline) in a double-blind, crossover design [1]. Despite this relatively simple design, several aspects need attention.

H.W. Schytz (✉)
Danish Headache Center, Department of Neurology, Nordre Ringvej 56,
Glostrup 2600, Denmark
e-mail: henrikschytz@dadlnet.dk

G.G. Schoonman, MD, PhD (✉)
Neurology, Leiden University Medical Centre and Elisabeth/Tweesteden Hospital Tilburg,
Leiden and Tilburg, The Netherlands
e-mail: g.g.schoonman@lumc.nl

M. Ashina, P. Geppetti (eds.), *Pathophysiology of Headaches: From Molecule to Man*, Headache, DOI 10.1007/978-3-319-15621-7_5

The first aspect is selection of patients and controls. The patients should also only have a moderate baseline attack frequency, as patients need to be headache-free no less than 5 days before the experimental days. Conversely, there is no strict limit on minimum attack frequency, as attack frequency does not seem to predict the susceptibility to a headache trigger [2]. The control population should not be susceptible to migraine, so one should exclude subjects with a first relative suffering from migraine.

The second aspect is to consider the experimental design. The blinding procedure is important in order for the investigator to be objective and the patient as unbiased as possible. It is important to be aware of negative patient expectations, which may induce and increase headache [3, 4]. Thus, Benedetti et al. [3] investigated hypobaric hypoxia headache provocation in two healthy groups, where the nocebo group received negative information about the risk of headache, whereas the control group did not know about the possible occurrence of headache. Interestingly, the study showed a significant increase in headache and salivary prostaglandins and thromboxane in the nocebo group compared to the control group, suggesting that negative expectations enhance headache and prostaglandins synthesis. Given that the patients attend on two different days, it is also important to account for a possible day-to-day variation. However, it has been demonstrated, using glyceryl trinitrate (GTN) provocation on 2 separate days, that the maximal headache response differs in less than 50 % and only one headache score in 40 % of healthy subjects [5]. Furthermore, headache characteristics are also reproducible [5].

The third aspect is the route of administration of the investigated pharmacological substance. It is preferable to use an administration form, which gives a precise amount of the investigated substance to the head, without imposing too many systemic effects. Thus, intracarotid injection could be the ideal way to infuse substances, but this administration form is technically demanding, creates unneeded stress and pain in the patients. Furthermore, it has also been shown that the intracarotid injection of saline may in itself induce migraine, likely due to vascular pressure alterations [6]. The intravenous route via a cubital vein seems to be a balanced compromise, whereas the oral route seems to induce a more variable and less frequent headache result [7].

The fourth aspect is documentation of headache intensity, which can be recorded on an ordinal verbal rating scale from 0 to 10 (0, no headache; 1, a very mild headache (including a feeling of pressing or throbbing); 5, moderate headache; 10, worst imaginable headache). Experimental headache induced by infusion of a neurotransmitter cannot fulfil the criteria for a primary headache. As an example, pharmacological provocation of migraine is obviously not a spontaneous event and therefore does not fulfil the International Classification of Headache Disorders (ICHD) criteria for migraine [8]. However, provoked migraine attacks phenotypically mimic spontaneous migraine attacks in the majority of patients [9, 10]. In addition, most spontaneous migraine attacks develop in a matter of hours and often go through a phase where they phenomenologically fulfil the criteria for tension-type headache before the headache intensifies, becomes unilateral and has the associated symptoms required for migraine [8]. Furthermore, patients in experimental provocation studies cannot be denied treatment of the induced attacks and often treat before all migraine

Table 5.1 Migraine-like attacks fulfilling either 1 or 2

1. Headache fulfilling criteria C[a] and D for MO
2. Headache described as mimicking usual migraine attack and treated with a triptan

[a]Moderate to severe pain intensity is considered ≥4 on the VRS

criteria are fulfilled. For this reason, attacks aborted by migraine-specific treatment before fulfilling all criteria for migraine were accepted in the new criteria for chronic migraine [8]. Based on these considerations, the following criteria for a migraine-like attack induced 0–12 h after infusion of an experimental drug have been used in several experimental models [11] (Table 5.1). These circumstances are similar in other primary headache, even though such experimental criteria have not yet been defined.

The fifth aspect is experimental outcome measures. In the Copenhagen Group, it is very common to use transcranial Doppler ultrasonography (TCD) [1] to investigate blood flow velocity in the middle cerebral artery (MCA), as this is an intracranial artery, which is known to be pain sensitive [12, 13]. Furthermore, velocity recordings of the MCA are very reproducible, with a day-to-day coefficient of variation of 16 % and a 5 min coefficient of variation of 7 % [14]. Using high-resolution ultrasonography, the diameter of the frontal branch of the superficial temporal artery (STA) and radial artery (RA) can also be obtained with a day-to-day coefficient of variation of 12 % [15] (Fig. 5.1). Magnetic resonance imaging (MRI) is increasingly used in human headache models, and is very valuable, as it can image multiple arterial compartments extra- and intracranial [15–17]. Furthermore, the use of functional MRI (fMRI) using blood-oxygen-level-dependent (BOLD) contrast and resting state fMRI can further elucidate functional changes during headache ictally and interictally. It is an ongoing debate whether cranial vessels are important for pain development [18, 19]. Nevertheless, the advantage of imaging vessels in human headache models is that vessels are easy accessible markers for how, when and where the pharmacological triggers may be affecting the brain. The limitations of imaging are that the researcher, obviously, can only find a possible effect in the imaged areas of interest, which may lead to an oversimplification of the complexities in primary headaches.

5.3 Migraine Models

There is a vast experimental experience on pharmacological migraine models (Table 5.2). In the following, we will focus on the most promising pharmacological models so far.

5.3.1 Nitric Oxide

In 1987, Sicuteri et al. [35] demonstrated how sublingual GTN, a nitric oxide (NO) donor, induced an initial mild headache in healthy subjects, whereas 29 % of healthy subjects with a first-degree relative having migraine reported a delayed headache, which was also found in 67 % of migraine patients. Later, the Copenhagen Headache

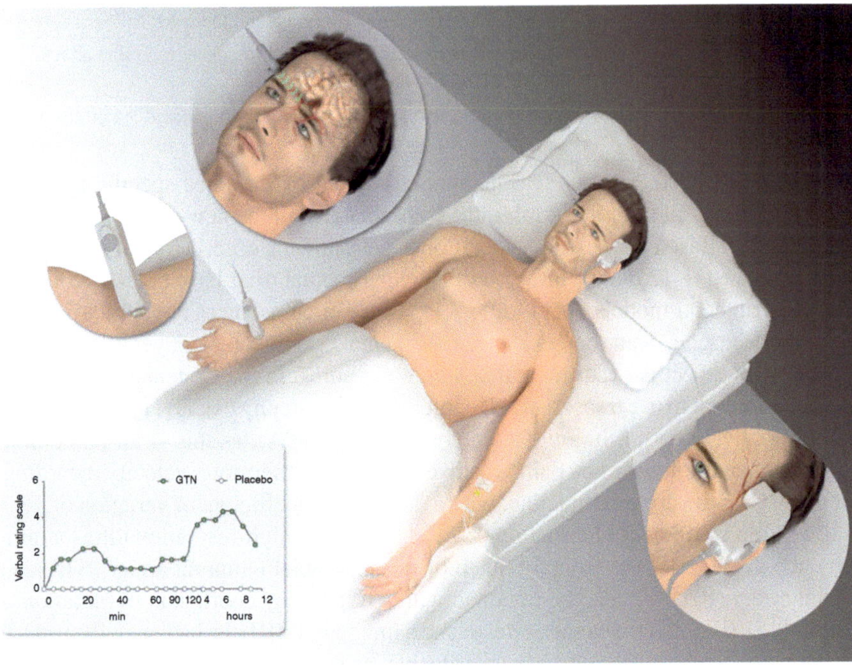

Fig. 5.1 Experimental design of a human model of migraine. In the main version of this model, patients with migraine are randomly allocated to receive intravenous infusion (25 min) of 'target substance' or placebo (isotonic saline) in a double-blind, crossover design. Headache intensity is recorded on a verbal rating scale from 0 to 10 (*0* no headache, *1* a very mild headache (including a feeling of pressing or throbbing), *5* moderate headache, *10* worst imaginable headache). The following haemodynamic variables are recorded at intervals: mean velocity of blood flow in the middle cerebral artery (MCA) by transcranial Doppler with hand-held probes, diameter of the frontal branch of the superficial temporal artery (With permission from Wiley-Blackwell)

Research Group developed, through systematic and extensive studies, a human GTN model of migraine [5]. Olesen et al. reported that GTN induced migraine-like attacks in migraine sufferers compared to patients with tension-type headache and healthy subjects [36]. Furthermore, Thomsen et al. demonstrated that 80 % of migraine without aura (MO) patients developed a delayed headache fulfilling IHS criteria for migraine peaking 5 h after end of the infusion compared to 10 % after placebo [22]. Other groups have validated the model and found it to be very reproducible and reliable [7, 21]. Furthermore, the ability of GTN to induce migraine does not seem to vary according to the frequency of attacks in migraine patients GTN and frequency of attacks [2]. The Copenhagen group uses intravenous GTN, which gives a robust response, but the sublingual route is also well studied [7], even though the results are more variable and the migraine rate is reduced. Interestingly, GTN induces MA attacks in a lower rate, and migraine aura is rarely induced [1], which shows that GTN causes specific alterations in each migraine phenotype.

NO is a vasodilator and arterial dilatation can cause headache [12, 13]. Interestingly, GTN causes a larger degree of dilatation of extra- and intracerebral

Table 5.2 Experimental experience on pharmacological migraine models

Compound		Dose	Migraine-like attacks (%)	Aura (%)	Ref.
Glyceryl trinitrate	Migraine with aura	Intravenous 0.5 μg/kg/min	50–67	0–10	[20, 21]
		Sublingual 0.9 mg	41	14	[7]
	Migraine without aura	Intravenous 0.5 μg/kg/min	80–83	0	[21, 22]
		Sublingual 0.9 mg	82	0	[7]
	Hemiplegic migraine	Intravenous 0.5 μg/kg/min	13–30	0–18	[23–25]
Sildenafil		100 mg per os	83	0	[26]
Histamine		Intravenous 0.5 μg/kg/min	70	0	[27]
CGRP	Migraine with aura	Intravenous 1.5 μg/min	57	28	[28]
	Migraine without aura	Intravenous 2 μg/min	67	0	[9]
	Hemiplegic migraine	Intravenous 1.5 μg/min	9–22	0	[29, 30]
Dipyridamole		Intravenous 0.142 mg/kg/min	50	0	[31]
Vasoactive intestinal peptide		Intravenous 8 pmol/kg/min	0	0	[32]
PACAP38		Intravenous 10 pmol/kg/min	66–73	0	[11, 33]
Prostaglandin I_2		Intravenous 10 ng/kg/min	50	0	[34]

arteries in migraine patient than healthy controls [10]. However, meta-analysis of human headache models in healthy subjects shows no relationship between maximal headache score and intra- and extracerebral artery changes [37], and the arterial dilatation occurs only in the first 60 min [25] and not during the delayed migraine phase [38]. Thus, the neurobiological mechanisms of GTN-induced migraine-like attacks are not fully clarified [19, 38]. Given that NO penetrates the blood–brain barrier, it is possible that NO may trigger migraine through peripheral and/or central modulation of the brain.

5.3.2 CGRP

The involvement of calcitonin gene-related peptide (CGRP) in the human trigeminovascular reflex was first demonstrated by Goadsby et al. [39] showing that thermocoagulation of the trigeminal ganglion leads to CGRP release into the extracerebral circulation of humans. During spontaneous MO attacks, one study has shown CGRP elevation sampled from the external jugular vein [40] in comparison to healthy controls, but later Tvedskov et al. found no changes in CGRP from the external jugular vein during attacks compared to outside attacks [41]. The same discrepancy was shown in three studies in migraine with aura (MA) patients [42–44]. However, using a human migraine model, Lassen et al. [9] infused CGRP or placebo for 20 min in 12 MO patients in a double-blind crossover study. All of the patients experienced headache, and 3 patients experienced headache according to the ICHD-2 criteria for MO. Using newer criteria for migraine-like headache in experimental models [11], 67 % experienced migraine-like attacks after CGRP compared to only one after placebo. The effect of CGRP in migraine with aura has later been explored by Hansen et al. [28], who revealed that CGRP induced MO attacks in 57 % of MA patients and aura in 28 %. Using MR angiography, the vascular effects of CGRP have been investigated in healthy subjects, where CGRP caused dilatation of extracranial middle meningeal artery (MMA), but not the MCA, which was reversed after subcutaneous sumatriptan parallel with abortion of headache [17]. In addition, CGRP-induced migraine-like attacks are associated with dilatation of extra- and intracranial arteries on the same side as the headache location [16]. Whether or how the perivascular space is important for the generation of CGRP-induced migraine remains to be further elucidated in future human and animal studies.

5.3.3 VIP and PACAP

Pituitary adenylate cyclase-activating polypeptide (PACAP) is distributed in human sensory [45] and parasympathetic nerve ganglia [46] with perivascular nerve fibre projections. PACAP coexists with vasoactive intestinal peptide (VIP) [47] and both belong to the secretin–glucagon family [48]. The $VPAC_1$ and $VPAC_2$ receptors bind both PACAP and VIP ligands with similar affinities, but the PACAP type 1 receptor preferentially binds PACAP [49]. In a recent study, Tuka et al. [50] demonstrated a low level of PACAP38 in the interictal plasma of migraineurs compared with healthy controls, but elevated PACAP38 levels during attacks compared to the overall population of migraineurs outside of attacks. The headache-eliciting effect of VIP and PACAP38, the most predominant PACAP form, has been systematically studied in human models in healthy volunteers [11, 51, 52] and in MO patients [11, 32]. The studies have shown that the systemic administration of VIP induces only a very mild and short-lasting immediate headache both in healthy subjects [32, 51] and migraineurs [32, 33]. Thus, VIP is the first vasoactive substance found not to robustly induce migraine. In contrast to VIP, PACAP38 infusion induced migraine attacks in 58–73 % of MO patients [11, 33]. Amin et al. [33] recently investigated VIP and PACAP38 provocation in MO patients and found using MRA that both peptides induced marked dilatation of the extracranial, but not intracranial arteries. However, PACAP38-induced vasodilatation was longer lasting (>2 h), whereas vasoactive intestinal polypeptide-induced dilatation was normalized after 2 h. Furthermore, explorative analysis revealed that PACAP38 concentration at 60 min was increased significantly in the group of patients who reported delayed migraine attacks than in the group of patients ($n=6$) who did not report delayed attacks. This finding is quite extraordinary, given that PACAP38 has a half-life of only 3.5 min

[52]. Whether this finding is caused by delayed elimination, de novo synthesis or endogenous release is unknown. It is likely that the PAC1 receptor plays an important role in the induction of migraine [53], given that PACAP38 has a much larger affinity than VIP for this receptor. However, other mechanisms may also be important, such as dural mast cell degranulation [54] or central release of CGRP from the TNC [55], which may be independent of the PAC1 receptor. Further study on the role of PACAP38 in inducing migraine is very much needed and anticipated.

5.3.4 *Non-pharmacological Triggers Such as Stress, Exercise, Visual Triggering and Hypoxia*

Several non-pharmacological triggers have been tested for their ability to induce migraine. When asking patients for potential triggers, they report stress (both mental and physical), visual stimulation, several food substances and atmospheric conditions [56]. Of these, a few have been tested in a randomized controlled trial. Mental stress is an important trigger according to patients and doctors alike, but experiments so far did not measure the clinical headache response after a stressor. Physical stress is also frequent identified trigger. Hougaard et al. [57] in a recent controlled study recruited 27 patients with MA, who reported that bright or flickering light or strenuous exercise would trigger their migraine attacks. The patients were then experimentally provoked by different types of photostimulation and strenuous exercise, but, surprisingly, only 3/27 (11 %) reported MA attacks following provocation. Of the different food substances, only chocolate, red wine and tyramine have been tested in a randomized clinical trial (RCT) for their migraine-provoking abilities. Red wine provoked migraine in 9 out of 11 migraine patients who were preselected on being sensitive for red wine [58]. Chocolate triggered migraine in 5 out of 12 'chocolate-sensitive' migraine patients, whereas in a second study, the headache response after chocolate did not differ from placebo [59, 60]. Tyramine (200 mg), a naturally occurring catecholamine analogue releasing agent found in fermented food, has also been tested in a provocation study in 80 migraine patients, and there was no difference in the occurrence of headache between tyramine and placebo [61]. Atmospheric factors are difficult to manipulate, but it is possible to position migraine patients in a hypoxic condition. After 5 h of normobaric hypoxia, 6 out of 16 patients had a full-blown migraine attack [62].

5.3.5 *Migraine Aura*

Migraine aura is likely caused by cortical spreading depression (CSD) [6], as the rate and progression of CSD and MA are similar. However, CSD has never been proven as the cause of MA, and, therefore, there is a great need to investigate MA mechanisms under controlled conditions. Using GTN, Christiansen et al. [20]

demonstrated that 50 % of the patients suffering exclusively from migraine with aura developed migraine headache with associated symptoms, but *none* of them developed migraine aura. Afridi et al. [21] reported that 1 out of 21 patients with MA had an aura triggered on 2 separate occasions by GTN, and only one during the second session. Following sublingual GTN provocation, Sances et al. [7] reported 3/22 (14 %) developed a visual aura. Interestingly, Hansen et al. [28] demonstrated that CGRP infusion in 14 MA patients caused aura in 4/14 (29 %). There are also two clinical reports of aura triggered by visual stimuli and vigorous physical activity in a few MA patients [63, 64]. However, the validity of triggering using light or exercise was recently challenged by Hougaard et al. [57] described above. In conclusion, so far, no valid experimental model exists to reproduce aura episodes in MA patients. In Table 5.2, percentages of patients reporting migraine-like attacks in experimental studies are shown.

5.4 Provocation Studies in Other Primary Headache Syndromes

Most experimental studies have been done in migraine patients. A few studies looked at other primary headache syndromes and, so far, only nitroglycerin is used to provoke tension-type headache (TTH), cluster headache (CH) and chronic paroxysmal hemicrania (CPH). Infusion of GTN in TTH patients in a randomized controlled trial caused both an immediate headache during infusion and delayed-type headache after 2–4 h. The delayed type of headache resembled TTH phenotype [65]. Measurements on muscle hardness, myofascial tenderness and pain thresholds during the immediate-type headache were not different after GTN compared to placebo [65]. This led to the conclusion that there is no peripheral or central sensitization during immediate-type headache. There was also no significant change in CGRP after GTN compared to placebo [66]. Unfortunately, no measurements were done during delayed-type headache. Comparing TTH and controls after GTN infusion, arginine was not different, but TTH patients showed an increase in citrulline after 60 min [67]. Possibly, GTN triggers the endogenous production of NO in TTH patient but not in healthy controls. In cluster headache, attacks can be triggered during an episode using GTN. The attacks started 30–50 min after the challenge of 1 mg sublingual [68]. The attacks did not differ from the spontaneous attacks. When cluster patients need therapeutic administration of nitrates for angina, some progress into a new cluster period [69]. Few studies examined the pathophysiological effect of GTN infusion in CH patient, but specific areas of the hypothalamus might be activated during a trigger attack [70]. Why GTN triggers different types of headaches in separate groups of primary headache patients is unknown. A prospective comparative trial to study the effect of GTN in different headache models will certainly increase our knowledge of the pathophysiology of the different headache types.

5.5 Application of Human Models in Drug Development

In view of the potentially important role of NO in primary headaches, surprisingly, few therapeutic studies have been performed with NOS inhibitors. Lassen et al. [71] investigated whether a non-selective nitric oxide synthase (NOS) inhibitor, *N*(G)-mono-methyl-L-arginine (L-NMMA), might have anti-migraine effects. This proof of concept study demonstrated that NOS inhibition is effective in treating spontaneous migraine attacks. Ashina et al. investigated in a series of studies patients with chronic tension-type headache, who received L-NMMA or placebo to test changes in pain scores, trapezius muscle hardness and pericranial myofascial tenderness [72, 73]. Compared to placebo, L-NMMA reduced pain scores and the summary score of trapezius muscle hardness, but not pericranial tenderness [72, 73]. The role in modulation of trigeminovascular nociception, and in particular migraine generating properties of CGRP, stimulated interest in CGRP antagonism as a potential target for anti-migraine drugs. The first proof of concept study showed that olcegepant, a selective CGRP antagonist, was effective in treating acute migraine attacks [74]. Later, a phase II trial demonstrated that a novel oral CGRP receptor antagonist, telcagepant, was effective and generally well tolerated for the acute treatment of migraine [75].

Currently, only the triptans are used as specific acute treatment in migraine and cluster headache [76, 77], but their exact mode of action is still unresolved. Preventive primary headache treatments are used for various disorders, such as hypertension, epilepsy and depression, and their anti-nociceptive effects primarily derive from serendipity and not neurobiological considerations [78]. The question is if experimental headache models may be a helpful tool to explore relevant neurobiological mechanisms of existing used drugs in primary headache.

The effect of sumatriptan on GTN-induced headache has been examined in several studies [79–81]. In a double-blind crossover study, Iversen and Olesen [79] injected sumatriptan 6 mg or placebo subcutaneously in ten healthy controls, followed by GTN infusion. This study demonstrated that sumatriptan significantly reduced the GTN-induced immediate headache and aborted cranial dilatation. Another study by Schmetterer et al. [80] confirmed the efficacy of sumatriptan to prevent GTN-induced headache and dilatation of the MCA. A recent study [82] tested the effect of zolmitriptan and aspirin to a 140 min infusion of GTN (0.125 μg/kg/min) in healthy subjects, where the drugs were given 20 min into the GTN infusion. The study showed no effect on aspirin or zolmitriptan, and the authors therefore suggested that NO might work later in the cascade of events that lead to headache than the anti-migraine drugs. However, there is a clear discrepancy between triptan response in the Iversen and Olesen study [79] compared to the Tvedskov et al. [82], which is likely to be caused by administration route and drug timing in relation to GTN infusion.

Tvedskov et al. [83] used the GTN model of migraine to test the effect of valproate, a well-known prophylactic drug in migraine treatment. This study showed that pretreatment with valproate was better than placebo in preventing GTN-induced migraine. In another study, Tvedskov et al. [83] observed no effect of the prophylactic drug propranolol on GTN-induced headache and migraine. Tfelt-Hansen et al. [84] explored the effect of 15 MO patients pretreated with 150 mg of prednisolone or placebo followed by GTN infusion in a double-blind placebo-controlled study. Pretreatment with prednisolone did not reduce the immediate GTN-induced headache or inhibit the frequency of delayed headache. However, the intensity of delayed GTN-induced headache was significantly decreased. This suggests that GTN causes induction of inflammatory mediators, which can be suppressed by prednisolone, and a likely mechanism of delayed GTN-induced migraine. These studies suggest that the GTN model of experimental headache may represent a powerful tool for testing anti-migraine drugs and thereby contribute to better understanding of the mechanism of action of existing and future migraine or other primary headache therapies.

5.6 Genetic Background and Provocation

The only known primary headache with an autosomal dominant inheritance is the MA subtype familial hemiplegic migraine (FHM) in which there is hemiparesis during the aura phase [8]. The identification of the mutated FHM genes [85–87] stimulated an interest to explore the link between genotype and phenotype [88]. Hansen et al. [23, 24] used intravenous infusion of GTN in a series of studies with FHM-1 and FHM-2 patients. Quite surprisingly, only 13–25 % of FHM patients reported migraine-like attacks [23, 24], which was a very different response than in the common types of migraine. Furthermore, Hansen et al. [25] showed that following GTN infusion patients with FHM with coexisting migraine with and without aura reported statistical significantly more migraine attacks than patients with pure FHM. These results might indicate that the pathophysiological mechanisms behind FHM are different than in the majority of patients with and without aura. Thus, the FHM results show that human headache models have a huge potential to explore possible links between genetic mutations and neurobiological pathways.

5.7 Concluding Remarks

Experimental human models are extremely useful to study the pathophysiology of primary headache disorders. Unlike many other neurological disorders, these primary headache disorders are fully reversible, and using different experimental conditions, we will be able to increase our knowledge of the function of the brain. Yet, there are limitations to human models. First, only one substance is investigated, which may not resemble a spontaneous migraine attack, where several substances may be released and have an effect simultaneously. Secondly, there is a limit to how much we can test and investigate in a human headache model, as blood samples, monitoring and imaging, etc. may induce stress in the subject investigated, which may bias the results. Thirdly, not all primary headache patients are sensitive to triggers, which highlights the complexity of headaches. In conclusion, the knowledge gained from human headache models continues to be enormous, and it is possible that these experiments will soon lead to the development of new prophylactic drugs that decrease the burden of headache.

References

1. Schytz HW, Schoonman GG, Ashina M (2010) What have we learnt from triggering migraine? Curr Opin Neurol 23:259–265
2. Christiansen I, Daugaard D, Lykke TL, Olesen J (2000) Glyceryl trinitrate induced headache in migraineurs – relation to attack frequency. Eur J Neurol 7:405–411
3. Benedetti F, Durando J, Vighetti S (2014) Nocebo and placebo modulation of hypobaric hypoxia headache involves the cyclooxygenase-prostaglandins pathway. Pain 155:921–928
4. Stovner LJ, Oftedal G, Straume A, Johnsson A (2008) Nocebo as headache trigger: evidence from a sham-controlled provocation study with RF fields. Acta Neurol Scand Suppl 188:67–71
5. Iversen HK, Olesen J, Tfelt-Hansen P (1989) Intravenous nitroglycerin as an experimental model of vascular headache. Basic characteristics. Pain 38:17–24
6. Olesen J, Larsen B, Lauritzen M (1981) Focal hyperemia followed by spreading oligemia and impaired activation of rCBF in classic migraine. Ann Neurol 9:344–352
7. Sances G, Tassorelli C, Pucci E, Ghiotto N, Sandrini G, Nappi G (2004) Reliability of the nitroglycerin provocative test in the diagnosis of neurovascular headaches. Cephalalgia 24:110–119
8. International Classification Committee of the International Headache Society (IHS) (2013) The International Classification of Headache Disorders, 3rd edition (beta version). Cephalalgia 33:629–808
9. Lassen LH, Haderslev PA, Jacobsen VB, Iversen HK, Sperling B, Olesen J (2002) CGRP may play a causative role in migraine. Cephalalgia 22:54–61
10. Thomsen LL, Iversen HK, Brinck TA, Olesen J (1993) Arterial supersensitivity to nitric oxide (nitroglycerin) in migraine sufferers. Cephalalgia 13:395–399
11. Schytz HW, Birk S, Wienecke T, Kruuse C, Olesen J, Ashina M (2009) PACAP38 induces migraine-like attacks in patients with migraine without aura. Brain 132(Pt 1):16–25
12. Nichols FT III, Mawad M, Mohr JP, Stein B, Hilal S, Michelsen WJ (1990) Focal headache during balloon inflation in the internal carotid and middle cerebral arteries. Stroke 21:555–559
13. Ray BS, Wolff HG (1940) Experimental studies on headache. Arch Surg 41:813–856
14. Thomsen LL, Iversen HK (1993) Experimental and biological variation of three-dimensional transcranial Doppler measurements. J Appl Physiol 75:2805–2810
15. Amin FM, Lundholm E, Hougaard A, Arngrim N, Wiinberg L, de Koning PJ, Larsson HB, Ashina M (2014) Measurement precision and biological variation of cranial arteries using automated analysis of 3 T magnetic resonance angiography. J Headache Pain 15:25
16. Asghar MS, Hansen AE, Amin FM, van der Geest RJ, Koning P, Larsson HB, Olesen J, Ashina M (2011) Evidence for a vascular factor in migraine. Ann Neurol 69:635–645
17. Asghar MS, Hansen AE, Kapijimpanga T, van der Geest RJ, van der Koning P, Larsson HB, Olesen J, Ashina M (2010) Dilation by CGRP of middle meningeal artery and reversal by sumatriptan in normal volunteers. Neurology 75:1520–1526
18. Goadsby PJ, Charbit AR, Andreou AP, Akerman S, Holland PR (2009) Neurobiology of migraine. Neuroscience 161:327–341
19. Olesen J, Burstein R, Ashina M, Tfelt-Hansen P (2009) Origin of pain in migraine: evidence for peripheral sensitisation. Lancet Neurol 8:679–690
20. Christiansen I, Thomsen LL, Daugaard D, Ulrich V, Olesen J (1999) Glyceryl trinitrate induces attacks of migraine without aura in sufferers of migraine with aura. Cephalalgia 19:660–667

21. Afridi SK, Kaube H, Goadsby PJ (2004) Glyceryl trinitrate triggers premonitory symptoms in migraineurs. Pain 110:675–680
22. Thomsen LL (1994) KCIHOJ: a nitric oxide donor triggers genuine migraine attacks. Eur J Neurol 1:73–80
23. Hansen JM, Thomsen LL, Olesen J, Ashina M (2008) Familial hemiplegic migraine type 1 shows no hypersensitivity to nitric oxide. Cephalalgia 28:496–505
24. Hansen JM, Thomsen LL, Marconi R, Casari G, Olesen J, Ashina M (2008) Familial hemiplegic migraine type 2 does not share hypersensitivity to nitric oxide with common types of migraine. Cephalalgia 28:367–375
25. Hansen JM, Thomsen LL, Olesen J, Ashina M (2010) Coexisting typical migraine in familial hemiplegic migraine. Neurology 74:594–600
26. Kruuse C, Thomsen LL, Birk S, Olesen J (2003) Migraine can be induced by sildenafil without changes in middle cerebral artery diameter. Brain 126:241–247
27. Lassen LH, Thomsen LL, Olesen J (1995) Histamine induces migraine via the H1-receptor. Support for the NO hypothesis of migraine. Neuroreport 6:1475–1479
28. Hansen JM, Hauge AW, Olesen J, Ashina M (2010) Calcitonin gene-related peptide triggers migraine-like attacks in patients with migraine with aura. Cephalalgia 30:1179–1186
29. Hansen JM, Thomsen LL, Olesen J, Ashina M (2011) Calcitonin gene-related peptide does not cause migraine attacks in patients with familial hemiplegic migraine. Headache 51:544–553
30. Hansen JM, Thomsen LL, Olesen J, Ashina M (2008) Calcitonin gene-related peptide does not cause the familial hemiplegic migraine phenotype. Neurology 71:841–847
31. Kruuse C, Lassen LH, Iversen HK, Oestergaard S, Olesen J (2006) Dipyridamole may induce migraine in patients with migraine without aura. Cephalalgia 26:925–933
32. Rahmann A, Wienecke T, Hansen JM, Fahrenkrug J, Olesen J, Ashina M (2008) Vasoactive intestinal peptide causes marked cephalic vasodilation, but does not induce migraine. Cephalalgia 28:226–236
33. Amin FM, Hougaard A, Schytz HW, Asghar MS, Lundholm E, Parvaiz AI, de Koning PJ, Andersen MR, Larsson HB, Fahrenkrug J, Olesen J, Ashina M (2014) Investigation of the pathophysiological mechanisms of migraine attacks induced by pituitary adenylate cyclase-activating polypeptide-38. Brain 137:779–794
34. Wienecke T, Olesen J, Ashina M (2010) Prostaglandin I2 (epoprostenol) triggers migraine-like attacks in migraineurs. Cephalalgia 30:179–190
35. Sicuteri F, Del BE, Poggioni M, Bonazzi A (1987) Unmasking latent dysnociception in healthy subjects. Headache 27:180–185
36. Olesen J, Iversen HK, Thomsen LL (1993) Nitric oxide supersensitivity: a possible molecular mechanism of migraine pain. Neuroreport 4:1027–1030
37. Ashina M, Tfelt-Hansen P, Dalgaard P, Olesen J (2011) Lack of correlation between vasodilatation and pharmacologically induced immediate headache in healthy subjects. Cephalalgia 31:683–690
38. Schoonman GG, van der Grond J, Kortmann C, van der Geest RJ, Terwindt GM, Ferrari MD (2008) Migraine headache is not associated with cerebral or meningeal vasodilatation – a 3T magnetic resonance angiography study. Brain 131:2192–2200
39. Goadsby PJ, Edvinsson L, Ekman R (1988) Release of vasoactive peptides in the extracerebral circulation of humans and the cat during activation of the trigeminovascular system. Ann Neurol 23:193–196
40. Goadsby PJ, Edvinsson L, Ekman R (1990) Vasoactive peptide release in the extracerebral circulation of humans during migraine headache. Ann Neurol 28:183–187
41. Tvedskov JF, Lipka K, Ashina M, Iversen HK, Schifter S, Olesen J (2005) No increase of calcitonin gene-related peptide in jugular blood during migraine. Ann Neurol 58:561–568
42. Stepien A, Jagustyn P, Trafny EA, Widerkiewicz K (2003) Suppressing effect of the serotonin 5HT1B/D receptor agonist rizatriptan on calcitonin gene-related peptide (CGRP) concentration in migraine attacks. Neurol Neurochir Pol 37:1013–1023
43. Gallai V, Sarchielli P, Floridi A, Franceschini M, Codini M, Glioti G, Trequattrini A, Palumbo R (1995) Vasoactive peptide levels in the plasma of young migraine patients with and without aura assessed both interictally and ictally. Cephalalgia 15:384–390

44. Friberg L, Olesen J, Olsen TS, Karle A, Ekman R, Fahrenkrug J (1994) Absence of vasoactive peptide release from brain to cerebral circulation during onset of migraine with aura. Cephalalgia 14:47–54
45. Tajti J, Uddman R, Moller S, Sundler F, Edvinsson L (1999) Messenger molecules and receptor mRNA in the human trigeminal ganglion. J Auton Nerv Syst 76:176–183
46. Uddman R, Hara H, Edvinsson L (1989) Neuronal pathways to the rat middle meningeal artery revealed by retrograde tracing and immunocytochemistry. J Auton Nerv Syst 26:69–75
47. Edvinsson L, Elsas T, Suzuki N, Shimizu T, Lee TJ (2001) Origin and Co-localization of nitric oxide synthase, CGRP, PACAP, and VIP in the cerebral circulation of the rat. Microsc Res Tech 53:221–228
48. Harmar AJ, Arimura A, Gozes I, Journot L, Laburthe M, Pisegna JR, Rawlings SR, Robberecht P, Said SI, Sreedharan SP, Wank SA, Waschek JA, International Union of Pharmacology (1998) XVIII. Nomenclature of receptors for vasoactive intestinal peptide and pituitary adenylate cyclase-activating polypeptide. Pharmacol Rev 50:265–270
49. Fahrenkrug J (2006) PACAP – a multifaceted neuropeptide. Chronobiol Int 23:53–61
50. Tuka B, Helyes Z, Markovics A, Bagoly T, Szolcsanyi J, Szabo N, Toth E, Kincses ZT, Vecsei L, Tajti J (2013) Alterations in PACAP-38-like immunoreactivity in the plasma during ictal and interictal periods of migraine patients. Cephalalgia 33:1085–1095
51. Hansen JM, Sitarz J, Birk S, Rahmann AM, Oturai PS, Fahrenkrug J, Olesen J, Ashina M (2006) Vasoactive intestinal polypeptide evokes only a minimal headache in healthy volunteers. Cephalalgia 26:992–1003
52. Birk S, Sitarz JT, Petersen KA, Oturai PS, Kruuse C, Fahrenkrug J, Olesen J (2007) The effect of intravenous PACAP38 on cerebral hemodynamics in healthy volunteers. Regul Pept 140:185–191
53. Schytz HW, Olesen J, Ashina M (2010) The PACAP receptor: a novel target for migraine treatment. Neurotherapeutics 7:191–196
54. Baun M, Pedersen MH, Olesen J, Jansen-Olesen I (2012) Dural mast cell degranulation is a putative mechanism for headache induced by PACAP-38. Cephalalgia 32:337–345
55. Jansen-Olesen I, Baun M, Amrutkar DV, Ramachandran R, Christophersen DV, Olesen J (2014) PACAP-38 but not VIP induces release of CGRP from trigeminal nucleus caudalis via a receptor distinct from the PAC1 receptor. Neuropeptides 48:53–64
56. Wober C, Brannath W, Schmidt K, Kapitan M, Rudel E, Wessely P, Wober-Bingol C (2007) Prospective analysis of factors related to migraine attacks: the PAMINA study. Cephalalgia 27:304–314
57. Hougaard A, Amin FM, Hauge AW, Ashina M, Olesen J (2013) Provocation of migraine with aura using natural trigger factors. Neurology 80:428–431
58. Littlewood JT, Gibb C, Glover V, Sandler M, Davies PT, Rose FC (1988) Red wine as a cause of migraine. Lancet 1:558–559
59. Marcus DA, Scharff L, Turk D, Gourley LM (1997) A double-blind provocative study of chocolate as a trigger of headache. Cephalalgia 17:855–862
60. Gibb CM, Davies PT, Glover V, Steiner TJ, Clifford RF, Sandler M (1991) Chocolate is a migraine-provoking agent. Cephalalgia 11:93–95
61. Ziegler DK, Stewart R (1977) Failure of tyramine to induce migraine. Neurology 27: 725–726
62. Schoonman GG, Sandor PS, Agosti RM, Siccoli M, Bartsch P, Ferrari MD, Baumgartner RW (2006) Normobaric hypoxia and nitroglycerin as trigger factors for migraine. Cephalalgia 26: 816–819
63. Cao Y, Welch KM, Aurora S, Vikingstad EM (1999) Functional MRI-BOLD of visually triggered headache in patients with migraine. Arch Neurol 56:548–554
64. Hadjikhani N, Sanchez del Rio M, Wu O, Schwartz D, Bakker D, Fischl B, Kwong KK, Cutrer FM, Rosen BR, Tootell RB, Sorensen AG, Moskowitz MA (2001) Mechanisms of migraine aura revealed by functional MRI in human visual cortex. Proc Natl Acad Sci U S A 98: 4687–4692
65. Ashina M, Bendtsen L, Jensen R, Olesen J (2000) Nitric oxide-induced headache in patients with chronic tension-type headache. Brain 123(Pt 9):1830–1837

66. Ashina M, Bendtsen L, Jensen R, Schifter S, Olesen J (2001) Calcitonin gene-related peptide levels during nitric oxide-induced headache in patients with chronic tension-type headache. Eur J Neurol 8:173–178
67. Ashina M, Simonsen H, Bendtsen L, Jensen R, Olesen J (2004) Glyceryl trinitrate may trigger endogenous nitric oxide production in patients with chronic tension-type headache. Cephalalgia 24:967–972
68. Ekbom K (1968) Nitroglycerin as a provocative agent in cluster headache. Arch Neurol 19:487–493
69. Ekbom K, Sjostrand C, Svensson DA, Waldenlind E (2004) Periods of cluster headache induced by nitrate therapy and spontaneous remission of angina pectoris during active clusters. Cephalalgia 24:92–98
70. May A, Bahra A, Buchel C, Frackowiak RS, Goadsby PJ (1998) Hypothalamic activation in cluster headache attacks. Lancet 352:275–278
71. Lassen LH, Ashina M, Christiansen I, Ulrich V, Grover R, Donaldson J, Olesen J (1998) Nitric oxide synthase inhibition: a new principle in the treatment of migraine attacks. Cephalalgia 18:27–32
72. Ashina M, Bendtsen L, Jensen R, Lassen LH, Sakai F, Olesen J (1999) Possible mechanisms of action of nitric oxide synthase inhibitors in chronic tension-type headache. Brain 122(Pt 9):1629–1635
73. Ashina M, Lassen LH, Bendtsen L, Jensen R, Olesen J (1999) Effect of inhibition of nitric oxide synthase on chronic tension-type headache: a randomised crossover trial. Lancet 353:287–289
74. Olesen J, Diener HC, Husstedt IW, Goadsby PJ, Hall D, Meier U, Pollentier S, Lesko LM (2004) Calcitonin gene-related peptide receptor antagonist BIBN 4096 BS for the acute treatment of migraine. N Engl J Med 350:1104–1110
75. Ho TW, Mannix LK, Fan X, Assaid C, Furtek C, Jones CJ, Lines CR, Rapoport AM (2008) Randomized controlled trial of an oral CGRP receptor antagonist, MK-0974, in acute treatment of migraine. Neurology 70:1304–1312
76. Derry CJ, Derry S, Moore RA (2014) Sumatriptan (all routes of administration) for acute migraine attacks in adults – overview of Cochrane reviews. Cochrane Database Syst Rev (5):CD009108
77. Law S, Derry S, Moore RA (2013) Triptans for acute cluster headache. Cochrane Database Syst Rev (7):CD008042
78. Vollbracht S, Rapoport AM (2013) The pipeline in headache therapy. CNS Drugs 27:717–729
79. Iversen HK, Olesen J (1996) Headache induced by a nitric oxide donor (nitroglycerin) responds to sumatriptan. A human model for development of migraine drugs. Cephalalgia 16:412–418
80. Schmetterer L, Wolzt M, Krejcy K, Graselli U, Findl O, Eichler HG, Singer EA (1996) Cerebral and ocular hemodynamic effects of sumatriptan in the nitroglycerin headache model. Clin Pharmacol Ther 60:199–205
81. Fullerton T, Komorowski-Swiatek D, Forrest A, Gengo FM (1999) The pharmacodynamics of sumatriptan in nitroglycerin-induced headache. J Clin Pharmacol 39:17–29
82. Tvedskov JF, Iversen HK, Olesen J, Tfelt-Hansen P (2010) Nitroglycerin provocation in normal subjects is not a useful human migraine model? Cephalalgia 30:928–932
83. Tvedskov JF, Thomsen LL, Iversen HK, Gibson A, Wiliams P, Olesen J (2004) The prophylactic effect of valproate on glyceryltrinitrate induced migraine. Cephalalgia 24:576–585
84. Tfelt-Hansen P, Daugaard D, Lassen LH, Iversen HK, Olesen J (2009) Prednisolone reduces nitric oxide-induced migraine. Eur J Neurol 16:1106–1111
85. De FM, Marconi R, Silvestri L, Atorino L, Rampoldi L, Morgante L, Ballabio A, Aridon P, Casari G (2003) Haploinsufficiency of ATP1A2 encoding the Na^+/K^+ pump alpha2 subunit associated with familial hemiplegic migraine type 2. Nat Genet 33:192–196
86. Dichgans M, Freilinger T, Eckstein G, Babini E, Lorenz-Depiereux B, Biskup S, Ferrari MD, Herzog J, van den Maagdenberg AM, Pusch M, Strom TM (2005) Mutation in the neuronal voltage-gated sodium channel SCN1A in familial hemiplegic migraine. Lancet 366:371–377

87. Ophoff RA, Terwindt GM, Vergouwe MN, van Ejik R, Oefner PJ, Hoffman SM, Lamerdin JE, Mohrenweiser HW, Bulman DE, Ferrari M, Haan J, Lindhout D, van Ommen GJ, Hofker MH, Ferrari MD, Frants RR (1996) Familial hemiplegic migraine and episodic ataxia type-2 are caused by mutations in the Ca2+ channel gene CACNL1A4. Cell 87:543–552
88. van de Ven RC, Kaja S, Plomp JJ, Frants RR, van den Maagdenberg AM, Ferrari MD (2007) Genetic models of migraine. Arch Neurol 64:643–646

Chapter 6
Imaging of Migraine

Michaela Andelova, David Borsook, and Till Sprenger

6.1 Introduction

Over the last two decades, neuroimaging studies have led to a reappraisal of central mechanisms involved in migraine pathophysiology. Neuroimaging studies clearly support the view of migraine being a primary brain disorder with altered sensory processing even in pain-free periods. In the future, neuroimaging has the potential to provide a noninvasive biomarker that will potentially facilitate headache diagnosis and aid physicians in treatment decisions and treatment monitoring. However, findings from current studies are still partially inconsistent. In this chapter, the findings of neuroimaging studies in migraine are summarized according to brain anatomy separately for (peri)ictal and interictal phase where applicable. Figure 6.1 and Table 6.1 summarize brain areas where activation has been reported during attacks, respectively, which have been suggested to play a role outside of attacks. Findings in medication-overuse headache are briefly outlined at the end of the chapter.

M. Andelova • T. Sprenger (✉)
Department of Neurology, University Hospital Basel,
Petersgraben 4, Basel 4031, Switzerland

Division of Neurology, DKD Helios Klinik Wiesbaden,
Aukammallee 33, Wiesbaden 65191, Switzerland
e-mail: till.sprenger@usb.ch

D. Borsook
Center for Pain and the Brain, Boston Children's, Massachusetts
General and McLean Hospitals, Harvard Medical School, Boston, MA, USA

M. Ashina, P. Geppetti (eds.), *Pathophysiology of Headaches: From Molecule to Man*, Headache, DOI 10.1007/978-3-319-15621-7_6

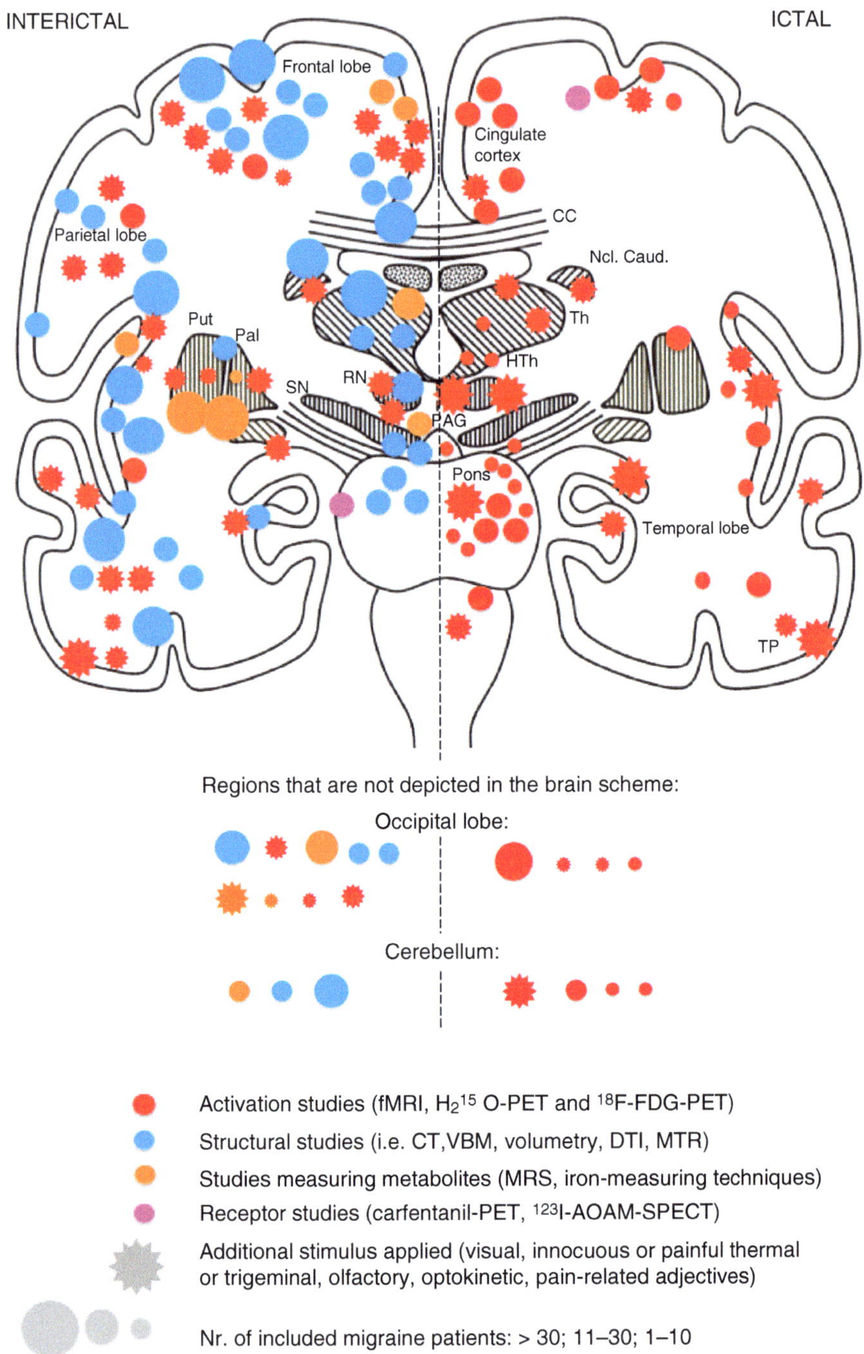

Fig. 6.1 Brain imaging findings in patients with migraine (with or without aura). The *left side* represents studies on interictal state in migraineurs, while on the *right side* results from ictal studies are shown. *Right*: during attacks, brain activations have been reported most consistently in the brainstem, but also in other brain areas implicated with pain processing (such as the cingulate cortex and insula). *Left*: interictally, several studies have shown a decrease in gray matter density in a distributed network across the entire brain

6.2 Brainstem

There is growing evidence that dysfunction of multiple brainstem nuclei may reflect an important part of the migraine pathophysiology.

Table 6.1 Summary of structural and functional brain imaging findings in the interictal and ictal migraine state; Abbreviation: TBD: to be determined

Brain structure	Ictal and peri-ictal studies	Interictal studies	Possible role in migraine pathogenesis
Spinal trigeminal nucleus	Preictal normalization	Gradient-like activity following nociceptive stimulation	Generation of migraine attack? Regulation of cortical excitability/dishabituation
Pons	Activation consistently observed in PET and fMRI studies	Hypometabolism in brainstem parallel to visual cortex hypersensitivity Increase in 5HT1A receptor and SERT availability	"Migraine generator"? Dysfunction of inhibitory pathways Serotonergic dysregulation
Substantia nigra	Activation	Reduced activity during executive task in pts. with MOH	Dopaminergic dysfunction, overlap between MOH and addiction
Red nucleus	Activation	Iron accumulation	TBD
PAG	Activation in premonitory phase and during headache	Reduced gray matter in migraine and MOH	Altered ascending inhibition/modulation Medication-overuse headache
Nucleus cuneiformis	Activation during headache	Hypoactivation in response to heat stimuli	Altered descending modulation/enhanced facilitation of ascending nociceptive pathways (→ trigeminal hyperexcitability) → migraine transformation?
Nucleus accumbens		Volume reduction correlating with disease duration, increased response to pain in high-frequency vs. low-frequency migraineurs	Dysfunction of endogenous opioid system? Altered reward processing?
Amygdala		Gray matter volume reduction	TBD
Hypothalamus	Activation in premonitory phase and during early headache phase		Generation of premonitory symptoms Potential association with migraine triggers (i.e., disrupted sleep)

(continued)

Table 6.1 (continued)

Brain structure	Ictal and peri-ictal studies	Interictal studies	Possible role in migraine pathogenesis
Thalamus	Activation during attack, DTI normalization during attack	Higher thalamic FA values, higher MTR values, iron increase, pulvinar hyperactivation as correlate to dishabituation, hyperactivity of pulvinar in response to thermal nox. stimuli in migraineurs with allodynia, increased GM in MOH	Pathophysiology of photophobia (LGN, Pulvinar) Transformation of trigeminal allodynia into whole-body allodynia Potential role in abnormal habituation MOH
Globus pallidus	Activation during attacks (nucleus lentiformis)	Iron increase, correlating with disease duration	Chronicity?
Putamen	Activation during attacks (nucleus lentiformis)	Iron increase, correlating with disease duration	Chronicity?
Nucleus caudatus		Volume reduction, lower responses to painful stimulation in high-frequency migraineurs, lower volume in high-frequency migraineurs, increased GM in MOH	Chronicity? Association with triptan use and MOH?
Corpus callosum		Decreased fractional anisotropy correlating with interhemispheric resting-state functional connectivity	Altered modulation of interhemispheric connectivity
Anterior cingulate cortex	Activation during attacks, probably not migraine specific, but pain related	Abnormal response to thermal and trigeminal noxious stimuli	Affective and attentive dimensions of pain, multiple mechanisms
Orbitofrontal cortex (OFC)		Increase of OFC gray matter volume in MOH, decrease of OFC gray matter volume predicts response to detoxification	In MOH/detoxification/ addiction Altered sensory integration, decision-making, and expectation and planning behavior associated with sensitivity to reward and punishment (not migraine specific)

Table 6.1 (continued)

Brain structure	Ictal and peri-ictal studies	Interictal studies	Possible role in migraine pathogenesis
Temporal cortex	Activation during headache phase, hyperexcitability to noxious heat	Greater responses to olfactory stimuli in migraineurs with interictal odor hypersensitivity, hyperexcitability to noxious heat, positive correlation between cortical thickness in a superior temporal/inferior parietal region and pain thresholds	Altered multisensory (noxious, olfactory, visual) processing, altered motion processing Inability to modulate pain via shifting attention
Visual cortex	Activation during headache and peri-ictally	Different responses to luminous stimulation with and without concomitant pain, hyperresponsiveness in visual areas activated by motion perception (V5 and V3) to optokinetic visual stimulation, decrease of NAA after photic stimulation, increased Glu/Gln ratio, lactate baseline increase in purely visual aura vs. lactate increase only after vis. stimulation in multimodal aura pts.	Attenuation of habituation of visual stimuli by pain stimuli Altered interictal motion processing Different metabolism alterations between migraineurs with purely visual and multimodal aura

6.2.1 *Ictal Findings*

As early as 1995, Weiller et al. conducted a positron emission tomography (PET) activation study ($H_2{}^{15}O$-PET) in which nine patients were examined during spontaneous migraine attacks [1]. The authors observed enhanced activity in a part of the brainstem (dorsal midbrain/dorsolateral pons) that was later reported to anatomically correspond to the nucleus cuneiformis, colliculus inferior, and reticular nuclei [2], although other authors believe it could reflect activation of the periaqueductal gray matter or locus coeruleus. This activation persisted after successful headache treatment with sumatriptan. The involvement of pontine nuclei in migraine attack processing and possibly generation was confirmed in subsequent studies in both spontaneous and provoked migraine attacks [3–5]. Further brainstem areas such as the red nucleus and substantia nigra have been shown to possess enhanced activity during visually triggered migraine attacks in an fMRI study [6]. More recently, activations of the bilateral ventral midbrain, dorsal midbrain, and dorsomedial pons were documented by Denuelle et al. in a PET study during spontaneous migraine attacks [7].

Regarding lateralization of the brainstem activations during migraine attacks, the results of the studies are somewhat inconsistent. In the abovementioned study by Weiller et al., activations were observed contralaterally to the headache side. In contrary, enhanced brainstem activity ipsilateral to the headache side was observed in PET during GTN-triggered migraine [8]. Left-sided activation of the dorsal pons (regardless of headache side) was observed by Afridi et al. during spontaneous migraine attacks [5]. Hence, lateralization or asymmetry of brainstem activation seems to play a role in migraine that is not yet fully elucidated.

Regarding the relationship between brainstem activation and cortical changes, animal studies have shown that stimulation of brainstem nuclei results in reduction of cortical blood flow with maximal changes observed in the occipital cortex [9, 10] as well as in caudate nucleus [11]. This finding is in line with findings of Cao et al., who showed that brainstem activation precedes activation of the occipital cortex in visually triggered headache attacks in 8 of 12 migraineurs (regardless of the presence of aura) [6]. Thus, these findings support the concept of migraine as a primary brainstem dysfunction with subsequent changes of cortical activity that may in susceptible patients result in CSD and full-blown migraine attacks [12].

Case reports describing symptomatic migraine-like headache with photophobia, nausea, and aggravation by exercise due to demyelinating or ischemic lesions and vascular malformation of the brainstem, e.g., in the PAG [13], dorsal midbrain [14], midpontine tegmentum [15], or middle cerebellar peduncle [16], also support the key role of brainstem in migraine pathophysiology. Furthermore, the presence of midbrain plaques in proximity to the PAG likelihood of headache with migrainous features in multiple sclerosis increases by four [17]. On the other hand, many patients with diffuse brainstem involvement do not experience symptomatic headache.

The efficacy of serotonergic drugs in migraine has driven research of potential underlying serotonergic mechanisms in migraine. Demarquay et al. stimulated migraine patients with known olfactory hypersensitivity and healthy controls with olfactory stimuli prior to PET scanning. Using the radiolabeled 5HT1A receptor antagonist [4-(2′-methoxyphenyl)-1-[2′-(N-2-pirydynyl)-p-fluorobenzamido]-ethylpiperazine ((18)F-MPPF)], differences between interictal migraineurs and controls were studied [18]. After the olfactory stimulation, 4 of 10 migraine patients experienced a migraine attack during the PET scanning, and in these patients, significant increases in 5HT1A receptor availability were observed in the pontine region when compared to headache-free migraineurs and healthy controls. Compared to headache-free migraineurs, patients who developed an attack also had significantly increased 5HT1A availability in serotonergic projection areas, i.e., the orbitofrontal cortex, precentral gyrus, and temporal pole. In a PET study using alpha-[(11) C]methyl-l-tryptophan as marker for brain serotonin synthesis, Sakai et al. demonstrated low interictal serotonergic activity in cortical regions of migraineurs with an increase of serotonergic synthetic activity during attacks which was potently reversed by sumatriptan [19]. Whether such sumatriptan-related changes reflect the mechanisms of action of the compound in migraine attacks remains open to debate. The same group later used tryptophan-PET to study response of migraineurs to another triptan, eletriptan. No difference was observed in baseline global cerebral 5-HT synthesis between migraine and control subjects. After administration of eletriptan, however, migraineurs had striking reductions in global cerebral 5-HT synthesis, suggesting alterations in interictal serotonin metabolism [20].

6.2.2 Interictal Findings

The role of brainstem in migraine is further supported by structural abnormalities, receptor changes, and altered function and connectivity of particular brainstem structures in between migraine attacks.

The nucleus cuneiformis seems to be less activated by thermal stimuli in interictal migraineurs than in healthy controls [21], suggesting a permanent dysfunction of descending pain modulatory pathways. Alterations in processing of the primary brainstem relay station that may be influenced by altered descending modulation have also been noted. Specifically, Stankewitz et al. have shown evidence of cycling responses of the spinal trigeminal nucleus to experimental painful trigeminal stimuli [22]. Repetitive nociceptive stimulation with ammonia resulted in significantly stronger activation of the lower pons corresponding to the location of the spinal trigeminal nucleus (STN) in interictal migraineurs compared to healthy controls. Unexpectedly, no activation differences were observed in other structures of the trigeminal pain pathway. The authors cleverly tested the association between the height of the STN activation and time to the next spontaneous migraine attack and found that the stronger the STN activation, the closer the next attack. This finding suggests that the excitability of the STN is oscillating over time. This probably reflects the changing susceptibility of the brain to generate migraine attack (in response to triggers such as lack of sleep). Moreover, STN activations in preictal migraineurs (scanned 12–48 h before migraine attack) did not differ from healthy controls, and patients scanned during headache attacks showed even lower STN activations than controls and preictal patients, suggesting normalization of STN activity in the preictal phase and during attacks. These findings provide an interesting link to the electrophysiological literature in migraine where dishabituation of sensory stimuli has been shown with a similar behavior of peri-ictal normalization.

The role of the PAG in pain modulation has been well described [23]. Mainero et al. have studied the functional connectivity of the PAG in migraineurs and observed stronger interictal resting-state connectivity between the PAG and several brain areas involved in nociceptive processing and pain modulation [24]. For example, migraineurs with higher attack frequency had reduced connectivity between the PAG, prefrontal regions, and the ACC compared to migraineurs with fewer attacks. Interestingly, the same pattern was observed in patients with a history of allodynia as compared to patients who do not experience allodynia. In contrast to this, however, the severity of ictal allodynia was correlated with the resting-state functional connectivity between key structures of brainstem descending modulatory regions (PAG and nucleus cuneiformis) and other brainstem structures, the thalamus, as well as frontal and temporal regions involved in pain modulation in a more recent study [25]. In patients with chronic migraine, significant correlations have been reported between the duration of the disorder and functional connectivity between anterior insula and PAG [26].

Schuh-Hofer et al. used the radioligand (123)I-ADAM[2-((2-((dimethylamino)methyl)phenyl)thio)-5-iodophenylamine] targeting the brain serotonin transport protein (SERT) and observed significantly increased SERT availability of mesopontine brainstem areas in interictal migraineurs compared to healthy controls, whereas in the thalamus, SERT availability did not differ between the two groups [27].

Finally, structural changes such as increased periaqueductal gray matter density (PAG) and abnormalities in iron metabolism in the PAG and red nucleus have also been reported in migraineurs [28, 29]. Interestingly, changes in the PAG and nucleus cuneiformis (i.e., reduction of preexisting pathological gray matter increase) were observed by means of VBM after successful detoxification in medication-overuse headache [30].

6.3 Hypothalamus

Hypothalamic symptoms are common in migraineurs [31]. Neuroimaging has further supported a role for the structure in migraine.

6.3.1 Ictal and Premonitory Findings

Evidence for an important role of the hypothalamus in migraine pathophysiology comes from a study evidencing activation of the hypothalamus already during the premonitory phase of triggered migraine attacks as shown by PET scanning [32]. Hypothalamic activation has also been shown during the headache phase of attacks [7]. However, there are also ictal migraine studies that did not show hypothalamic activation [3, 5, 8]. Probably, hypothalamic activation is limited to the early phases of attacks, and indeed hypothalamic dysfunction would be well suited to explain premonitory symptoms (i.e., changes of alertness and fatigue, yawning, appetite changes, or thirst that occurs up to 48 h before the attack). Again, the temporal and functional relationship between brainstem and hypothalamic activations remains to be elucidated.

6.3.2 Interictal Findings

Interictal changes in hypothalamic connectivity (i.e., increased connectivity with the parahippocampal gyrus, locus coeruleus, caudate nucleus, and temporal pole) may reflect some of the autonomic symptoms that trigger/precede and accompany migraine attacks [33]. In animal models, direct neuronal projections between thalamic trigeminovascular neurons (posterior and lateral posterior thalamic nuclei) and ventromedial and ventral tuberomammillary hypothalamic nuclei were found [34]. These studies confirm the view of the (posterior) hypothalamus as a potential key area in the initial phase of primary headache syndromes such as in the development of premonitory symptoms and/or its association with some migraine triggers as disrupted sleep, altered eating patterns, and emotional reactions.

Regarding PET for studying hypothalamic abnormalities, new radiotracers targeting dopamine, orexin, or somatostatin receptor could in the future help in studying hypothalamic ictal, peri-ictal, and interictal changes and potentially guide the development of new migraine drugs.

6.4 Thalamus

The general role of the thalamus in the integration and modulation of nociceptive inputs is well established. In migraine imaging studies, specifically abnormalities of the posterior thalamus/pulvinar nucleus have been reported.

6.4.1 Ictal Findings

Activation of the thalamus contralateral to the head pain has been observed during migraine attacks [3, 5]. In an animal model, sensitization of trigeminovascular thalamic neurons from cranial meninges as well as from the extracephalic skin has been shown to be associated with the transformation of headache into whole-body allodynia [35]. In line with this, fMRI blood oxygen level-dependent (BOLD) responses in the posterior thalamus (roughly corresponding to the pulvinar nucleus) induced by brush and innocuous heat stimulation were significantly stronger during an attack than in the pain-free period in migraineurs, who regularly experienced extracephalic allodynia [35]. The authors concluded that third-order thalamic neurons receiving inputs from the meninges and facial and body skin areas may be responsible for the spreading of allodynia beyond the area where migraine headache is located.

6.4.2 Interictal Findings

A putative role of the posterior thalamus in migraine-associated central sensitization was also suggested by Stankewitz et al. In their fMRI study, interictal migraineurs sensitized, whereas control subjects habituated in terms of ratings to nociceptive trigeminal stimuli (i.e., pain ratings were increasing in migraineurs while being attenuated in controls with repetitive stimulation). This behavior was reflected by BOLD increases in the amygdala, the cingulate cortex, and the pulvinar in migraineurs, in contrast to BOLD decreases in the same structures of controls [36].

Abnormal resting-state functional connectivity between the pulvinar and affective pain-processing regions has been reported in patients with chronic migraine [26]. The pulvinar receives inputs from both the trigeminal and optical nerve [37], and hence, it is a potential site of interaction between the processing of different sensory stimuli possibly explaining the bidirectional [38] and dysfunctional relationship between pain and visual processing in migraine. By using DTI and fMRI, increased structural and functional connectivity between the pulvinar and temporal pole was found by Moulton et al., who proposed a potential trigeminothalamic pathway through the pulvinar sending nociceptive signals to the temporal pole [39].

Regarding structural changes in the thalamus, Granziera et al. have studied thalamic microstructure and its dependence on attack frequency in 37 migraineurs both with and without aura by means of MTR, DTI, and relaxation mapping techniques and found significantly shorter T1 relaxation times, higher MTR values, as well as shorter T2* relaxation times suggesting increased cellularity and/or relatively increased iron content, respectively, in migraineurs with aura compared to migraineurs without aura and healthy controls [40].

6.5 Basal Ganglia

The basal ganglia have been shown to play a significant role in acute and chronic pain processing [41].

6.5.1 Interictal Findings

In line with findings of increased iron content in the thalamus [40] and red nucleus [29], increased iron deposition has also been observed in the putamen and caudate nucleus more recently [42]. Among T2, T2*, and T2′ relaxation times, only T2 values in the globus pallidus were able to distinguish episodic from chronic headache patients [43], suggesting that iron metabolism as measured by T2 may be an objective marker of migraine frequency.

Comparison of high- and low-frequency migraineurs revealed significantly lower BOLD responses to noxious thermal stimuli bilaterally throughout the caudate, putamen, and pallidum. Interestingly, increased caudate volume was observed in the high-frequency group [44]. However, in another recent study, migraineurs without aura had a relatively reduced volume of the caudate nucleus and nucleus accumbens (NAc), which correlated with the disease duration [45].

In an FDG-PET study in chronic migraineurs, relatively reduced metabolism was shown bilaterally in the caudate [46]. Additionally, the study by Yuan et al.

mentioned above evidenced that attack frequency and disease duration are associated with increased functional connectivity between the caudate and insula and between NAc and ACC [45]. Interestingly, in a recent animal KCl-induced model of CSD, reduced neuronal activity in the caudate was observed about 45 min after CSD [11]. Changes in caudate volume and connectivity may therefore represent another link between CSD and migraine.

A short-term longitudinal study in migraineurs without aura showed progressively dysfunctional connectivity between the putamen and brainstem, thalamus, and secondary somatosensory and orbitofrontal cortex [47].

6.6 Amygdala

The amygdala is involved in numerous aspects of pain processing [48].

6.6.1 Interictal Findings

The amygdala, especially the laterocapsular part, may play an important role in migraine as it is connected to both the mediodorsal thalamus and the trigeminal nucleus caudalis. Moreover, amygdalar activity can be disrupted by cortical spreading depression in rats [49]. One VBM study demonstrated structural changes in the amygdala and other structures involved in affective pain processing (e.g., ACC, insula) in patients with chronic migraine when compared to episodic migraine patients [50], indicating that repeated migraine attacks induce structural alterations in affective parts of the pain matrix that may subsequently lead to dysfunctional connectivity/affective pain processing. Indeed, resting-state functional connectivity of the amygdala and other affective pain regions is altered in chronic migraine [26]. The finding of altered connectivity between the amygdala and PAG in migraineurs has been mentioned above [24]. Interestingly, an increased connectivity between the amygdala and several viscerosensitive cortical areas including the anterior insula, parietal operculum, thalamus, and temporal pole was found in migraineurs, whereas patients with trigeminal neuralgia or carpal tunnel syndrome did not differ from healthy control, suggesting a relatively specific dysfunction of neurolimbic networks in migraine [51].

6.7 Visual Pathway

Patients with migraine very often complain of photophobia (either as light worsening their pain, so-called photic allodynia, or light itself being unusually unpleasant, i.e., pure photophobia). The pathophysiology of such abnormal sensations and its association to pain, CSD, and aura has yet to be fully elucidated. Whereas the relationship between CSD and aura is quite well established, a potential relationship between (silent) cortical spreading depression and migraine headache remains controversial.

6.7.1 *Ictal Findings and Abnormalities During Aura*

It is generally accepted that migraine aura most likely reflects cortical spreading depression (CSD), a slowly propagating wave of depolarization followed by suppression of neuronal activity [52]. Many imaging studies showed aura correlates with CSD-like events. In one of the first human imaging studies, Olesen applied the Xenon method [53] and observed hyperemia in frontal and parietal areas followed by occipital spreading oligemia during episodes of aura. In a subsequent SPECT study, frontal, temporal, and parietal but not occipital hypoperfusion was reported by Lauritzen and Olesen [54]. The study which provided the most convincing evidence for the occurrence of CSD-like events in humans as the correlate of migraine aura was performed by Hadjikhani et al., who applied fMRI and studied retinotopic changes of the BOLD signal during five episodes of migraine aura in three male patients. The BOLD changes were indicative of an initial phase with cortical hyperemia, with duration and velocity characteristic for CSD that was not crossing prominent sulci, followed by cortical hypoperfusion with attenuated responses to visual stimulation and subsequent spontaneous recovery to baseline levels and recovery of the stimulus-driven brain activation. The progression of the BOLD signal changes paralleled the retinotopy of the perceived visual aura [55].

By means of $H_2{}^{15}O$-PET, Denuelle et al. investigated brain responses of migraineurs to luminous stimulation in three conditions: during spontaneous migraine attacks, after headache resolution through sumatriptan application, and in the interictal period. Greatest blood flow in the visual cortex was observed during the headache phase, suggesting that pain plays a role in the cortical response. The fact that blood flow increases persisted even after headache and photophobia subsided after sumatriptan treatment supports the view that pain could be seen as one possibly separable aspect of migraine pathogenesis [56].

6.7.2 *Interictal Findings*

Optical coherence tomography (OCT), a relatively novel high-resolution method to visualize the retina, could be a new technique for studying migraine and evaluating its progression. The retinal nerve fiber layer thickness (RNFL) in the temporal quadrant of the eye was found to be reduced in migraineurs, and average RNFL thickness was strongly correlated with migraine severity as evidenced by OCT [57]. Moreover, RNFL was thinner in chronic migraineurs than healthy controls [58]. The most recent OCT studies found thinning of RNLF thickness and ganglion cell layer thickness in migraine patients with aura as compared to both migraineurs without aura and healthy controls [59]. Longitudinal OCT studies are needed to elucidate whether RNFL changes are associated with progressive axonal loss or whether these changes are reversible.

Boulloche et al. provided another piece of evidence on abnormalities of cerebral processing of luminous/visual stimuli when showing a facilitation of visual responses to luminous stimuli in the visual cortex of migraineurs. Concomitant trigeminal pain stimulation increased the activation of the visual cortex (BA 7) and precuneus in both groups, although the pattern of activation differed [60]. Interestingly, it has been shown that patients experiencing photophobia in the premonitory phase have significantly greater activation of the extrastriate visual cortex (BA 18) as compared to baseline scans and to patients without photophobia [61].

Datta et al. have reported interictal visual discomfort in both migraineurs with and without aura. However, only migraineurs with aura demonstrated enhanced geniculostriate (LGN and V1) BOLD responses to visual stimuli as opposed to migraineurs without aura and control subjects [62]. In another small fMRI study, enhanced interictal responsiveness to visual stimulation (incongruent lines) was observed in five migraineurs, who experienced typical visual aura. This stimulation was used to successively activate distinct visual cortex areas with distinct line preferences that are believed to play a role in the genesis of typical zigzag-like patterns as seen in visual aura [63].

In response to visual stimuli, migraineurs with and without aura had stronger activations of the medial superior temporal area, suggesting involvement of higher visual processing areas in migraine [64]. Another study combined fMRI and transcranial sonography to study hyperresponsiveness of visual areas activated by motion perception (V5 and V3) to optokinetic visual stimulation and further supported the concept of an interictal motion-processing deficit in migraine [65]. However, interictal alterations in the occipital cortex and LGN of migraineurs with aura were not reproduced in another well-conducted recent fMRI study in a meticulously phenotyped population of migraineurs with side-fixed visual aura [66]. The authors described hyperresponsiveness of frontoparietal visually driven networks involved in oculomotor control, movement guidance, motion perception, visual attention, and spatial memory in the symptomatic hemisphere in these patients. There was no difference between the asymptomatic hemispheres of migraineurs and healthy controls.

By using voxel- and surface-based morphometry, the same group did not found any association between gray matter structure and aura symptoms in migraineurs. Comparison of cortical thickness between the hemisphere usually affected by migraine headaches and the unaffected side showed a difference in cortical thickness in the inferior frontal gyrus, suggesting a potential structural reorganization of inhibitory pain pathways [67].

No difference between patients with and without aura was observed in another similar study by Granziera et al., who reported increased cortical thickness in occipital areas MT+/V3A in migraineurs with and without aura compared to healthy controls [68]. Moreover, the same two groups of migraineurs had significantly lower fractional anisotropy in other parts of the visual pathway (WM adjacent to V3A, colliculus superior, left LGN). The authors stressed the possible bidirectional role of an altered visual motion processing network on cortical hyperexcitability. Thickening of the temporo-occipital incisure, an area related to visual motion processing, was observed by Messina et al. [69]; Using diffusion tensor imaging, Rocca et al. showed reduced fractional anisotropy in the optic radiation in migraineurs with aura compared to migraineurs without aura and controls [70].

In an animal model, it was showed that non-image-forming retinal pathway modulate the activity of dura-sensitive thalamocortical neurons [71]. Bright light can activate nociceptive neurons in the trigeminal nucleus caudalis through intraocular mechanisms by luminance-responsive circuits and increases parasympathetic outflow in rodents [72].

Sarchielli et al. used MR spectroscopy to measure metabolic changes in the occipital cortex during photic stimulation in interictal migraineurs and healthy subjects. N-acetyl-aspartate (NAA), a marker of neuronal, in particular axonal, integrity and mitochondrial function, was decreased more after photic stimulation in migraineurs with aura, and its subsequent recovery was less pronounced in migraineurs with aura as compared to migraineurs without aura and healthy controls [73]. One other MR spectroscopy study revealed a significantly higher Glu/Gln ratio in migraine patients than in healthy controls, suggesting that either altered neuronal–glial coupling of glutamatergic metabolism or an increased neuron/astrocyte ratio in the OC [74] may be part of migraine pathophysiology.

Sandor et al. combined magnetic spectroscopy with a functional paradigm using sustained visual stimulation in migraineurs with purely visual aura and migraineurs who experienced sensory, motor, or dysphasic symptoms in addition to visual aura. Two different metabolic patterns were observed: in the group with purely visual aura, resting lactate was high and did not further increase with visual stimulation. In the second group, lactate increased during stimulation, only in visual cortex. This may reflect a mitochondrial dysfunction in migraineurs with purely visual aura and dishabituation in migraineurs with complex aura as habituation of evoked potentials is known to be correlated with lactate decreases in the visual cortex during sustained visual stimuli [75, 76].

6.8 Additional Cortical Structures

6.8.1 Ictal Findings

In addition to the pain matrix structures that are consistently reported to be activated/affected in a multitude of different pain conditions (predominantly ACC, insula, and prefrontal cortex), there is increasing evidence that the temporal lobe and specifically the temporal pole (TP), which has not usually been associated with the processing of pain, may play an important role in migraine pathophysiology. Activation of the temporal lobe has previously been reported during migraine attacks [4, 5, 43]. The temporal pole is a multisensory (visual, olfactory, auditory) integration area with extraordinary connectivity (described in a review on interoception [77]), sometimes referred to as part of the paralimbic network. It relates complex highly processed perceptual inputs to visceral emotional responses [78]. Therefore, temporal pole dysfunction may explain sensory and behavioral changes in migraine.

6.8.2 Interictal/Habituation Findings

TP activation with painful heat was exacerbated during migraine, suggesting that repeated migraines may sensitize TP [39]. In an interictal PET study, a group of migraine patients with self-reported olfactory hypersensitivity had stronger activations of the TP [79]. The TP also showed enhanced functional connectivity with other pain-processing areas, e.g., the thalamus (especially pulvinar), ACC, anterior insula, amygdala, and nucleus caudatus [39]. In another study, the right temporal pole was one of the areas that showed atypical age-related cortical thinning in migraine patients [80].

Greater activations in the ACC in response to moderately painful heat stimulation and reduced activity of SII in response to more intense noxious heat stimuli were observed in migraineurs when compared to healthy controls. When the two noxious conditions were compared, migraine patients showed a greater response to moderately painful heat compared to the highly painful condition, whereas the reverse pattern was observed in healthy controls [81]. These findings are consistent with a study of Aderjan et al., who applied repetitive trigeminal-nociceptive stimuli to migraineurs and healthy controls. Behavioral attenuation of pain ratings did not differ between both groups. However, gradual increases of ACC responses were paralleled by decreases in SII responses in healthy volunteers, whereas in migraineurs, responses in the ACC decreased over time. The authors suggested dysfunctional pain inhibitory circuits, which may be associated with a lack of habituation [82], which is one of the most consistent psychophysiological and electrophysiological findings in migraineurs. In a study performed by the same group and aiming to explore habituation of painful trigeminal and olfactory stimuli, pain ratings in healthy controls decreased, whereas pain ratings remained unchanged in patients. This pattern of habituation respectively lack of habituation was reflected by increased BOLD responses in the insula, cingulate cortex, and thalamus in migraineurs, but decreases in the control group. Interestingly, in contrary to the headache-free period, ictal patients did not differ from control subjects [36]. This is in line with electrophysiological findings where interictal abnormalities of habituation have been shown to normalize in the premonitory phase and during migraine attacks, possibly reflecting an increase in the cortical pre-activation level due to enhanced activity in raphe–cortical serotonergic pathways [83].

Multiple VBM studies have examined gray matter changes in migraine. In these studies, gray matter reductions were consistently reported in the ACC, insula, and prefrontal cortex [28, 50, 84–87].

However, gray matter alterations in these regions have been also observed in an ever-increasing number of VBM studies on other types of chronic pain conditions (i.e., chronic back pain or fibromyalgia). These alterations are probably unspecific and rather a reversible fingerprint of pain.

6.9 Medication-Overuse Headache

Glucose hypometabolism within the orbitofrontal cortex (OFC), ACC, insula, ventral striatum, right inferior parietal lobule, and thalamus was observed in patients with medication-overuse headache (all migraineurs) in a PET study [88]. Three weeks after medication withdrawal, these metabolic changes normalized with the exception of the OFC that remained hypometabolic. The hypometabolism was more pronounced in patients using combination analgesics than those with single analgesics [88]. As OFC hypofunction is known to play an important role in substance abuse, medication-overuse headache seems to share pathophysiological features. Fronto-striatal dysfunction may reflect a predisposing psychobiological susceptibility to medication overuse [89] and may potentially serve as a future marker for risk of medication overuse.

In line with these PET findings, Riederer et al. observed gray matter increases in the bilateral thalamus and ventral striatum and gray matter decreases in frontal regions including the orbitofrontal cortex, ACC, insula, and precuneus in patients with both migraine and medication-overuse headache relative to healthy controls in a cross-sectional MRI study [90]. In a subsequent longitudinal study performed by the same group, only patients with clinical improvement after detoxification showed a significant decrease of previously increased gray matter in the midbrain including periaqueductal gray matter and nucleus cuneiformis. Strikingly, decreases of OFC gray matter volume predicted the clinical response to the detoxification [30].

Furthermore, dysfunctional mesocorticolimbic dopamine circuits have been shown in MOH patients. In particular, changes were seen in the ventromedial prefrontal cortex and in the substantia nigra/ventral tegmental area. The authors again stressed the apparent overlap between dopaminergic mechanisms in medication overuse and other types of substance abuse/overuse [91].

6.10 Conclusions

Taken together, neuroimaging techniques have helped define processes that take place in the brains of migraineurs. As such they have helped support prior clinical or preclinical findings and have opened a new understanding of the human condition.

References

1. Weiller C et al (1995) Brain stem activation in spontaneous human migraine attacks. Nat Med 1(7):658–660

2. Borsook D, Burstein R (2012) The enigma of the dorsolateral pons as a migraine generator. Cephalalgia 32(11):803–812
3. Bahra A et al (2001) Brainstem activation specific to migraine headache. Lancet 357(9261): 1016–1017
4. Matharu MS et al (2004) Central neuromodulation in chronic migraine patients with suboccipital stimulators: a PET study. Brain 127(Pt 1):220–230
5. Afridi SK et al (2005) A positron emission tomographic study in spontaneous migraine. Arch Neurol 62(8):1270–1275
6. Cao Y et al (2002) Functional MRI-BOLD of brainstem structures during visually triggered migraine. Neurology 59(1):72–78
7. Denuelle M et al (2007) Hypothalamic activation in spontaneous migraine attacks. Headache 47(10):1418–1426
8. Afridi SK et al (2005) A PET study exploring the laterality of brainstem activation in migraine using glyceryl trinitrate. Brain 128(Pt 4):932–939
9. Lance JW et al (1983) Brainstem influences on the cephalic circulation: experimental data from cat and monkey of relevance to the mechanism of migraine. Headache 23(6):258–265
10. Goadsby PJ, Duckworth JW (1989) Low frequency stimulation of the locus coeruleus reduces regional cerebral blood flow in the spinalized cat. Brain Res 476(1):71–77
11. Seghatoleslam M et al (2014) Cortical spreading depression modulates the caudate nucleus activity. Neuroscience 267:83–90
12. Denuelle M et al (2008) Posterior cerebral hypoperfusion in migraine without aura. Cephalalgia 28(8):856–862
13. Haas DC, Kent PF, Friedman DI (1993) Headache caused by a single lesion of multiple sclerosis in the periaqueductal gray area. Headache 33(8):452–455
14. Goadsby PJ (2002) Neurovascular headache and a midbrain vascular malformation: evidence for a role of the brainstem in chronic migraine. Cephalalgia 22(2):107–111
15. Rubin MN, Garza I (2012) A discrete lesion of the midpontine tegmentum causing migrainous headache and numbness. Headache 52(9):1428–1429
16. Mariotti P et al (2012) Chronic migraine-like headache caused by a demyelinating lesion in the brain stem. Pain Med 13(4):610–612
17. Gee JR et al (2005) The association of brainstem lesions with migraine-like headache: an imaging study of multiple sclerosis. Headache 45(6):670–677
18. Demarquay G et al (2011) Brainstem changes in 5-HT1A receptor availability during migraine attack. Cephalalgia 31(1):84–94
19. Sakai Y et al (2008) Sumatriptan normalizes the migraine attack-related increase in brain serotonin synthesis. Neurology 70(6):431–439
20. Sakai Y et al (2014) alpha-[11C] methyl-L tryptophan-PET as a surrogate for interictal cerebral serotonin synthesis in migraine without aura. Cephalalgia 34(3):165–173
21. Moulton EA et al (2008) Interictal dysfunction of a brainstem descending modulatory center in migraine patients. PLoS One 3(11):e3799
22. Stankewitz A et al (2011) Trigeminal nociceptive transmission in migraineurs predicts migraine attacks. J Neurosci 31(6):1937–1943
23. Basbaum AI, Fields HL (1984) Endogenous pain control systems: brainstem spinal pathways and endorphin circuitry. Annu Rev Neurosci 7:309–338
24. Mainero C, Boshyan J, Hadjikhani N (2011) Altered functional magnetic resonance imaging resting-state connectivity in periaqueductal gray networks in migraine. Ann Neurol 70(5):838–845
25. Schwedt TJ et al (2014) Allodynia and descending pain modulation in migraine: a resting state functional connectivity analysis. Pain Med 15(1):154–165
26. Schwedt TJ et al (2013) Atypical resting-state functional connectivity of affective pain regions in chronic migraine. Headache 53(5):737–751
27. Schuh-Hofer S et al (2007) Increased serotonin transporter availability in the brainstem of migraineurs. J Neurol 254(6):789–796

28. Rocca MA et al (2006) Brain gray matter changes in migraine patients with T2-visible lesions: a 3-T MRI study. Stroke 37(7):1765–1770
29. Welch KM et al (2001) Periaqueductal gray matter dysfunction in migraine: cause or the burden of illness? Headache 41(7):629–637
30. Riederer F et al (2013) Decrease of gray matter volume in the midbrain is associated with treatment response in medication-overuse headache: possible influence of orbitofrontal cortex. J Neurosci 33(39):15343–15349
31. Geraud G, Donnet A (2013) Migraine and hypothalamus. Rev Neurol (Paris) 169(5): 372–379
32. Maniyar FH et al (2014) Brain activations in the premonitory phase of nitroglycerin-triggered migraine attacks. Brain 137(Pt 1):232–241
33. Moulton EA et al (2014) Altered hypothalamic functional connectivity with autonomic circuits and the locus coeruleus in migraine. PLoS One 9(4):e95508
34. Kagan R et al (2013) Hypothalamic and basal ganglia projections to the posterior thalamus: possible role in modulation of migraine headache and photophobia. Neuroscience 248:359–368
35. Burstein R et al (2010) Thalamic sensitization transforms localized pain into widespread allodynia. Ann Neurol 68(1):81–91
36. Stankewitz A, Schulz E, May A (2013) Neuronal correlates of impaired habituation in response to repeated trigemino-nociceptive but not to olfactory input in migraineurs: an fMRI study. Cephalalgia 33(4):256–265
37. Maleki N et al (2012) Direct optic nerve pulvinar connections defined by diffusion MR tractography in humans: implications for photophobia. Hum Brain Mapp 33(1):75–88
38. Drummond PD, Woodhouse A (1993) Painful stimulation of the forehead increases photophobia in migraine sufferers. Cephalalgia 13(5):321–324
39. Moulton EA et al (2011) Painful heat reveals hyperexcitability of the temporal pole in interictal and ictal migraine States. Cereb Cortex 21(2):435–448
40. Granziera C et al (2014) Structural abnormalities in the thalamus of migraineurs with aura: a multiparametric study at 3 T. Hum Brain Mapp 35(4):1461–1468
41. Borsook D et al (2010) A key role of the basal ganglia in pain and analgesia–insights gained through human functional imaging. Mol Pain 6:27
42. Kruit MC et al (2009) Iron accumulation in deep brain nuclei in migraine: a population-based magnetic resonance imaging study. Cephalalgia 29(3):351–359
43. Tepper SJ et al (2012) Iron deposition in pain-regulatory nuclei in episodic migraine and chronic daily headache by MRI. Headache 52(2):236–243
44. Maleki N et al (2011) Migraine attacks the basal ganglia. Mol Pain 7:71
45. Yuan K et al (2013) Altered structure and resting-state functional connectivity of the basal ganglia in migraine patients without aura. J Pain 14(8):836–844
46. Aurora SK et al (2007) Brainstem dysfunction in chronic migraine as evidenced by neurophysiological and positron emission tomography studies. Headache 47(7):996–1003; discussion 1004–1007
47. Zhao L et al (2014) Abnormal brain activity changes in patients with migraine: a short-term longitudinal study. J Clin Neurol 10(3):229–235
48. Simons LE et al (2014) The human amygdala and pain: evidence from neuroimaging. Hum Brain Mapp 35(2):527–538
49. Dehbandi S et al (2008) Cortical spreading depression modulates synaptic transmission of the rat lateral amygdala. Eur J Neurosci 27(8):2057–2065
50. Valfre W et al (2008) Voxel-based morphometry reveals gray matter abnormalities in migraine. Headache 48(1):109–117
51. Hadjikhani N et al (2013) The missing link: enhanced functional connectivity between amygdala and visceroceptive cortex in migraine. Cephalalgia 33(15):1264–1268
52. Leao AA (1951) The slow voltage variation of cortical spreading depression of activity. Electroencephalogr Clin Neurophysiol 3(3):315–321

53. Olesen J, Larsen B, Lauritzen M (1981) Focal hyperemia followed by spreading oligemia and impaired activation of rCBF in classic migraine. Ann Neurol 9(4):344–352
54. Lauritzen M, Olesen J (1984) Regional cerebral blood flow during migraine attacks by Xenon-133 inhalation and emission tomography. Brain 107(Pt 2):447–461
55. Hadjikhani N et al (2001) Mechanisms of migraine aura revealed by functional MRI in human visual cortex. Proc Natl Acad Sci U S A 98(8):4687–4692
56. Denuelle M et al (2011) A PET study of photophobia during spontaneous migraine attacks. Neurology 76(3):213–218
57. Martinez A, Proupim N, Sanchez M (2008) Retinal nerve fibre layer thickness measurements using optical coherence tomography in migraine patients. Br J Ophthalmol 92(8): 1069–1075
58. Kirbas S et al (2013) Evaluation of the retinal changes in patients with chronic migraine. Acta Neurol Belg 113(2):167–172
59. Ekinci M et al (2014) Retinal nerve fibre layer, ganglion cell layer and choroid thinning in migraine with aura. BMC Ophthalmol 14:75
60. Boulloche N et al (2010) Photophobia in migraine: an interictal PET study of cortical hyperexcitability and its modulation by pain. J Neurol Neurosurg Psychiatry 81(9):978–984
61. Maniyar FH et al (2014) Photic hypersensitivity in the premonitory phase of migraine–a positron emission tomography study. Eur J Neurol 21(9):1178–1183
62. Datta R et al (2013) Interictal cortical hyperresponsiveness in migraine is directly related to the presence of aura. Cephalalgia 33(6):365–374
63. Vincent M et al (2003) Enhanced interictal responsiveness of the migraineous visual cortex to incongruent bar stimulation: a functional MRI visual activation study. Cephalalgia 23(9):860–868
64. Antal A et al (2011) Differential activation of the middle-temporal complex to visual stimulation in migraineurs. Cephalalgia 31(3):338–345
65. Griebe M et al (2014) Multimodal assessment of optokinetic visual stimulation response in migraine with aura. Headache 54(1):131–141
66. Hougaard A et al (2014) Interhemispheric differences of fMRI responses to visual stimuli in patients with side-fixed migraine aura. Hum Brain Mapp 35(6):2714–2723
67. Hougaard A et al (2015) Structural gray matter abnormalities in migraine relate to headache lateralization, but not aura. Cephalalgia 35(1):3–9
68. Granziera C et al (2006) Anatomical alterations of the visual motion processing network in migraine with and without aura. PLoS Med 3(10):e402
69. Messina R et al (2013) Cortical abnormalities in patients with migraine: a surface-based analysis. Radiology 268(1):170–180
70. Rocca MA et al (2008) Selective diffusion changes of the visual pathways in patients with migraine: a 3-T tractography study. Cephalalgia 28(10):1061–1068
71. Noseda R et al (2010) A neural mechanism for exacerbation of headache by light. Nat Neurosci 13(2):239–245
72. Okamoto K et al (2010) Bright light activates a trigeminal nociceptive pathway. Pain 149(2):235–242
73. Sarchielli P et al (2005) Functional 1H-MRS findings in migraine patients with and without aura assessed interictally. Neuroimage 24(4):1025–1031
74. Gonzalez de la Aleja J et al (2013) Higher glutamate to glutamine ratios in occipital regions in women with migraine during the interictal state. Headache 53(2):365–375
75. Sandor PS et al (2005) MR-spectroscopic imaging during visual stimulation in subgroups of migraine with aura. Cephalalgia 25(7):507–518
76. Sappey-Marinier D et al (1992) Effect of photic stimulation on human visual cortex lactate and phosphates using 1H and 31P magnetic resonance spectroscopy. J Cereb Blood Flow Metab 12(4):584–592
77. Craig AD (2002) How do you feel? Interoception: the sense of the physiological condition of the body. Nat Rev Neurosci 3(8):655–666

78. Olson IR, Plotzker A, Ezzyat Y (2007) The enigmatic temporal pole: a review of findings on social and emotional processing. Brain 130(Pt 7):1718–1731
79. Demarquay G et al (2008) Olfactory hypersensitivity in migraineurs: a H(2)(15)O-PET study. Cephalalgia 28(10):1069–1080
80. Chong CD et al (2014) Atypical age-related cortical thinning in episodic migraine. Cephalalgia 34(14):1115–1124
81. Russo A et al (2012) Pain processing in patients with migraine: an event-related fMRI study during trigeminal nociceptive stimulation. J Neurol 259(9):1903–1912
82. Aderjan D, Stankewitz A, May A (2010) Neuronal mechanisms during repetitive trigemino-nociceptive stimulation in migraine patients. Pain 151(1):97–103
83. Judit A, Sandor PS, Schoenen J (2000) Habituation of visual and intensity dependence of auditory evoked cortical potentials tends to normalize just before and during the migraine attack. Cephalalgia 20(8):714–719
84. Obermann M et al (2014) Central vestibular system modulation in vestibular migraine. Cephalalgia 34(13):1053–1061
85. Schmitz N et al (2008) Attack frequency and disease duration as indicators for brain damage in migraine. Headache 48(7):1044–1055
86. Kim JH et al (2008) Regional grey matter changes in patients with migraine: a voxel-based morphometry study. Cephalalgia 28(6):598–604
87. Jin C et al (2013) Structural and functional abnormalities in migraine patients without aura. NMR Biomed 26(1):58–64
88. Fumal A et al (2006) Orbitofrontal cortex involvement in chronic analgesic-overuse headache evolving from episodic migraine. Brain 129(Pt 2):543–550
89. Biagianti B et al (2012) Orbitofrontal dysfunction and medication overuse in patients with migraine. Headache 52(10):1511–1519
90. Riederer F et al (2012) Grey matter changes associated with medication-overuse headache: correlations with disease related disability and anxiety. World J Biol Psychiatry 13(7):517–525
91. Ferraro S et al (2012) In medication-overuse headache, fMRI shows long-lasting dysfunction in midbrain areas. Headache 52(10):1520–1534

Chapter 7
Imaging of Other Primary Headaches

Sarah Miller and Manjit S. Matharu

The use of structural and functional imaging of the human brain has led to significant advances in our understanding of pain processing and headache. The application of such techniques to primary headache conditions, especially migraine and the trigeminal autonomic cephalalgias, has provided major advances in the understanding of their underlying pathophysiology. This chapter will focus on trigeminal autonomic cephalalgias but will also briefly explore the findings of neuroimaging studies in tension-type and hypnic headache.

7.1 Tension-Type Headache

Tension-type headache (TTH) is a common condition with lifetime prevalence ranging from 30 to 80 %. The condition is characterized by episodes of bilateral, pressing or tight pain which is mild to moderate in intensity and which has no associated features [1]. It was previously thought to be psychogenic, but the current evidence base no longer supports this view [1].

7.2 Hypnic Headache

Hypnic headaches (HH) are frequent recurring headache attacks occurring only during sleep, generally without any cranial autonomic symptoms [1].

S. Miller, MBBS, BSc, MRCP(Neuro) • M.S. Matharu, MBChB, FRCP, PhD (✉)
Headache Group, Institute of Neurology and The National Hospital for Neurology and Neurosurgery, Queen Square, London WC1N 3BG, UK
e-mail: sarah.miller.12@ucl.ac.uk; m.matharu@uclmail.net

M. Ashina, P. Geppetti (eds.), *Pathophysiology of Headaches: From Molecule to Man*, Headache, DOI 10.1007/978-3-319-15621-7_7

7.3 Trigeminal Autonomic Cephalalgias

Trigeminal autonomic cephalalgias (TACs) are a group of primary headache disorders, which share distinctly similar phenotypes. The 3rd edition (beta version) of the International Classification of Headache Disorders (ICHD-IIIb) currently lists the TACs as cluster headache (CH), paroxysmal hemicrania (PH), short-lasting unilateral neuralgiform headache attacks (comprising both short-lasting unilateral neuralgiform headaches with conjunctival injection and tearing [SUNCT] and short-lasting unilateral neuralgiform headaches with cranial autonomic features [SUNA]) and hemicrania continua (HC) [1]. The TACs are characterized by intense, unilateral trigeminal distribution pain with concomitant cranial autonomic features. The common clinical presentation of these disorders has raised the possibility of a shared pathophysiological mechanism.

7.4 Diagnostic Imaging of Primary Headaches

Diagnosis of all primary headaches is based on careful clinical phenotyping. The European Federation of Neurological Sciences guidelines state that neuroimaging should only be considered for those with atypical headache patterns or focal neurological signs [2]. Likewise, the National Institute for Health and Care Excellence (NICE) guidelines in the United Kingdom state that imaging should be reserved for those with atypical clinical features or other conditions making them high risk for secondary headaches (such as malignancy and immunodeficiency). The NICE guidelines specifically state that imaging should not be conducted purely for reassurance [3].

Recent reviews of the literature on symptomatic TACs have shown a number of secondary causes [4, 5]. The most striking finding to emerge from these reviews is the number of symptomatic TACs associated with pituitary lesions [4, 5]. A large observational study performed in a tertiary referral centre reported that 4 % of patients with pituitary tumours had CH; however, the objective of this study was to describe the phenotypes of the headaches that occur in patients with pituitary tumours and not the prevalence of the different headache types as the patient group was highly selected and therefore not representative of the general pituitary tumour cohort [6]. A causal link between pituitary tumours and trigeminal autonomic cephalalgias cannot be assumed on the basis of these observational findings, and there is no place for routine pituitary imaging in clinical practice until further data from population-based studies is available. Furthermore, there is considerable risk of incidental findings with 1 in 10 of the general population having a microadenoma and 1 in 500 a macroadenoma on routine MRI [5].

Recent evidence has suggested that a significant proportion of patient with SUNCT and SUNA have trigemino-vascular conflict and these patients respond well to trigeminal microvascular decompression [7]. It is therefore recommended that all patients with short-lasting neuralgiform headache attacks undergo dedicated trigeminal nerve imaging.

We suggest a routine MRI brain scan in CH, PH and HC and MRI brain scan with dedicated trigeminal nerve imaging in SUNCT and SUNA.

7.5 Functional Neuroimaging of Experimental Head and Facial Pain

Functional neuroimaging studies have helped to establish the brain structures involved in nociception. Two major studies on experimental facial and head pain, alongside a broad literature on experimentally induced pain, have shown a widespread brain network activated during nociceptive processing [8–11]. Positron emission tomography (PET) studies in experimental head and facial pain have demonstrated significant activations were recorded in the insulae, thalamus, anterior cingulate cortex (ACC), prefrontal cortex, periaqueductal grey and the cerebellum during the acute pain state when compared to the pain-free state [9, 10]. Findings are summarized in Table 7.1.

7.6 Structural Neuroimaging in Tension-Type Headache

A voxel-based morphometry (VBM) study investigated 20 patients with chronic TTH compared to subjects with medication-overuse headache (and migraine) and headache-free controls [12]. A significant decrease in grey matter density within the pain-processing networks was observed only in those with TTH thereby providing evidence that TTH is a different disorder from migraine. Areas involved in TTH included the dorsal rostral and ventral pons, ACC, bilateral insulae, orbitofrontal cortex, bilateral parahippocampal regions and cerebellum (Table 7.1).

7.7 Structural Neuroimaging in Hypnic Headache

VBM of 14 HH patients revealed decreased grey matter in areas known to be involved in cortical pain processing [13]. A reduction in grey matter was also seen in the left posterior hypothalamus, lateralized to the left independent of headache side (Table 7.1).

7.8 Structural and Functional Neuroimaging in Trigeminal Autonomic Cephalalgias

7.8.1 Cluster Headache

Typical features of cluster headache (CH) include a trigeminal distribution of pain, circadian and circannual rhythmicity and ipsilateral cranial autonomic features [1].

Table 7.1 Table showing patterns of activation or involvement in advanced neuroimaging studies of tension-type headache, hypnic headache and trigeminal autonomic cephalalgias and experimental head pain

Study	Imaging modality	Number imaged	Specific structures activated in headache syndromes						Structures of general pain matrix										
			Dorsal pons	Hy	RN	SN	PAG	PMJ	ACC	PCC	Ins	Th	BG	Frontal cortex	Parietal cortex	Temporal cortex	Occipital cortex	Cerebellum	Others
Tension-type headache																			
Schmidt-Wicke et al. (2005) [12]	VBM	20	✓						✓	✓	✓			✓		✓	✓	✓	
Hypnic headache																			
Holle et al. (2011) [13]	VBM	14		✓					✓		✓			✓		✓		✓	
Cluster headache																			
Di Piero et al. (1997) [22]	SPECT	7 ECH							✓			✓		✓	✓	✓			
Hsieh et al. (1996) [23]	PET	7 ECH							✓		✓			✓	✓	✓	✓		
May et al. (1998) [9, 27]	PET	9 CCH		✓					✓		✓	✓	✓	✓					

May et al. (2000) [28]	PET	18 ECH		✓															
Sprenger et al. (2004) [29]	PET	1 ECH		✓					✓			✓	✓	✓					
Sprenger et al. (2007) [25]	PET	11 ECH		✓					✓		✓	✓	✓	✓		✓		✓	
Morelli et al. (2009) [31]	BOLD-fMRI	4 ECH		✓					✓		✓	✓	✓	✓				✓	
May et al. (1999) [33]	VBM	25 ECH		✓															
Absinta et al. (2012) [34]	VBM	15 ECH								✓	✓	✓	✓	✓	✓	✓			
Yang et al. (2013) [26]	VBM	49 ECH							✓		✓			✓		✓	✓		

(continued)

Table 7.1 (continued)

Study	Imaging modality	Number imaged	Specific structures activated in headache syndromes						Structures of general pain matrix										
			Dorsal pons	Hy	RN	SN	PAG	PMJ	ACC	PCC	Ins	Th	BG	Frontal cortex	Parietal cortex	Temporal cortex	Occipital cortex	Cerebellum	Others
Teepker et al. (2011)	DTI	7 ECH										✓		✓		✓	✓	✓	Brainstem, internal capsule
Szabó et al. (2013) [39]	DTI	13 ECH												✓	✓	✓	✓		
Chou et al. (2014) [37]	DTI	17 ECH		✓							✓		✓	✓		✓		✓	
Rocca et al. (2010) [40]	RsfMRI	13 ECH		✓					✓		✓	✓		✓	✓	✓	✓		
Qiu et al. (2012) [42]	RsfMRI	12 ECH		✓					✓	✓	✓			✓	✓	✓	✓		
Yang et al. (2014) [41]	RsfMRI	18 ECH		✓										✓		✓	✓	✓	

Paroxysmal hemicrania																			
Matharu et al. (2005) [51]	PET	7		✓	✓	✓			✓	✓	✓	✓	✓	✓	✓	✓		✓	
Schlake et al. (1990) [44]	SPECT	1												✓	✓				
SUNCT/SUNA																			
May et al. (1999) [33]	BOLD-fMRI	1		✓							✓	✓							
Sprenger et al. (2005) [47]	BOLD-fMRI	1		✓				✓	✓		✓	✓		✓				✓	
Cohen et al. (2006) [45]	BOLD-fMRI	2		✓														✓	
Auer et al. (2009) [49]	BOLD-fMRI	1												✓					Brainstem
Hemicrania continua																			
Matharu et al. (2005) [51]	PET	7	✓	✓	✓	✓		✓	✓	✓	✓	✓		✓	✓	✓	✓	✓	

(continued)

Table 7.1 (continued)

Study	Imaging modality	Number imaged	Specific structures activated in headache syndromes						Structures of general pain matrix										
			Dorsal pons	Hy	RN	SN	PAG	PMJ	ACC	PCC	Ins	Th	BG	Frontal cortex	Parietal cortex	Temporal cortex	Occipital cortex	Cerebellum	Others
Experimental head pain																			
May et al. (1998) [9, 27]	PET	7							✓		✓	✓						✓	
Kupers et al. (2004) [10]	PET	10					✓		✓		✓		✓	✓	✓			✓	

ACC anterior cingulate gyrus, *BG* basal ganglia, *BOLD-fMRI* blood-oxygenation-level-dependent functional magnetic resonance imaging, *CCH* chronic cluster headache, *DTI* diffusion tensor imaging, *ECH* episodic cluster headache, *Hy* posterior hypothalamus, *Ins* insula, *PAG* periaqueductal grey, *PCC* posterior cingulate gyrus, *PET* positron emission tomography, *PMJ* pontomedullary junction, *RsfMRI* resting state functional magnetic resonance imaging, *RN* red nucleus, *SN* substantia nigra, *SPECT* single-photon emission computed tomography, *SUNCT* short-lasting unilateral neuralgiform headache attacks with conjunctival injection and tearing, *SUNA* short-lasting unilateral neuralgiform headache attacks with autonomic features, *Th* thalamus, *VBM* voxel-based morphometry

Neuroimaging has made a substantial contribution to the understanding of this condition (Table 7.1).

Recent experimental work into the pathophysiology of CH has led to the understanding that the severe unilateral pain is likely mediated by activation of the first division of the trigeminal nerve, whilst autonomic symptoms are due to the activation of the cranial parasympathetic outflow via the seventh cranial nerve [14]. The circannual and circadian periodicity of CH led to the concept of a central origin of the headache condition and implicated the hypothalamus, in particular, as this is the structure where the body clock is located in the brain [15, 16]. Functional neuroimaging involving blood flow studies, magnetic resonance spectroscopy (MRS) and more recently connectivity studies have all been studied in CH and have led to advances in pathophysiology constructs and clinical treatments.

The current findings of studies into TACs can be summarized by three major abnormalities:

1. Involvement of the pain matrix
2. Posterior hypothalamic dysfunction
3. Involvement of the central opioid system

7.8.1.1 Pain Neuromatrix

Early studies on cerebral blood flow in CH were few in number and used single-photon emission computed tomography (SPECT) techniques. This semiquantitative technique taken together with the methodological differences led to heterogeneous results, with some studies reporting increases, some decreases and some no detectable difference in cortical blood flow in CH [17–21]. A more recent study investigating the cerebral blood flow changes using Xenon-133 SPECT in CH patients outside of a bout and healthy controls [22] demonstrated differences in cerebral blood flow in the contralateral primary sensorimotor and thalamic regions of CH sufferers compared to controls. The presence of alterations in pain processing outside of an active cluster bout suggested a possible involvement of central tonic pain mechanisms in the pathogenesis of cluster headache.

The first PET study on CH was performed examining seven patients (four in and three out of a cluster bout) during nitroglycerine evoked pain [23]. The authors reported a significant increase in regional cerebral blood flow (rCBF) in the right caudal and rostrocaudal ACC, temporopolar region, supplementary motor area, bilateral primary motor and premotor areas, bilateral opercula region, bilateral insula and bilateral inferior frontal cortex. A reduction in rCBF bilaterally in the posterior parietal cortex, occipitotemporal region and prefrontal cortex was observed in the pain state. The authors concluded that this work supported their earlier findings suggesting a preference of the nondominant hemisphere, especially the ACC, in affective processing of chronic ongoing pain [24]. Sprenger and colleagues conducted fluoro-D-glucose PET (FDG-PET) in 11 episodic CH subjects both during and out of a cluster bout [25]. In a bout compared to out of a bout scans showed increased metabolism in the ACC, posterior cingulate gyrus, insula, thalamus and temporal cortex. Decreased metabolism was observed in the cerebellopontine area. The authors surmised that the structures activated are involved in descending pain control and hypothesized a deficient top-down modulation of the antinociceptive circuits in CH patients.

Using VBM analysis, Yang and co-workers studied CH patients in and out of a bout [26]. They reported that in bout, CH patients had significantly reduced grey matter volume in the middle frontal and in the superior and medial frontal gyri when compared to controls. A significant increase in grey matter volume outside a bout compared to in a bout was reported in the anterior cingulate, insula and fusiform gyrus. The affected regions were all frontal pain modulation areas and may reflect an insufficient pain modulation capacity in CH patients.

7.8.1.2 Hypothalamic Dysfunction

May and colleagues conducted PET imaging in nine chronic CH subjects using $H_2{}^{15}O$ PET during nitroglycerine-induced attacks and were the first to demonstrate ipsilateral hypothalamic grey matter activation during cluster attacks [27]. Increased rCBF was also observed in the areas known to be involved with pain processing such as the contralateral ventroposterior thalamus, the ACC, bilateral insulae, basal ganglia and anterior frontal cortex and extracerebral areas consistent with large intracranial blood vessels. The significant activation of the ipsilateral hypothalamus was not seen when patients were out of a bout [28]. Findings were reproduced during $H_2{}^{15}O$ PET studies of patients during a spontaneous attack [28, 29]. Given that hypothalamic activation had not been observed in migraine or in experimental facial pain, it was concluded that the hypothalamus is involved in the underlying pathogenesis of CH rather than activation being due to a secondary response to first division trigeminal pain [30]. In contrast to migraine, none of these studies reported brainstem activation during an acute attack compared to resting state [28, 30].

Morelli and colleagues were the first to use functional magnetic resonance imaging (fMRI) employing the blood-oxygenation-level-dependent effect (BOLD-fMRI) techniques to study cerebral activation in CH patients during a bout, both in and out of attacks [31]. In the pain state compared to pain-free states, significant activation was reported in the ipsilateral hypothalamus. Trends towards activation were also reported in areas involved in pain processing.

Further evidence for hypothalamic involvement in CH has also emerged from other neuroimaging techniques. A study of magnetic resonance spectroscopy ($^{1H\text{-}MRS}$) on 26 patients (18 ECH, 10 in a bout and eight out of a bout; eight CCH) showed that N-acetyl aspartate levels (a marker of neuronal density) were reduced in the hypothalamus of CH patients compared to healthy controls [32]. The reduction of this neuronal marker was surmised to be consistent with persistent hypothalamic dysfunction in CH patients.

VBM analysis of CH patients compared to healthy subjects has provided some data to suggest that posterior hypothalamic grey matter is increased in volume, both in patients examined during and outside a bout [33]. This study was highly flawed with poor age and sex matching of subjects and controls as well as errors in the software used to analyse the data. More recent VBM studies have failed to show any grey matter changes in the hypothalamus but did show grey matter volume changes within the pain matrix previously described in a number of other chronic pain syndromes [26, 34].

7.8.1.3 Opioidergic System

Opioid receptor binding in CH patients has been studied during a cluster bout but out of an attack [35]. Decreased opioid receptor binding was observed in the pineal gland. The pineal gland is known to have functional connections to the trigeminal system – mainly ophthalmic division projections from the trigeminal ganglion [36]. The authors suggest that the findings are due to receptor downregulation or an increased release of endogenous opioids. Opioids are known to act on melatonin release, and these alterations of opioidergic function may relate to the therapeutic effect of melatonin in CH. The same study also reported decreased opioid activity in the ipsilateral hypothalamus and ACC, which were inversely related to the duration of disease.

7.8.1.4 Connectivity Studies in CH

A number of recent studies have reported on white matter microstructure abnormalities or functional connectivity changes in CH subjects. Diffusion tensor imaging (DTI) techniques have been applied by Teepker and colleagues, Szabó and colleagues and Chou and colleagues [37–39]. All groups reported significant differences in white matter microstructure in areas of the pain matrix (frontal, parietal and temporal lobes). In addition all described involvement of areas of the traditional pain matrix, Chou et al. in the limbic system, Szabo and colleagues in the occipital lobes and Teepker and colleagues in the occipital lobe and cerebellum [37–39]. Chou and colleagues also used probabilistic tractography to identify highly consistent and direct anatomical connections between the altered areas of diffusivity on DTI and the hypothalamus and thalamus [37].

Resting state functional MRI (RsfMRI) has also shown significant differences in the functional connectivity of white matter networks in CH patients compared to controls. Both Rocca and colleagues and Yang and colleagues reported increased functional connectivity within networks related to the ipsilateral hypothalamus and to areas of the pain matrix [40, 41]. Another RsfMRI study compared CH patients in and out of attacks compared to controls [42]. In an attack, there was significant increase of functional connection to the ipsilateral hypothalamus when compared to out of an attack. Further alterations in connectivity were seen in areas involved with pain processing and the emotional modulation of pain.

7.8.2 *Paroxysmal Hemicrania*

Paroxysmal hemicrania (PH) is a rare syndrome characterized by severe unilateral paroxysms of pain localized to the ophthalmic division of the trigeminal distribution accompanied by autonomic features. A diagnostic feature of the headache is an absolute response to indometacin [1].

Matharu and colleagues are the only group to have performed PET imaging in PH. $H_2{}^{15}O$ PET scanning during acute attacks and pain-free states revealed that during the headache phase compared to pain-free state, significant activation occurred in the contralateral posterior hypothalamus [43]. Other areas of the general pain matrix also showed activation in the headache state (Table 7.1). Indometacin administration was found to reverse this activation.

The only other functional imaging work in PH is of an HMPAO SPECT scan conducted on a single patient in 1990 by Schlake and colleagues [44]. Bilateral hypoperfusion in the frontoparietal region was noted between attacks with normalization of the perfusion pattern during a headache.

7.8.3 *Short-Lasting Unilateral Neuralgiform Headache Attacks (SUNCT and SUNA)*

SUNCT is a rare disorder with distinctive clinical similarities to CH and PH thus suggesting a shared pathophysiology [1]. SUNCT and SUNA are characterized by very brief, unilateral, severe, neuralgic attacks involving the ophthalmic distribution of the trigeminal nerve associated with conjunctival injection and lacrimation [45].

As with CH, May and colleagues observed activation of the ipsilateral inferior posterior hypothalamus on fMRI scanning during spontaneous SUNCT attacks when compared to the pain-free state [46]. However, although other groups have also identified activation of the hypothalamus in SUNCT and SUNA attacks using fMRI techniques, both bilateral and also contralateral activations are described [47, 48]. Auer and colleagues used fMRI imaging to study three attacks in a single patient and reported strong activation in the brainstem region, which were suggested to represent activation of the trigeminal autonomic reflex [49]. These studies are summarized in Table 7.1.

7.8.4 *Hemicrania Continua*

Hemicrania continua is a primary headache condition that has clinical similarities to both migraine and TACs. HC is characterized by a strictly unilateral headache of moderate intensity with superimposed exacerbations of severe intensity accompanied by autonomic features and migrainous symptoms [50]. Similar to PH, it has an absolute response to indometacin [1].

Only one functional imaging study has been conducted in HC, and that is from Matharu and colleagues [51]. PET scans of seven patients in the pain state showed significant activations of the contralateral posterior hypothalamus and ipsilateral dorsal rostral pons (Table 7.1).

7.9 Conclusion

Neuroimaging has made substantial contributions to the understanding of these rare but important primary headache syndromes.

For chronic TTH and HH, the single publications on each subject report a decrease in grey matter in areas involved in pain processing. It is postulated that these findings may be a sign of neural plasticity in response to prolonged nociceptive input and generation of central sensitization. However, little conclusion can be drawn on the basis of a single publication, so more studies are needed into these disorders.

The hypothesis of a common pathophysiological background for all TACs has been strengthened by the use of functional neuroimaging. Hypothalamic involvement has been shown in CH, SUNCT, PH and HC (Table 7.1). The pathophysiological importance of the hypothalamus would appear to be robust on the basis of the neuroimaging studies reviewed here. However, it is important to consider any contradictory evidence before concluding a causal link between hypothalamic dysfunction and TACs. Many positive studies have focused on the hypothalamus, and other data has been considered insignificant. Hypothalamic activation and structural changes have now been reported in other primary headache conditions such as migraine and HH [10, 13, 52]. In fact, hypothalamic changes have been observed in a wide range of pain conditions such as angina and irritable bowel syndrome but also non-pain-related conditions such as narcolepsy and autism [52–55]. Although the majority of neuroimaging studies on non-CH pain do not report hypothalamic dysfunction, most of these would not be using the hypothalamic area as a target region thus making them less likely to detect any subtle changes below the set threshold for significance.

The limited spatial resolution of VBM, PET and fMRI techniques has led some groups to suggest that the observed areas of activation are not in the hypothalamus but actually within the midbrain tegmentum [56]. This observation again challenges the conclusions made from neuroimaging studies that the hypothalamus is the key region of importance in TACs.

Consistent findings of involvement of various areas belonging to the pain matrix (e.g. prefrontal cortex, ACC, thalamus, insula and cerebellum) are seen across imaging techniques (Table 7.1). These areas are not specific to TACs but are seen across a very broad range of acute and chronic pain conditions including migraine [8] and TTH [12] and are believed to be involved in descending pain modulation. Therefore, in TACs, activation of these areas is likely due to a response to acute pain and not indicating areas of attack generation. This view is supported by the observations that abnormal activation patterns return to normal in CCH when treated with neuromodulation or when indometacin is used in PH or HC [43, 51, 57].

Opioidergic system involvement in TACs is suggested by the observations of decreased activity in the ACC in ECH compared to controls [25]. This area is thought to play a major role in the central descending opioidergic pain control mechanisms, and dysfunction in this area may therefore predispose to CH and or its recurrence. Further evidence of the importance of this system is the decreased receptor binding seen in the rostral ACC and hypothalamus related to CH disease duration [35]. The observation that those who respond to ONS for CCH have increased metabolism in their ACC compared to nonresponders also supports the concept and suggests that restoration of a normal opioidergic system is important in treatment mechanisms [57].

To conclude, the rapid advancements in functional and structural imaging techniques will continue to advance our understanding of the complex nature of brain dysfunction in primary headaches. On balanced reflection, neuroimaging studies of primary headaches, especially the TACs, appear to suggest a complex neural pain network dysfunction. Although the hypothalamus is of definite importance in the pathophysiology of TACs, neuroimaging studies cannot be used to indicate that it acts on a region of pain generation. Challenges for the future include defining the importance of the hypothalamus and its associated pain pathways in TACs and the possible mechanisms of treatment effects.

References

1. Headache Classification Committee of the International Headache S (2013) The international classification of headache disorders, 3rd edition (beta version). Cephalalgia Int J Headache 33(9):629–808. PubMed PMID: 23771276
2. Sandrini G, Friberg L, Coppola G, Janig W, Jensen R, Kruit M et al (2011) Neurophysiological tests and neuroimaging procedures in non-acute headache (2nd edition). Eur J Neurol Off J Eur Fed Neurol Soc 18(3):373–381. PubMed PMID: 20868464. Epub 2010/09/28. eng
3. National Institute for Health and Clinical Excellence (2012) Headaches: diagnosis and management of headaches in young people and adults. NICE Clin Guideline 150. 8:82–96
4. Edvardsson B (2014) Symptomatic cluster headache: a review of 63 cases. SpringerPlus 3:64. PubMed PMID: 24570848. Pubmed Central PMCID: PMC3928394. Epub 2014/02/27. eng
5. Cittadini E, Matharu MS (2009) Symptomatic trigeminal autonomic cephalalgias. Neurologist 15(6):305–312. PubMed PMID: 19901708. Epub 2009/11/11. eng
6. Levy MJ, Matharu MS, Meeran K, Powell M, Goadsby PJ (2005) The clinical characteristics of headache in patients with pituitary tumours. Brain J Neurol 128(Pt 8):1921–1930. PubMed PMID: 15888539. Epub 2005/05/13. eng
7. Sebastian S, Schweitzer D, Tan L, Broadley SA (2013) Role of trigeminal microvascular decompression in the treatment of SUNCT and SUNA. Curr Pain Headache Rep 17(5):332. PubMed PMID: 23564233
8. Tracey I (2005) Nociceptive processing in the human brain. Curr Opin Neurobiol 15(4): 478–487. PubMed PMID: 16019203. Epub 2005/07/16. eng
9. May A, Kaube H, Buchel C, Eichten C, Rijntjes M, Juptner M et al (1998) Experimental cranial pain elicited by capsaicin: a PET study. Pain 74(1):61–66. PubMed PMID: 9514561

10. Kupers RC, Svensson P, Jensen TS (2004) Central representation of muscle pain and mechanical hyperesthesia in the orofacial region: a positron emission tomography study. Pain 108(3):284–293. PubMed PMID: 15030948
11. Derbyshire SW (2000) Exploring the pain "neuromatrix". Curr Rev Pain 4(6):467–477. PubMed PMID: 11060593
12. Schmidt-Wilcke T, Leinisch E, Straube A, Kampfe N, Draganski B, Diener HC et al (2005) Gray matter decrease in patients with chronic tension type headache. Neurology 65(9): 1483–1486. PubMed PMID: 16275843. Epub 2005/11/09. eng
13. Holle D, Naegel S, Krebs S, Gaul C, Gizewski E, Diener HC et al (2011) Hypothalamic gray matter volume loss in hypnic headache. Ann Neurol 69(3):533–539. PubMed PMID: 21446025. Epub 2011/03/30. eng
14. Goadsby PJ, Edvinsson L (1994) Human in vivo evidence for trigeminovascular activation in cluster headache. Neuropeptide changes and effects of acute attacks therapies. Brain J Neurol 117(Pt 3):427–434. PubMed PMID: 7518321
15. Kudrow L (1987) The cyclic relationship of natural illumination to cluster period frequency. Cephalalgia Int J Headache 7(Suppl 6):76–78. PubMed PMID: 3442810
16. Strittmatter M, Hamann GF, Grauer M, Fischer C, Blaes F, Hoffmann KH et al (1996) Altered activity of the sympathetic nervous system and changes in the balance of hypophyseal, pituitary and adrenal hormones in patients with cluster headache. Neuroreport 7(7):1229–1234. PubMed PMID: 8817538
17. Sakai F, Meyer JS, Ishihara N, Naritomi H, Deshmukh VD (1977) Noninvasive 133XE inhalation measurements of regional cerebral blood flow in migraine and related headaches. Acta Neurol Scand Suppl 64:196–197. PubMed PMID: 268785
18. Henry PY, Vernhiet J, Orgogozo JM, Caille JM (1978) Cerebral blood flow in migraine and cluster headache. Compartmental analysis and reactivity to anaesthetic depression. Res Clin Stud Headache 6:81–88. PubMed PMID: 725260
19. Nelson RF, du Boulay GH, Marshall J, Russell RW, Symon L, Zilkha E (1980) Cerebral blood flow studies in patients with cluster headache. Headache 20(4):184–189. PubMed PMID: 7390799
20. Krabbe AA, Henriksen L, Olesen J (1984) Tomographic determination of cerebral blood flow during attacks of cluster headache. Cephalalgia Int J Headache 4(1):17–23. PubMed PMID: 6424945
21. Norris JW, Hachinski VC, Cooper PW (1976) Cerebral blood flow changes in cluster headache. Acta Neurol Scand 54(4):371–374. PubMed PMID: 973557
22. Di Piero V, Fiacco F, Tombari D, Pantano P (1997) Tonic pain: a SPET study in normal subjects and cluster headache patients. Pain 70(2–3):185–191. PubMed PMID: 9150292
23. Hsieh JC, Hannerz J, Ingvar M (1996) Right-lateralised central processing for pain of nitroglycerin-induced cluster headache. Pain 67(1):59–68. PubMed PMID: 8895232
24. Hsieh JC, Belfrage M, Stone-Elander S, Hansson P, Ingvar M (1995) Central representation of chronic ongoing neuropathic pain studied by positron emission tomography. Pain 63(2): 225–236. PubMed PMID: 8628589
25. Sprenger T, Ruether KV, Boecker H, Valet M, Berthele A, Pfaffenrath V et al (2007) Altered metabolism in frontal brain circuits in cluster headache. Cephalalgia Int J Headache 27(9):1033–1042. PubMed PMID: 17666083. Epub 2007/08/02. eng
26. Yang FC, Chou KH, Fuh JL, Huang CC, Lirng JF, Lin YY et al (2013) Altered gray matter volume in the frontal pain modulation network in patients with cluster headache. Pain 154(6):801–807. PubMed PMID: 23582154. Epub 2013/04/16. eng
27. May A, Bahra A, Büchel C et al (1998) Hypothalamic activation in cluster headache attacks. Lancet 352:275–278
28. May A, Bahra A, Buchel C, Frackowiak RS, Goadsby PJ (2000) PET and MRA findings in cluster headache and MRA in experimental pain. Neurology 55(9):1328–1335. PubMed PMID: 11087776. Epub 2000/11/23. eng
29. Sprenger T, Boecker H, Tolle TR, Bussone G, May A, Leone M (2004) Specific hypothalamic activation during a spontaneous cluster headache attack. Neurology 62(3):516–517. PubMed PMID: 14872051

30. Weiller C, May A, Limmroth V, Juptner M, Kaube H, Schayck RV et al (1995) Brain stem activation in spontaneous human migraine attacks. Nat Med 1(7):658–660. PubMed PMID: 7585147. Epub 1995/07/01. eng
31. Morelli N, Pesaresi I, Cafforio G, Maluccio MR, Gori S, Di Salle F et al (2009) Functional magnetic resonance imaging in episodic cluster headache. J Headache Pain 10(1):11–14. PubMed PMID: 19083151. Pubmed Central PMCID: Pmc3451754. Epub 2008/12/17. eng
32. Lodi R, Pierangeli G, Tonon C, Cevoli S, Testa C, Bivona G et al (2006) Study of hypothalamic metabolism in cluster headache by proton MR spectroscopy. Neurology 66(8):1264–1266. PubMed PMID: 16636250. Epub 2006/04/26. eng
33. May A, Ashburner J, Buchel C, McGonigle DJ, Friston KJ, Frackowiak RS et al (1999) Correlation between structural and functional changes in brain in an idiopathic headache syndrome. Nat Med 5(7):836–838. PubMed PMID: 10395332. Epub 1999/07/08. eng
34. Absinta M, Rocca MA, Colombo B, Falini A, Comi G, Filippi M (2012) Selective decreased grey matter volume of the pain-matrix network in cluster headache. Cephalalgia Int J Headache 32(2):109–115. PubMed PMID: 22174349. Epub 2011/12/17. eng
35. Sprenger T, Willoch F, Miederer M, Schindler F, Valet M, Berthele A et al (2006) Opioidergic changes in the pineal gland and hypothalamus in cluster headache: a ligand PET study. Neurology 66(7):1108–1110. PubMed PMID: 16606930. Epub 2006/04/12. eng
36. Reuss S (1999) Trigeminal innervation of the mammalian pineal gland. Microsc Res Tech 46(4–5):305–309. PubMed PMID: 10469466. Epub 1999/09/01. eng
37. Chou KH, Yang FC, Fuh JL, Huang CC, Lirng JF, Lin YY et al (2014) Altered white matter microstructural connectivity in cluster headaches: a longitudinal diffusion tensor imaging study. Cephalalgia Int J Headache 25. PubMed PMID: 24668118. Epub 2014/03/29. Eng
38. Teepker M, Menzler K, Belke M, Heverhagen JT, Voelker M, Mylius V et al (2012) Diffusion tensor imaging in episodic cluster headache. Headache 52(2):274–282. PubMed PMID: 22082475. Epub 2011/11/16. eng
39. Szabo N, Kincses ZT, Pardutz A, Toth E, Szok D, Csete G et al (2013) White matter disintegration in cluster headache. J Headache Pain 14:64. PubMed PMID: 23883140. Pubmed Central PMCID: Pmc3728007. Epub 2013/07/26. eng
40. Rocca MA, Valsasina P, Absinta M, Colombo B, Barcella V, Falini A et al (2010) Central nervous system dysregulation extends beyond the pain-matrix network in cluster headache. Cephalalgia Int J Headache 30(11):1383–1391. PubMed PMID: 20959433. Epub 2010/10/21. eng
41. Yang FC, Chou KH, Fuh JL, Lee PL, Lirng JF, Lin YY et al (2014) Altered hypothalamic functional connectivity in cluster headache: a longitudinal resting-state functional MRI study. J Neurol Neurosurg Psychiatry 30. PubMed PMID: 24983632. Epub 2014/07/02. Eng
42. Qiu EC, Yu SY, Liu RZ, Wang Y, Ma L, Tian LX (2012) Altered regional homogeneity in spontaneous cluster headache attacks: a resting-state functional magnetic resonance imaging study. Chin Med J (Engl) 125(4):705–709. PubMed PMID: 22490500 .Epub 2012/04/12. eng
43. Matharu MS, Cohen AS, Frackowiak RS, Goadsby PJ (2006) Posterior hypothalamic activation in paroxysmal hemicrania. Ann Neurol 59(3):535–545. PubMed PMID: 16489610
44. Schlake HP, Bottger IG, Grotemeyer KH, Husstedt IW, Schober O (1990) Single photon emission computed tomography (SPECT) with 99mTc-HMPAO (hexamethyl propylenamino oxime) in chronic paroxysmal hemicrania–a case report. Cephalalgia Int J Headache 10(6):311–315. PubMed PMID: 2289232. Epub 1990/12/01. eng
45. Cohen AS, Matharu MS, Goadsby PJ (2006) Short-lasting unilateral neuralgiform headache attacks with conjunctival injection and tearing (SUNCT) or cranial autonomic features (SUNA)–a prospective clinical study of SUNCT and SUNA. Brain J Neurol 129(Pt 10): 2746–2760. PubMed PMID: 16905753
46. May A, Bahra A, Buchel C, Turner R, Goadsby PJ (1999) Functional magnetic resonance imaging in spontaneous attacks of SUNCT: short-lasting neuralgiform headache with conjunctival injection and tearing. Ann Neurol 46(5):791–794. PubMed PMID: 10554000

47. Sprenger T, Valet M, Platzer S, Pfaffenrath V, Steude U, Tolle TR (2005) SUNCT: bilateral hypothalamic activation during headache attacks and resolving of symptoms after trigeminal decompression. Pain 113(3):422–426. PubMed PMID: 15661452. Epub 2005/01/22. eng
48. Cohen AS (2007) Short-lasting unilateral neuralgiform headache attacks with conjunctival injection and tearing. Cephalalgia Int J Headache 27(7):824–832. PubMed PMID: 17598764. Epub 2007/06/30. eng
49. Auer T, Janszky J, Schwarcz A, Doczi T, Trauninger A, Alkonyi B et al (2009) Attack-related brainstem activation in a patient with SUNCT syndrome: an ictal fMRI study. Headache 49(6):909–912. PubMed PMID: 19220497. Epub 2009/02/18. eng
50. Cittadini E, Goadsby PJ (2010) Hemicrania continua: a clinical study of 39 patients with diagnostic implications. Brain J Neurol 133(Pt 7):1973–1986. PubMed PMID: 20558416. Epub 2010/06/19. eng
51. Matharu MS, Cohen AS, McGonigle DJ, Ward N, Frackowiak RS, Goadsby PJ (2005) Posterior hypothalamic and brainstem activation in hemicrania continua. Headache 44(8):747–761. PubMed PMID: 15330820. Epub 2004/08/28. eng
52. Rosen SD, Paulesu E, Frith CD, Frackowiak RS, Davies GJ, Jones T et al (1994) Central nervous pathways mediating angina pectoris. Lancet 344(8916):147–150. PubMed PMID: 7912763
53. Blankstein U, Chen J, Diamant NE, Davis KD (2010) Altered brain structure in irritable bowel syndrome: potential contributions of pre-existing and disease-driven factors. Gastroenterology 138(5):1783–1789. PubMed PMID: 20045701
54. Kim SJ, Lyoo IK, Lee YS, Lee JY, Yoon SJ, Kim JE et al (2009) Gray matter deficits in young adults with narcolepsy. Acta Neurol Scand 119(1):61–67. PubMed PMID: 18624787
55. Kurth F, Narr KL, Woods RP, O'Neill J, Alger JR, Caplan R et al (2011) Diminished gray matter within the hypothalamus in autism disorder: a potential link to hormonal effects? Biol Psychiatry 70(3):278–282. PubMed PMID: 21531390. Pubmed Central PMCID: 3134572
56. Matharu MS, Zrinzo L (2010) Deep brain stimulation in cluster headache: hypothalamus or midbrain tegmentum? Curr Pain Headache Rep 14(2):151–159. PubMed PMID: 20425205. Epub 2010/04/29. eng
57. Magis D, Bruno MA, Fumal A, Gerardy PY, Hustinx R, Laureys S et al (2011) Central modulation in cluster headache patients treated with occipital nerve stimulation: an FDG-PET study. BMC Neurol 11:25. PubMed PMID: 21349186. Pubmed Central PMCID: 3056751. Epub 2011/02/26. eng

Chapter 8
Neurophysiology of Migraine

Gianluca Coppola, Francesco Pierelli, Petter M. Omland, and Trond Sand

Abbreviations

BAEP	Brainstem auditory evoked potential
CAP	Cyclic alternating pattern
CM	Chronic migraine
CNV	Contingent negative variation
CR	Corneal reflex
CSP	Cortical silent period
EEG	Electroencephalography
EMG	Electromyography
ERP	Event-related potential
ES	Exteroceptive suppression
HR	H-response-increased photic driving amplitude
IDAP	Intensity-dependent auditory evoked cortical potentials
LEP	Laser evoked potential
MA	Migraine with aura

G. Coppola (✉)
Department of Neurophysiology of Vision and Neuroophthalmology,
G.B. Bietti Foundation IRCCS, Rome, Italy
e-mail: gianluca.coppola@gmail.com

F. Pierelli
Department of Medico-Surgical Sciences and Biotechnologies,
Sapienza University of Rome Polo Pontino, Latina, Italy
e-mail: francesco.pierelli@uniroma1.it

P.M. Omland • T. Sand (✉)
Department of Neuroscience, Norwegian University of Science and Technology,
Trondheim, Norway
e-mail: petter.m.omland@ntnu.no; trond.sand@ntnu.no

M. Ashina, P. Geppetti (eds.), *Pathophysiology of Headaches: From Molecule to Man*, Headache, DOI 10.1007/978-3-319-15621-7_8

MEP	Motor evoked potentials
MO	Migraine without aura
MOH	Medication overuse headache
nBR	Nociceptive blink reflex
NSM	Non-sleep-related migraine
PAS	Paired associative stimulation
PD	Photic driving
PR	Pattern reversal
PSG	Polysomnography
PT	Phosphene threshold
qEEG	Quantitative electroencephalography
rTMS	Repetitive transcranial magnetic stimulation
SM	Sleep-related migraine
SSEP	Somatosensory evoked potential
sTMS	Single-pulse transcranial magnetic stimulation
TCR	Trigemino-cervical reflex
tDCS	Transcranial direct-current stimulation
TMS	Transcranial magnetic stimulation
TSR	Trigemino-spinal reflex
VEP	Visual evoked potential

8.1 Introduction

Migraine is one of the most common and disabling neurological disorders. It is characterised by recurrent attacks of headache that are widely variable in duration (i.e. between 4 and 72 h), intensity and frequency and are accompanied by nausea/vomiting and/or photo-/phonophobia. In some cases, migraine attacks are preceded by, or associated with, focal neurological symptoms, i.e. aura. A proportion of episodic migraine patients experiences a progressive increase in attack frequency leading to chronic migraine (CM), defined as 15 or more headache days with eight or more migraine attacks per month. Owing to the lack of interictal sequelae after transient ictal dysfunction, migraine is commonly considered as a prototype of functional disorders of the brain.

The clinical manifestations of the more common migraine types, with and without aura, probably depend upon a complex relationship between genetic, environmental and endogenous cognitive and emotive factors, the so-called migraine susceptibility. These factors should also be studied during the interictal period to search for the underlying dysfunctions that are able to cyclically ignite migraine attacks, probably involving both neuronal and vascular components within the head. These components include the cerebral cortex, the brainstem (e.g. periaqueductal grey matter and the monoaminergic nuclei), the thalamus and the peripheral and central components of the trigemino-cervico-vascular complex. The relative importance and the exact sequence of activation of these structures during a migraine attack remain elusive and are still under extensive investigation.

Many atraumatic methods are currently available for assessing neural functions in humans, contributing to the recent advances that have been made in understanding the pathophysiological facets of migraine. These methods of clinical neurophysiology have allowed the in vivo measurement of the migraineurs' responses to the application of various cephalic or extracephalic stimuli. Several interesting changes have been reported using multimodal evoked potentials, with both noxious and innocuous stimuli, brainstem and spinal withdrawal reflexes and transcranial neuromodulatory techniques such as magnetic or direct-current stimulations. Results have suggested that neurophysiological states undergo fluctuations related either to the development of a migraine disorder, to the cyclical recurrence of attacks or to the eventual migraine chronification. Changes induced by the use or the overuse of certain pharmacological treatments have also been reported.

This evidence indicates, on the one hand, a relationship between electrophysiological changes and migraine, in spite of some remaining controversies. On the other hand, it suggests that the methodologies of clinical neurophysiology seem to be suitable for acquiring further knowledge on the generation of the migraine attack, its recurrence as well as the transition from an episodic to a chronic state. The intent of this chapter is to provide a comprehensive overview of the results provided by different neurophysiological techniques that have been used to study migraine pathophysiology.

8.2 Electroencephalography

Electroencephalography (EEG) was the first electrophysiological method used to study brain function of migraine patients. Although EEG was not recognised as useful for the diagnosis of non-acute primary headache disorders, it may provide useful information in a research setting. Researchers have observed changes regarding three major EEG features: increased photic driving (PD) amplitude to trains above 18 Hz, often called 'H-response' (HR); alpha activity abnormalities; and the presence of slowing. In addition, EEG studies have been used to investigate the still ongoing discussion about a proposed relationship between EEG and epilepsy.

Several blinded studies have shown slight excess of various EEG features in migraine (for a review of the early EEG literature, see [1]). Definitely abnormal EEG has seldom been reported: focal slowing in 0–15 % and spiking in 0.2–9 % of patients, generally rather similar to prevalences in healthy controls [2]. Consistent EEG changes during a visual aura have not been reported, although patients with brainstem aura (previously termed basilar-type migraine) may have severe clinically relevant EEG slowing that may last for several days [3–6].

During the last 50 years of publication, the increased photic driving response has rather consistently been reported in migraineurs [7]. The specificity of HR is low however, because it may also occur in healthy subjects. In a recent study by Fogang et al. [8], moderate specificity (69 %), but good sensitivity (82 %) was observed using visual EEG inspection for PD at 20 Hz in a large cohort of migraine patients, but spectral analysis showed that visual sensitivity was considerably overestimated. A higher incidence of PD responses has been observed in patients with migraine-associated vertigo [9], as well as a positive correlation between PD and some of the clinical features of migraine, such as autonomic symptoms [10]. However, a major limitation of these studies was the inclusion of patients without taking into account the period in their migraine cycle. Bjørk et al. addressed this issue and compared recordings between attacks with those performed before, during and after an attack, as well as with the EEGs of healthy subjects in a fully blinded study [11]. PD was in fact depressed both during and between attacks in migraineurs without aura (MO), while it was increased immediately before an attack. This pattern was also related to the increased severity of symptoms [12].

Multichannel EEG during repetitive flash stimulation in MO patients who were between attacks identified phase hypersynchronisation of the alpha rhythm in all regions of the scalp, especially in migraine without aura [13]. Migraine with aura (MA) patients seem to behave differently since a distinct decrease was observed in the power of the beta frequency band compared with both migraineurs without aura and controls [13].

Quantitative electroencephalographic techniques have shown two parameters to be particularly significant in migraine: alpha activity and slowing (excess theta or delta activity). Alpha rhythm asymmetries and alpha total power abnormalities have been reported [14]. Alpha total power contralateral to the visually affected hemifield was decreased, within 3 days of an attack [15]. Alpha power was also reduced in MO on the headache side and in patients with menstrual migraine up to 24 h before the attack [15]. In some studies an increase in alpha power was observed [14, 16], while decreased alpha power was seen bilaterally in medial parts of the frontal cortex with the LORETA localisation method [17]. However, many past studies did not control for proximity to the next [14] or even the last attack [16]. In a recent blinded study, controlling for migraine phase, occipital alpha was normal interictally, while alpha rhythm variability increased in the pre-ictal phase and alpha power increased during the attack [18], suggesting that observed changes in the alpha rhythm may be caused by the temporal proximity to the next migraine attack. The changes in alpha rhythm seem to be related to increased migraine load and clinical photophobia [18].

Migraine patient groups may also have increased slow activity mostly over the temporo-occipital areas [14, 19]. An increase in theta power has been observed in all cortical regions and an increase in delta activity in the (painful) frontocentral region of adult migraineurs [18]. In another blinded paired qEEG study, these abnormalities in the frequency domain seemed to vary according to the time of examination: right before an attack (pre-ictal), during the attack (ictal) or between attacks (interictal) [20]. Interestingly, it was suggested that migraineurs are most susceptible to an attack when the anterior qEEG delta power and posterior alpha and theta asymmetry values are high [20].

8.3 Polysomnography and Sleep Dysfunction in Migraine

Migraine and sleep are connected, as sleep-related problems may act as a migraine trigger, drowsiness often precede an attack, and sleeping often relieves the attack. However, few objective polysomnographic (PSG) studies have been performed.

During the night before an attack, Göder et al. [21] found decreased cortical activation reflected by decreased number of EEG arousals and decreased REM density. Della Marca et al. [22] did also find reduced arousal index in REM and decreased cyclic alternating pattern (CAP) rate in NREM, suggesting reduced arousability interictally in patients with sleep-related migraine. In recent fully blinded and controlled studies, Engstrøm et al. [23, 24] found that patients with sleep-related migraine (attack onset during night or upon waking) had increased number of awakenings while those with non-sleep-related migraine had increased slow-wave sleep, compatible with a foregoing relative sleep deprivation. The authors hypothesised that a relative lack of sleep in non-sleep-related migraine also might explain reduced pain thresholds in this group. Thus, hypoarousal seems to characterise migraine patient groups before attacks, while the presence of nightly hyperarousal might determine if the migraineurs tend to experience attacks with nightly onset.

Using the method of nonlinear multi-electrode sleep EEG analysis, the maximum change in dimensional complexity was observed in the pre-ictal period over the scalp area where the migraine headache would subsequently be perceived [25].

8.4 Evoked Potentials

8.4.1 Visual Evoked Potentials

The amplitude of flash or pattern reversal visual evoked potentials (VEPs) was normal in the majority of studies, but in some it was increased and in others decreased compared with controls. Interhemispheric VEP amplitude asymmetry has been reported several times [26].

Amplitudes of VEPs normally decline during repeated stimulation, often referred to as habituation. Several pattern reversal VEP studies have shown an interictal habituation deficit between the first and the following blocks of responses [27–30] (see [31] for a review).

Other VEP studies have not managed to reproduce lack of habituation in migraine [32–34]. The discrepant findings may be caused by methodological differences between studies. A lack of habituation measured by VEP has not been reproduced in fully blinded studies [35]. Further studies are therefore needed in order to accept it as a reliable biomarker for migraine.

However, a lack of habituation was also recently found for visual evoked magneto-encephalographic responses [36] and motion-onset (M-VEP) visual evoked potentials [37]. Altered visual processing in migraine may be related to short-range lateral inhibition in the visual cortex [38], and it is possibly under abnormal thalamic and thalamocortical control [39]. In fact, experimental paradigms able to positively modulate thalamocortical activity, such as 3-min hyperventilation [40] and 1-h light deprivation [41], have re-established interictal VEP habituation. Finally, the degree of habituation may depend on where patients are in the migraine cycle, since some cross-sectional studies indicated that VEP habituation is minimal between attacks and more prominent during an attack [38, 42, 43]. However, this was not confirmed in a longitudinal blinded study [44].

8.4.2 Auditory Evoked Potentials

In the majority of studies, researchers were not able to find interictal abnormalities in the baseline parameters of short-latency brainstem auditory evoked potentials (BAEP). Sand et al. [45] provide a review of older BAEP studies in their Table 5. Similar to the results of VEP studies, a significant increase in side asymmetries has been reported for BAEP [46]. A lack of habituation of waves IV–V dispersion was found in migraineurs to 40 dB clicks (but not to 55 and 70 dB clicks) in a longitudinal blinded study, in which a direct relationship between BAEP amplitudes and blood 5HT levels was also reported in controls, but not in migraineurs [45]. A lack of habituation has also been reported for cortical auditory evoked responses for 70 dB, but not 40 dB stimuli [47].

Stronger stimulus intensity dependence of auditory evoked cortical potentials (IDAP) was found between attacks in migraine sufferers compared with control subjects in one study [47]. This may be another feature of lack of habituation, as another study reported a negative correlation between amplitude habituation and IDAP [48]. In common with VEP, IDAP has been reported to normalise during an attack [42]. One study, however, did not confirm this phenomenon [33].

Evidence that the thalamus in migraine abnormally controls the cortex between attacks is further supported by analysis of sensory gating, defined as a filtering of external stimuli by central sensory pathways, in which the thalamus seems to play a major role. In an auditory P50 event-related potential (ERP) paradigm, sensory gating was markedly reduced in migraine patients compared with controls [49, 50], probably in a way that is related to reduced short-term habituation.

8.4.3 Somatosensory Evoked Potentials

The amplitude and latency of standard somatosensory evoked potentials (SSEPs) after median nerve stimulation were normal between attacks in the majority of studies, although increases in amplitude were reported in the only study that used magnetoencephalography [26]. During a hemiparaesthetic migraine aura, the parietal N20 SSEP component was significantly delayed and reduced in amplitude, and both anomalies progressively returned to values within the normal range during the following headache phase [51].

In concordance with VEP and BAEP studies, a significant increase has been observed in interhemispheric asymmetries for the amplitude of the N30 SSEP component [52]. Deficient habituation has also been confirmed interictally for the SSEP components. In fact, both the cervical N13 [53] and the sensorimotor N20 [53–55] component have shown an increasing, instead of a decreasing, response during continuous electric stimulus repetition.

The application of a specific bandpass digital filter to broadband SSEP recordings permits the extraction of a series of high-frequency oscillations (HFOs). Multichannel source localisation analyses and pharmacological manipulation studies have shown that the separate analysis of the early (before N20 peak) and late (after N20 peak) HFO components enables the measurement of thalamocortical fibre activity and primary cortical activation, respectively. Between attacks, the early component of the HFOs, but not the late component, was significantly smaller in migraineurs [55–58], and this reduction was associated with a worsening in the clinical evolution of migraine [59].

8.4.4 Event-Related Potentials and P300

Contingent negative variation (CNV) is a long-latency EEG surface negative potential with cognitive and motor components and is considered to be an index of cortical arousal during orientation and attention. Two groups have consistently found increased CNV amplitude interictally in migraine, which is more pronounced for the early component (iCNV). The iCNV component seems to be enhanced during stress and the premenstrual phase of the ovarian cycle, but not during pregnancy [26, 60]. These iCNV changes correlated inversely with disorder duration [61], while the late component of CNV correlated inversely with depressive symptoms [62].

In concordance with some studies of VEPs, AEPs and SSEPs, CNV have showed a lack of habituation between attacks [63, 64]. This abnormal information processing has been observed only for the early and not for the late CNV component [64–67]. This phenomenon seems to have familial characteristics [68], increase just before an attack. Habituation seems to normalise during an attack [65, 68], after drug treatment [69, 70] and after non-pharmacological interventions [71].

A loss of habituation in migraine has also been described for cognitive functions, as measured by the event-related P300 potential. This was seen interictally using a visual or auditory oddball paradigm and was found to correlate inversely with platelet 5HT content [26]. However, the habituation deficit could not be confirmed in a recent study on menstrual migraine patients [72]. In addition, several authors, including the latter, did not control for proximity to the next attack.

8.4.5 Pain-Related Evoked Potentials

A reliable and objective way to study nociceptive evoked brain responses in the trigeminal or extracranial systems is by using brief laser pulses, which are able to excite Aδ and C nociceptors in the superficial skin layers.

Between migraine attacks, the N2-P2 laser evoked potential (LEP) is normal after cephalic and extracephalic stimulation. Remote heterotopic capsaicin application [73] as well as a distraction task [74] reduced LEP amplitude in healthy subjects, but not in migraineurs, probably because of defective brainstem inhibitory control. In migraine, the N2-P2 amplitude increased during the premenstrual phase [75], whereas it is decreased after interfering stimulation by images with different affective content [76], after excitability-enhancing 5 Hz-repetitive transcranial magnetic stimulation (rTMS) [77] and after visual and verbal suggestion, especially in patients with more severe migraine [78]. Moreover, LEP amplitudes increased [79], and their distribution was shifted rostrally during an attack in one study [80].

LEP studies have confirmed that the reduced habituation seen during repetitive stimulation between migraine attacks also can be found for the noxious stimulus modality during short [81] as well as long periods of painful stimulation [82]. Moreover, a lack of habituation of LEP amplitude has been found in patients for both cephalic (supraorbital zone) and extracephalic (hand dorsum) stimulation [82]. Interestingly, a persistent lack of LEP N2-P2 habituation was observed during an attack [82], which contrasts with the response normalisation found by some authors with non-noxious EPs and in the premenstrual phase [75].

8.5 Neuromodulation Methods

8.5.1 Transcranial Magnetic Stimulation

Transcranial magnetic stimulation (TMS) is a non-invasive method used to study the excitability of the underlying cortical area. Both single-pulse TMS (sTMS) and repetitive rTMS have been performed in migraine studies, the latter capable of durably modifying the excitability of the stimulated cortical area.

8.5.1.1 Single-Pulse TMS

With sTMS, both phosphene thresholds (PT) and motor thresholds have been assessed in migraine but with discrepant results. sTMS has the advantage of relying on an objective measure, the amplitude of motor evoked potential (MEP) recorded from a muscle. Briefly, both increased and decreased thresholds for MEP have been reported in migraineurs. However, most studies found no significant differences compared to controls [83, 84]. MEP thresholds were significantly increased in migraine after light deprivation, an experimental way to modulate subcortical and cortical activities, whereas they remained stable in controls [84]. Using paired TMS pulses, intracortical facilitation was found in one study [85], but not in another [83]. The cortical silent period was normal [85, 86] or reduced [87, 88] in the interictal period of episodic migraine. Cerebellar conditioning TMS showed a significant reduction in cerebellar inhibition on the motor cortex in migraine patients compared with controls [89].

With sTMS over the visual cortex, both decreased [86, 90] and increased [91] PT have been reported in migraine. Several studies also found no differences compared to controls [92, 93]. A recent meta-analysis found decreased PTs in migraineurs with aura interictally, but it was emphasised that the results of PT studies varied greatly because of methodological differences and that their results therefore should be interpreted with caution [94].

8.5.1.2 Repetitive Transcranial Magnetic Stimulation

Studies using repetitive transcranial magnetic stimulation (rTMS) have reported abnormal cortical excitability, manifesting as paradoxical effects, in response to both depressing and enhancing rTMS paradigms, particularly in MA. Brighina and coworkers observed that 1 Hz-rTMS at 90 % of the RMT over the motor cortex of MA patients significantly activates rather than inhibits intracortical facilitatory circuits [95]. More recently, two independent research groups provided evidence that 5 Hz-rTMS over the motor cortex induced short-term synaptic potentiation more easily in MA patients than those without aura and controls, in whom they did not find significant variations in MEP size [96, 97]. On the other hand, excitatory 5 Hz-rTMS induced a significant decrease in MEP size in MA patients rather than the clear MEP facilitation seen in controls [97]. The authors interpreted these paradoxical responses as being due to a compensatory cortical homeostatic metaplastic mechanism in response to a forced increase in cortical excitability. Consistent with evidence from several EP studies, the MEP response to 5 Hz-rTMS strongly depends on when patients are studied during the migraine cycle and on attack frequency [98]. In MO patients, Pierelli et al. [99] used a paired associative stimulation (PAS) paradigm, a protocol coupling a peripheral nerve and cortical TMS in order to study long-term associative learning mechanisms. The authors found that (the presumably inhibiting) PAS paradigm paradoxically increased MEP amplitudes, while the enhancing part of the PAS protocol did not induce potentiation [99]. More interestingly, the same authors observed in a subgroup that the PAS-induced plastic changes were inversely related to thalamocortical activation, as assessed by early somatosensory HFOs, suggesting a possible explanation for the observed paradoxical effects.

8.5.1.3 Transcranial Direct-Current Stimulation

Transcranial direct-current stimulation (tDCS) is another non-invasive method that can modify the excitability of the underlying cortex: cathodal tDCS is inhibitory and anodal tDCS excitatory. Chadaide et al. [92] studied the effect of tDCS on TMS-elicited phosphene thresholds (PT). While baseline PTs and the anodal tDCS-induced PT decrease were similar between migraine patients and control subjects, cathodal stimulation, that increased the PT in healthy subjects, did not affect the patient group. In accordance with the latter paper, Siniatchkin et al. [100] showed that VEP amplitude can increase under anodal and decrease under cathodal tDCS in healthy subjects, while neither affected VEPs in MA patients. Viganò et al. reported *that* N1-P1 and P1-N2 VEP amplitude habituations increased immediately after anodal tDCS applied over the visual area in migraineurs and controls [101]. Cathodal tDCS, but not anodal tDCS, restored the normal facilitatory response to 5 Hz-rTMS trains in MO and MA patients, as both groups showed a paradoxically inhibited response at baseline [98].

8.6 Electromyographic (EMG)-Recorded Reflexes

8.6.1 Brainstem Reflexes

Several research groups have reported that the exteroceptive suppression (ES) of temporalis muscle activity is unchanged between attacks in patients with migraine in comparison with controls, particularly the multisynaptic ES2 and its recovery curve, which are markers of the excitability of interneuronal networks in the ponto-medullary reticular formation [102].

The conventional blink reflex is used to explore the trigeminal system and has produced inconsistent results. Some studies have reported normal values for R1 and R2 latencies and amplitudes [103, 104], whereas some have found increased R2 latency during the pain-free period [105], and others have found lower values of R2 amplitude and size only during the headache phase of migraine, which returned to the normal range after sumatriptan injection [106]. Opposite results have been obtained from analysis of the BR recovery curve after supraorbital conditioning, since it was reported to be normal in one study [102] and slightly faster in another [107], especially when patients report allodynia during migraine [108].

Using a stimulation electrode that mainly activates Aδ-fibres, Katsarava et al. [109] found normal latencies and areas under the curve for the nociceptive blink reflex (nBR) R2 component in migraineurs interictally, but reduced habituation, as another group observed in a later study [110]. During the migraine attack, nBR-R2 amplitude and habituation were shown to be increased [109, 111, 112], which suggests temporary ictal sensitisation of the reflex pathway. Recovery curves for the nBR-R2 component both after supraorbital or peripheral conditioning were within normal limits between migraine attacks [113].

In a study of the corneal reflex (CR), which is mediated by small nociceptive fibres, a lower reflex threshold was found between attacks in migraineurs compared with controls [114]. This contrasts with the results of a subsequent study where baseline response areas under the curve and latencies of the CR- R2 components did not reach the level of significance [115].

Knowing that the trigeminal pathways and motor neurons in both the neck and upper limb muscles are functionally and anatomically connected, neurophysiological abnormalities in the pain-free phase have been similarly revealed by means of trigemino-cervical reflex (TCR) recordings [116], but no significant differences were observed for the trigemino-spinal reflex (TSR) [117]. Between attacks, the recovery cycle for the TCRs was markedly faster in migraine patients than in controls, while no significant differences were observed for the TSRs. A cold pressor stimulus reduced the TCR and TSR areas equally in both migraine patients and controls [117].

8.6.2 *Spinal Reflexes*

Within the nervous system, one of the most typical abnormalities resulting from dysfunction of pain processing is represented by activity-dependent changes in the excitability of central neurons resulting in an abnormal temporal summation (TS) of pain stimuli. In humans, the functional activity of the TS of pain can be tested using the temporal summation threshold (TST) of the nociceptive withdrawal reflex (NWR) method.

An increased NWR area with a normal reflex threshold has been reported in episodic migraine between attacks. A reduced TST of pain [118], accompanied by an increase in pain perception [119], was demonstrated in episodic migraine in between attacks. Interestingly, administration of a nitric oxide donor, glyceryl trinitrate, induced a transitory facilitation of TS-NWR only in those patients who developed a full-blown migraine attack [120].

8.7 Chronic Migraine

Some migraine patients experience a progressive increase in attack frequency, leading to headache chronification, i.e. they have 15 or more headache days per month with eight or more migraine attacks per month. The majority of these patients have CM, mostly associated with the excessive intake of acute medications, defining medication overuse headache (MOH). The precise pathophysiological mechanisms are not yet understood. Among various possible explanations, central sensitisation and defective central pain control systems are the most widely accepted causative factors.

An increase in the amplitude of pain-related cortical responses has been detected in chronic migraine both with [121] and without medication overuse [122]. Similar to the situation during an attack, the LEP brain distribution is shifted rostrally within the anterior cingulate cortex in CM [123].

Excessive cortical activation has also been reported in non-painful SSEP studies of CM or MOH [54, 58].

In CM, a neurophysiological pattern quite similar to that of episodic migraineurs recorded during an attack, including habituation, ictal thalamocortical activity (early HFO) normalisation and increased amplitude of the primary cortical component (late HFOs) [58].

In MOH, the initially higher SSEP amplitudes that reflect sensitisation were further increased during stimulus repetition, resulting in a persistent sensitisation proportional to the duration of the headache chronification phase [54]. In addition, SSEP amplitudes may differ according to the overused drug, being smaller in triptan overusers than in patients overusing nonsteroidal anti-inflammatory drugs (NSAIDs) [54]. These abnormalities in the cortical responses to somatosensory stimulation seem to be influenced by genetic factors [124]. Moreover, MOH patients still showed deficient habituation mechanisms during CNV [125] and LEP [126] recordings, the latter normalising after withdrawal of the overused medication. In agreement with the SSEP study mentioned above [54, 58], and as usually happens during an attack, the VEP P100m amplitude habituated normally on stimulus repetition in CM and in controls [36].

Using a test of cortical inhibition known as transcranial magnetic suppression of perceptual accuracy, researchers observed that CM patients (with or without medication overuse) had the lowest suppression index in comparison with healthy controls, with episodic migraineurs falling in between [127].

By further exploring inhibitory circuits, Currà et al. measured the TMS-induced cortical silent period (CSP) in a group of MOH patients. CSP duration was significantly shorter in triptan overusers than in the NSAID or triptan-plus-NSAID overuser subgroups [88]. Cosentino et al. reported an inhibitory response in CM patients during trains of TMS at 5 Hz, instead of a progressive facilitation, similar to results obtained by the same researchers during an attack of episodic migraine [98].

In CM, sensitisation phenomena might manifest either at the brainstem [128] or spinal levels. A significantly lower withdrawal reflex threshold, higher amplitude and lower TST were found in MOH patients before detoxification in comparison with episodic migraine and controls [118]. All these neurophysiological abnormalities tended to improve after a detoxification programme [118].

8.8 Conclusions

The diagnosis of migraine is still based on medical interviews and an objective neurological examination, while paraclinical tests mainly are useful to exclude secondary headache and to evaluate comorbid disorders or severe hemiplegic or brainstem aura [129]. The search for biomarkers that may predispose individuals to recurrent migraine attacks has provided a range of interesting bioelectrical parameters that correlate with migraine and seem to change during the migraine cycle on the group level. Notably, many neurophysiological studies have disclosed changes in the spinal, brainstem and cortical responsivity to external innocuous or noxious stimuli in migraine. These results can be summarised as follows (Table 8.1):

- Enhanced interictal photic driving in EEG seemed to be rather consistently reported, but this ‘H-response’ is not specific, and it was not confirmed in a fully blinded study.
- EEG power mapping (qEEG) studies have shown variable changes in two parameters alpha activity and excess slowing, but the influence of unspecific factors, like drowsiness, has not been settled.
- In EP studies of episodic migraine, a lack of habituation on recordings performed between attacks and sensitisation during the attack have been found, especially with somatosensory stimuli. Habituation tends to normalise during attacks. In the pre-ictal phase, both sensitisation and deficient habituation may variably co-exist in response to non-noxious and painful stimuli. In patients who evolve into CM, the cortical response pattern could be locked in a state combining both initial sensitisation and late habituation. The usability of VEP habituation as a neurophysiological biomarker in migraine is limited by the lack of replication in fully blinded VEP studies.

Table 8.1 Synoptic table of electrophysiological changes comparing episodic migraine between attacks, during an attack and chronic migraine with or without medication overuse

	Episodic migraine between attacks	Episodic migraine before or during an attack	Chronic migraine/MOH
Technique			
EMG-recorded reflexes	↓ Habituation of BR	Normal habituation	Persistent sensitisation at the trigeminal and spinal level
	↑ Recovery cycle of TCR	Transient sensitisation	
EEG	Normal or increased photic driving. Increased alpha variability and excess slowing	EEG activity changes shortly before and during the attack. Increased photic driving before attack	
PSG	↓ Arousals in NSM and preserved arousability in SM	↓ Arousals before attack in SM	
		↓ Sleep latency before attack	
TMS and tDCS	Paradoxical effects	Changes shortly before, during and immediately after the attack	Paradoxical prevalence of inhibitory responses
EP and ERP	↓ Thalamocortical activity	Normal thalamocortical activity	Normal thalamocortical activity
	Normal or ↓ habituation	Normal habituation or transient sensitisation	Normal (CM) or ↓ habituation (MOH)
			Persistent sensitisation

Arrows indicate the direction of change
CM chronic migraine, *EEG* electroencephalography, *EMG* electromyography, *EP*, evoked potentials (visual, sensory and auditory), *ERP* event-related potentials, *MOH* medication overuse headache, *TMS* transcranial magnetic stimulation, *tDCS* transcranial direct-current stimulation, *PSG* polysomnography, *SM* sleep-related migraine, *NSM* non-sleep-related migraine

- Only subtle abnormalities in the processing of noxious information have been revealed between migraine attacks, while more prominent changes seem to occur during an attack and when migraine becomes chronic.
- Studies with rTMS or tDCS have reported abnormal cortical excitability manifesting as paradoxical effects in response to both depressing and enhancing paradigms, particularly in MA. These paradoxical effects might be a consequence of an abnormal thalamocortical drive that impairs short- and longer-term changes in cortical synaptic effectivity, finally leading to maladaptive responses.
- Studies with EEG and visual and somatosensory evoked high-frequency oscillations suggest that an abnormal rhythmic activity between thalamus and cortex, namely, thalamocortical dysrhythmia, may be the pathophysiological mechanism underlying abnormal information processing in migraine.

Future research in this subject area in oncoming years should be devoted to understand the precise anatomical structures involved in the recurrence of migraine susceptibility and to the development of new target pharmacological and non-pharmacological interventions that are able to improve temporal information processing.

In order to reduce divergences between studies, more attention should be paid to performing blind studies. In confirmatory research, blinding both recording and analysis would be helpful. Accurate clinical data and a headache diary should be recorded before, during and after the day of testing, in order to prospectively monitor the patients' clinical fluctuations. The effects of homeostatic factors including sleep, arousal, attention and drowsiness should be explored, both with polysomnographic studies and more sophisticated ERP protocols.

Better insight into the nature of interictal cortical dysfunction will hopefully enable us to solve the mystery of migraine recurrence, a phenomenon so enthralling that the writer Oliver Sacks considered it to be 'not only an elemental activity of the cerebral cortex, but an entire self-organising system, a universal behaviour, at work....the creative heart of Nature itself' [130].

References

1. Sand T (1991) EEG in migraine: a review of the literature. Funct Neurol 6(1):7–22
2. Sand T (2003) Electroencephalography in migraine: a review with focus on quantitative electroencephalography and the migraine vs. epilepsy relationship. Cephalalgia 1:5–11
3. Parrino L, Pietrini V, Spaggiari M, Terzano M (1986) Acute confusional migraine attacks resolved by sleep: lack of significant abnormalities in post-ictal polysomnograms. Cephalalgia 6(2):95–100
4. Ganji S (1986) Basilar artery migraine: EEG and evoked potential patterns during acute stage. Headache 26(5):220–223
5. Pietrini V, Terzano M, D'Andrea G, Parrino L, Cananzi A, Ferro-Milone F (1987) Acute confusional migraine: clinical and electroencephalographic aspects. Cephalalgia 7(1): 29–37
6. Haan J, Ferrari M, Brouwer O (1988) Acute confusional migraine. Case report and review of literature. Clin Neurol Neurosurg 90(3):275–278
7. Golla FL, Winter AL (1959) Analysis of cerebral responses to flicker in patients complaining of episodic headache. Electroencephalogr Clin Neurophysiol 11(3):539–549
8. Fogang Y, Gérard P, De P, Pepin J, Ndiaye M, Magis D et al (2014) Analysis and clinical correlates of 20 Hz photic driving on routine EEG in migraine. Acta Neurol Belg, DOI 10.1007/s13760-014-0309-8
9. Goto F, Oishi N, Tsutsumi T, Ito T, Arai M, Ogawa K (2013) Characteristic electroencephalographic findings by photic driving in patients with migraine-associated vertigo. Acta Otolaryngol 133(3):253–256
10. Puca FM, de Tommaso M, Tota P, Sciruicchio V (1996) Photic driving in migraine: correlations with clinical features. Cephalalgia 16(4):246–250
11. Bjørk M, Hagen K, Stovner L, Sand T (2011) Photic EEG-driving responses related to ictal phases and trigger sensitivity in migraine: a longitudinal, controlled study. Cephalalgia 31(4):444–455
12. Bjørk M, Stovner L, Hagen K, Sand T (2011) What initiates a migraine attack? Conclusions from four longitudinal studies of quantitative EEG and steady-state visual-evoked potentials in migraineurs. Acta Neurol Scand Suppl 191:56–63
13. de Tommaso M, Stramaglia S, Marinazzo D, Trotta G, Pellicoro M (2013) Functional and effective connectivity in EEG alpha and beta bands during intermittent flash stimulation in migraine with and without aura. Cephalalgia 33(11):938–947
14. Facchetti D, Marsile C, Faggi L, Donati E, Kokodoko A, Poloni M (1990) Cerebral mapping in subjects suffering from migraine with aura. Cephalalgia 10(6):279–284

15. Schoenen J, Jamart B, Delwaide P (1987) Topographic EEG mapping in common and classic migraine during and between attacks. In: Rose FC (ed) Advances in headache research. Smith Gordon, London, pp 25–33
16. Hughes J, Robbins L (1990) Brain mapping in migraine. Clin Electroencephalogr 21(1):14–24
17. Clemens B, Bánk J, Piros P, Bessenyei M, Veto S, Tóth M et al (2008) Three-dimensional localization of abnormal EEG activity in migraine: a low resolution electromagnetic tomography (LORETA) study of migraine patients in the pain-free interval. Brain Topogr 21(1):36–42
18. Bjørk MH, Stovner LJ, Nilsen BM, Stjern M, Hagen K, Sand T (2009) The occipital alpha rhythm related to the "migraine cycle" and headache burden: a blinded, controlled longitudinal study. Clin Neurophysiol 120(3):464–471
19. Sauer S, Schellenberg R, Hofmann H, Dimpfel W (1997) Functional imaging of headache – first steps in an objective quantitative classification of migraine. Eur J Med Res 2(9): 367–376
20. Bjørk M, Sand T (2008) Quantitative EEG power and asymmetry increase 36 h before a migraine attack. Cephalalgia 28(9):960–968
21. Göder R, Fritzer G, Kapsokalyvas A, Kropp P, Niederberger U, Strenge H et al (2001) Polysomnographic findings in nights preceding a migraine attack. Cephalalgia 21(1):31–37
22. Della Marca G, Vollono C, Rubino M, Di Trapani G, Mariotti P, Tonali PA (2006) Dysfunction of arousal systems in sleep-related migraine without aura. Cephalalgia 26(7):857–864
23. Engstrøm M, Hagen K, Bjørk M, Gravdahl G, Sand T (2013) Sleep-related and non-sleep-related migraine: interictal sleep quality, arousals and pain thresholds. J Headache Pain 14:68
24. Engstrøm M, Hagen K, Bjørk M, Stovner L, Gravdahl G, Stjern M et al (2013) Sleep quality, arousal and pain thresholds in migraineurs: a blinded controlled polysomnographic study. J Headache Pain 14(1):12
25. Fritzer G, Strenge H, Göder R, Gerber WD, Aldenhoff J (2004) Changes in cortical dynamics in the preictal stage of a migraine attack. J Clin Neurophysiol 21(2):99–104
26. Coppola G, Pierelli F, Schoenen J (2007) Is the cerebral cortex hyperexcitable or hyperresponsive in migraine? Cephalalgia 27(12):1427–1439
27. Schoenen J, Wang W, Albert A, Delwaide P (1995) Potentiation instead of habituation characterizes visual evoked potentials in migraine patients between attacks. Eur J Neurol 2:115–122
28. Afra J, Cecchini AP, De Pasqua V, Albert A, Schoenen J (1998) Visual evoked potentials during long periods of pattern-reversal stimulation in migraine. Brain 121(Pt 2):233–241
29. Ozkul Y, Bozlar S (2002) Effects of fluoxetine on habituation of pattern reversal visually evoked potentials in migraine prophylaxis. Headache 42(7):582–587
30. Fumal A, Coppola G, Bohotin V, Gérardy PY, Seidel L, Donneau AF et al (2006) Induction of long-lasting changes of visual cortex excitability by five daily sessions of repetitive transcranial magnetic stimulation (rTMS) in healthy volunteers and migraine patients. Cephalalgia 26(2):143–149
31. Coppola G, Di Lorenzo C, Schoenen J, Pierelli F (2013) Habituation and sensitization in primary headaches. J Headache Pain 14(1):65
32. Oelkers R, Grosser K, Lang E, Geisslinger G, Kobal G, Brune K et al (1999) Visual evoked potentials in migraine patients: alterations depend on pattern spatial frequency. Brain 122(Pt 6):1147–1155
33. Sand T, Vingen JV (2000) Visual, long-latency auditory and brainstem auditory evoked potentials in migraine: relation to pattern size, stimulus intensity, sound and light discomfort thresholds and pre-attack state. Cephalalgia 20(9):804–820
34. Sand T, Zhitniy N, White LR, Stovner LJ (2008) Visual evoked potential latency, amplitude and habituation in migraine: a longitudinal study. Clin Neurophysiol 119(5):1020–1027
35. Omland P, Nilsen K, Uglem M, Gravdahl G, Linde M, Hagen K et al (2013) Visual evoked potentials in interictal migraine: no confirmation of abnormal habituation. Headache 53(7):1071–1086
36. Chen W, Wang S, Fuh J, Lin C, Ko Y, Lin Y (2011) Persistent ictal-like visual cortical excitability in chronic migraine. Pain 152(2):254–258

37. Bednář M, Kubová Z, Kremláček J (2014) Lack of visual evoked potentials amplitude decrement during prolonged reversal and motion stimulation in migraineurs. Clin Neurophysiol 125(6):1223–1230
38. Coppola G, Parisi V, Di Lorenzo C, Serrao M, Magis D, Schoenen J et al (2013) Lateral inhibition in visual cortex of migraine patients between attacks. J Headache Pain 14:20
39. Coppola G, Ambrosini A, Di Clemente L, Magis D, Fumal A, Gérard P et al (2007) Interictal abnormalities of gamma band activity in visual evoked responses in migraine: an indication of thalamocortical dysrhythmia? Cephalalgia 27(12):1360–1367
40. Coppola G, Currà A, Sava SL, Alibardi A, Parisi V, Pierelli F et al (2010) Changes in visual-evoked potential habituation induced by hyperventilation in migraine. J Headache Pain 11(6):497–503
41. Coppola G, Crémers J, Gérard P, Pierelli F, Schoenen J (2011) Effects of light deprivation on visual evoked potentials in migraine without aura. BMC Neurol 11:91
42. Judit A, Sándor PS, Schoenen J (2000) Habituation of visual and intensity dependence of auditory evoked cortical potentials tends to normalize just before and during the migraine attack. Cephalalgia 20(8):714–719
43. Chen W, Wang S, Fuh J, Lin C, Ko Y, Lin Y (2009) Peri-ictal normalization of visual cortex excitability in migraine: an MEG study. Cephalalgia 29(11):1202–1211
44. Sand T, White L, Hagen K, Stovner L (2009) Visual evoked potential and spatial frequency in migraine: a longitudinal study. Acta Neurol Scand Suppl 189:33–37
45. Sand T, Zhitniy N, White LR, Stovner LJ (2008) Brainstem auditory-evoked potential habituation and intensity-dependence related to serotonin metabolism in migraine: a longitudinal study. Clin Neurophysiol 119(5):1190–1200
46. Schlake HP, Grotemeyer KH, Hofferberth B, Husstedt IW, Wiesner S (1990) Brainstem auditory evoked potentials in migraine–evidence of increased side differences during the pain-free interval. Headache 30(3):129–132
47. Wang W, Timsit-Berthier M, Schoenen J (1996) Intensity dependence of auditory evoked potentials is pronounced in migraine: an indication of cortical potentiation and low serotonergic neurotransmission? Neurology 46(5):1404–1409
48. Ambrosini A, Rossi P, De Pasqua V, Pierelli F, Schoenen J (2003) Lack of habituation causes high intensity dependence of auditory evoked cortical potentials in migraine. Brain 126(Pt 9):2009–2015
49. Ambrosini A, De Pasqua V, Afra J, Sandor PS, Schoenen J (2001) Reduced gating of middle-latency auditory evoked potentials (P50) in migraine patients: another indication of abnormal sensory processing? Neurosci Lett 306(1–2):132–134
50. Siniatchkin M, Kropp P, Gerber WD (2003) What kind of habituation is impaired in migraine patients? Cephalalgia 23(7):511–518
51. Chayasirisobhon S (1995) Somatosensory evoked potentials in acute migraine with sensory aura. Clin Electroencephalogr 26(1):65–69
52. de Tommaso M, Sciruicchio V, Tota P, Megna M, Guido M, Genco S et al (1997) Somatosensory evoked potentials in migraine. Funct Neurol 12(2):77–82
53. Ozkul Y, Uckardes A (2002) Median nerve somatosensory evoked potentials in migraine. Eur J Neurol 9(3):227–232
54. Coppola G, Currà A, Di Lorenzo C, Parisi V, Gorini M, Sava SL et al (2010) Abnormal cortical responses to somatosensory stimulation in medication-overuse headache. BMC Neurol 10:126
55. Coppola G, De Pasqua V, Pierelli F, Schoenen J (2012) Effects of repetitive transcranial magnetic stimulation on somatosensory evoked potentials and high frequency oscillations in migraine. Cephalalgia 32(9):700–709
56. Sakuma K, Takeshima T, Ishizaki K, Nakashima K (2004) Somatosensory evoked high-frequency oscillations in migraine patients. Clin Neurophysiol 115(8):1857–1862
57. Coppola G, Vandenheede M, Di Clemente L, Ambrosini A, Fumal A, De Pasqua V et al (2005) Somatosensory evoked high-frequency oscillations reflecting thalamo-cortical activity are decreased in migraine patients between attacks. Brain 128(Pt 1):98–103

58. Coppola G, Iacovelli E, Bracaglia M, Serrao M, Di Lorenzo C, Pierelli F (2013) Electrophysiological correlates of episodic migraine chronification: evidence for thalamic involvement. J Headache Pain 14(1):76
59. Restuccia D, Vollono C, Virdis D, del Piero I, Martucci L, Zanini S (2014) Patterns of habituation and clinical fluctuations in migraine. Cephalalgia 34(3):201–210
60. Darabaneanu S, Kropp P, Niederberger U, Strenge H, Gerber W (2008) Effects of pregnancy on slow cortical potentials in migraine patients and healthy controls. Cephalalgia 28(10):1053–1060
61. Kropp P, Siniatchkin M, Gerber WD (2000) Contingent negative variation as indicator of duration of migraine disease. Funct Neurol 15(Suppl 3):78–81
62. Kropp P, Brecht I, Niederberger U, Kowalski J, Schröder D, Thome J et al (2012) Time-dependent post-imperative negative variation indicates adaptation and problem solving in migraine patients. J Neural Transm 119(10):1213–1221
63. Schoenen J, Maertens A, Timsit-Berthier M, Timsit M (1985) Contingent negative variation (CNV) as a diagnostic and physiopathologic tool in headache patients. In: Rose F (ed) Migraine. Clinical and research advances. Karger, Basel, pp 17–25
64. Kropp P, Gerber WD (1993) Contingent negative variation–findings and perspectives in migraine. Cephalalgia 13(1):33–36
65. Siniatchkin M, Gerber WD, Kropp P, Voznesenskaya T, Vein AM (2000) Are the periodic changes of neurophysiological parameters during the pain-free interval in migraine related to abnormal orienting activity? Cephalalgia 20(1):20–29
66. Siniatchkin M, Kropp P, Gerber WD (2001) Contingent negative variation in subjects at risk for migraine without aura. Pain 94(2):159–167
67. Siniatchkin M, Andrasik F, Kropp P, Niederberger U, Strenge H, Averkina N et al (2007) Central mechanisms of controlled-release metoprolol in migraine: a double-blind, placebo-controlled study. Cephalalgia 27(9):1024–1032
68. Siniatchkin M, Kropp P, Gerber WD, Stephani U (2000) Migraine in childhood–are periodically occurring migraine attacks related to dynamic changes of cortical information processing? Neurosci Lett 279(1):1–4
69. Schoenen J, Maertens de Noordhout A, Timsit-Berthier M, Timsit M (1986) Contingent negative variation and efficacy of beta-blocking agents in migraine. Cephalalgia 6(4):229–233
70. Tommaso M, Guido M, Sardaro M, Serpino C, Vecchio E, De S et al (2008) Effects of topiramate and levetiracetam vs placebo on habituation of contingent negative variation in migraine patients. Neurosci Lett 442(2):81–85
71. Overath C, Darabaneanu S, Evers M, Gerber W, Graf M, Keller A et al (2014) Does an aerobic endurance programme have an influence on information processing in migraineurs? J Headache Pain 15(1):11
72. Morlet D, Demarquay G, Brudon F, Fischer C, Caclin A (2014) Attention orienting dysfunction with preserved automatic auditory change detection in migraine. Clin Neurophysiol 125(3):500–511
73. de Tommaso M, Difruscolo O, Sardaro M, Libro G, Pecoraro C, Serpino C et al (2007) Effects of remote cutaneous pain on trigeminal laser-evoked potentials in migraine patients. J Headache Pain 8(3):167–174
74. de Tommaso M, Baumgartner U, Sardaro M, Difruscolo O, Serpino C, Treede RD (2008) Effects of distraction versus spatial discrimination on laser-evoked potentials in migraine. Headache 48(3):408–416
75. de Tommaso M, Valeriani M, Sardaro M, Serpino C, Fruscolo OD, Vecchio E et al (2009) Pain perception and laser evoked potentials during menstrual cycle in migraine. J Headache Pain 10(6):423–429
76. de Tommaso M, Calabrese R, Vecchio E, De Vito Francesco V, Lancioni G, Livrea P (2009) Effects of affective pictures on pain sensitivity and cortical responses induced by laser stimuli in healthy subjects and migraine patients. Int J Psychophysiol 74(2): 139–148

77. de Tommaso M, Brighina F, Fierro B, Francesco V, Santostasi R, Sciruicchio V et al (2010) Effects of high-frequency repetitive transcranial magnetic stimulation of primary motor cortex on laser-evoked potentials in migraine. J Headache Pain 11(6):505–512
78. de Tommaso M, Federici A, Franco G, Ricci K, Lorenzo M, Delussi M et al (2012) Suggestion and pain in migraine: a study by laser evoked potentials. CNS Neurol Disord Drug Targets 11(2):110–126
79. de Tommaso M, Guido M, Libro G, Losito L, Sciruicchio V, Monetti C et al (2002) Abnormal brain processing of cutaneous pain in migraine patients during the attack. Neurosci Lett 333(1):29–32
80. de Tommaso M, Guido M, Libro G, Losito L, Difruscolo O, Puca F et al (2004) Topographic and dipolar analysis of laser-evoked potentials during migraine attack. Headache 44(10):947–960
81. de Tommaso M, Libro G, Guido M, Losito L, Lamberti P, Livrea P (2005) Habituation of single CO2 laser-evoked responses during interictal phase of migraine. J Headache Pain 6(4):195–198
82. de Tommaso M, Lo Sito L, Di Fruscolo O, Sardaro M, Pia Prudenzano M, Lamberti P et al (2005) Lack of habituation of nociceptive evoked responses and pain sensitivity during migraine attack. Clin Neurophysiol 116(6):1254–1264
83. Afra J, Mascia A, Gérard P, Maertens de Noordhout A, Schoenen J (1998) Interictal cortical excitability in migraine: a study using transcranial magnetic stimulation of motor and visual cortices. Ann Neurol 44(2):209–215
84. Conforto A, Moraes M, Amaro E, Young W, Lois L, Gonçalves A et al (2012) Increased variability of motor cortical excitability to transcranial magnetic stimulation in migraine: a new clue to an old enigma. J Headache Pain 13(1):29–37
85. Siniatchkin M, Kröner-Herwig B, Kocabiyik E, Rothenberger A (2007) Intracortical inhibition and facilitation in migraine–a transcranial magnetic stimulation study. Headache 47(3):364–370
86. Gunaydin S, Soysal A, Atay T, Arpaci B (2006) Motor and occipital cortex excitability in migraine patients. Can J Neurol Sci 33(1):63–67
87. Curra A, Pierelli F, Coppola G, Barbanti P, Buzzi MG, Galeotti F et al (2007) Shortened cortical silent period in facial muscles of patients with migraine. Pain 132(1–2):124–131
88. Currà A, Coppola G, Gorini M, Porretta E, Bracaglia M, Di Lorenzo C et al (2011) Drug-induced changes in cortical inhibition in medication overuse headache. Cephalalgia 31(12):1282–1290
89. Brighina F, Palermo A, Panetta M, Daniele O, Aloisio A, Cosentino G et al (2009) Reduced cerebellar inhibition in migraine with aura: a TMS study. Cerebellum 8(3):260–266
90. Aurora SK, Ahmad BK, Welch KM, Bhardhwaj P, Ramadan NM (1998) Transcranial magnetic stimulation confirms hyperexcitability of occipital cortex in migraine. Neurology 50(4):1111–1114
91. Bohotin V, Fumal A, Vandenheede M, Bohotin C, Schoenen J (2003) Excitability of visual V1-V2 and motor cortices to single transcranial magnetic stimuli in migraine: a reappraisal using a figure-of-eight coil. Cephalalgia 23(4):264–270
92. Chadaide Z, Arlt S, Antal A, Nitsche M, Lang N, Paulus W (2007) Transcranial direct current stimulation reveals inhibitory deficiency in migraine. Cephalalgia 27(7):833–839
93. Omland P, Uglem M, Engstrøm M, Linde M, Hagen K, Sand T (2014) Modulation of visual evoked potentials by high-frequency repetitive transcranial magnetic stimulation in migraineurs. Clin Neurophysiol 125(10):2090–9
94. Brigo F, Storti M, Tezzon F, Manganotti P, Nardone R (2013) Primary visual cortex excitability in migraine: a systematic review with meta-analysis. Neurol Sci 34(6):819–830
95. Brighina F, Giglia G, Scalia S, Francolini M, Palermo A, Fierro B (2005) Facilitatory effects of 1 Hz rTMS in motor cortex of patients affected by migraine with aura. Exp Brain Res 161(1):34–38
96. Conte A, Barbanti P, Frasca V, Iacovelli E, Gabriele M, Giacomelli E et al (2010) Differences in short-term primary motor cortex synaptic potentiation as assessed by repetitive transcranial magnetic stimulation in migraine patients with and without aura. Pain 148(1):43–48

97. Brighina F, Cosentino G, Vigneri S, Talamanca S, Palermo A, Giglia G et al (2011) Abnormal facilitatory mechanisms in motor cortex of migraine with aura. Eur J Pain 15(9):928–935
98. Cosentino G, Fierro B, Vigneri S, Talamanca S, Paladino P, Baschi R et al (2014) Cyclical changes of cortical excitability and metaplasticity in migraine: evidence from a repetitive transcranial magnetic stimulation study. Pain 155(6):1070–1078
99. Pierelli F, Iacovelli E, Bracaglia M, Serrao M, Coppola G (2013) Abnormal sensorimotor plasticity in migraine without aura patients. Pain 154(9):1738–1742
100. Siniatchkin M, Sendacki M, Moeller F, Wolff S, Jansen O, Siebner H et al (2012) Abnormal changes of synaptic excitability in migraine with aura. Cereb Cortex 22(10):2207–2216
101. Viganò A, D'Elia T, Sava S, Auvé M, De P, Colosimo A et al (2013) Transcranial Direct Current Stimulation (tDCS) of the visual cortex: a proof-of-concept study based on interictal electrophysiological abnormalities in migraine. J Headache Pain 14(1):23
102. Aktekin B, Yaltkaya K, Ozkaynak S, Oguz Y (2001) Recovery cycle of the blink reflex and exteroceptive suppression of temporalis muscle activity in migraine and tension-type headache. Headache 41(2):142–149
103. Sand T, Zwart J (1994) The blink reflex in chronic tension-type headache, migraine, and cervicogenic headache. Cephalalgia 14(6):447–450
104. Sand T, Møll-Nilsen B, Zwart J (2006) Blink reflex R2 amplitudes in cervicogenic headache, chronic tension-type headache and migraine. Cephalalgia 26(10):1186–1191
105. Bánk J, Bense E, Király C (1992) The blink reflex in migraine. Cephalalgia 12(5):289–292
106. Avramidis T, Podikoglou D, Anastasopoulos I, Koutroumanidis M, Papadimitriou A (1998) Blink reflex in migraine and tension-type headache. Headache 38(9):691–696
107. de Tommaso M, Murasecco D, Libro G, Guido M, Sciruicchio V, Specchio L et al (2002) Modulation of trigeminal reflex excitability in migraine: effects of attention and habituation on the blink reflex. Int J Psychophysiol 44(3):239–249
108. Shibata K, Yamane K, Iwata M (2006) Change of excitability in brainstem and cortical visual processing in migraine exhibiting allodynia. Headache 46(10):1535–1544
109. Katsarava Z, Giffin N, Diener HC, Kaube H (2003) Abnormal habituation of 'nociceptive' blink reflex in migraine–evidence for increased excitability of trigeminal nociception. Cephalalgia 23(8):814–819
110. Di Clemente L, Coppola G, Magis D, Fumal A, De Pasqua V, Di Piero V et al (2007) Interictal habituation deficit of the nociceptive blink reflex: an endophenotypic marker for presymptomatic migraine? Brain 130(Pt 3):765–770
111. Kaube H, Katsarava Z, Przywara S, Drepper J, Ellrich J, Diener HC (2002) Acute migraine headache: possible sensitization of neurons in the spinal trigeminal nucleus? Neurology 58(8):1234–1238
112. Katsarava Z, Limmroth V, Baykal O, Akguen D, Diener H, Kaube H (2004) Differences of anti-nociceptive mechanisms of migraine drugs on the trigeminal pain processing during and outside acute migraine attacks. Cephalalgia 24(8):657–662
113. Coppola G, Di Clemente L, Fumal A, Magis D, De Pasqua V, Pierelli F et al (2007) Inhibition of the nociceptive R2 blink reflex after supraorbital or index finger stimulation is normal in migraine without aura between attacks. Cephalalgia 27(7):803–808
114. Sandrini G, Proietti C, Milanov I, Tassorelli C, Buzzi M, Nappi G (2002) Electrophysiological evidence for trigeminal neuron sensitization in patients with migraine. Neurosci Lett 317(3):135–138
115. Busch V, Kaube S, Schulte-Mattler W, Kaube H, May A (2007) Sumatriptan and corneal reflexes in headache-free migraine patients: a randomized and placebo-controlled crossover study. Cephalalgia 27(2):165–172
116. Nardone R, Ausserer H, Bratti A, Covi M, Lochner P, Marth R et al (2008) Trigemino-cervical reflex abnormalities in patients with migraine and cluster headache. Headache 48(4): 578–585
117. Serrao M, Perrotta A, Bartolo M, Fiermonte G, Pauri F, Rossi P et al (2005) Enhanced trigemino-cervical-spinal reflex recovery cycle in pain-free migraineurs. Headache 45(8):1061–1068

118. Perrotta A, Serrao M, Sandrini G, Burstein R, Sances G, Rossi P et al (2010) Sensitisation of spinal cord pain processing in medication overuse headache involves supraspinal pain control. Cephalalgia 30(3):272–284
119. Weissman-Fogel I, Sprecher E, Granovsky Y, Yarnitsky D (2003) Repeated noxious stimulation of the skin enhances cutaneous pain perception of migraine patients in-between attacks: clinical evidence for continuous sub-threshold increase in membrane excitability of central trigeminovascular neurons. Pain 104(3):693–700
120. Perrotta A, Serrao M, Tassorelli C, Arce-Leal N, Guaschino E, Sances G et al (2011) Oral nitric-oxide donor glyceryl-trinitrate induces sensitization in spinal cord pain processing in migraineurs: a double-blind, placebo-controlled, cross-over study. Eur J Pain 15(5):482–490
121. Ayzenberg I, Obermann M, Nyhuis P, Gastpar M, Limmroth V, Diener HC et al (2006) Central sensitization of the trigeminal and somatic nociceptive systems in medication overuse headache mainly involves cerebral supraspinal structures. Cephalalgia 26(9):1106–1114
122. de Tommaso M, Valeriani M, Guido M, Libro G, Specchio LM, Tonali P et al (2003) Abnormal brain processing of cutaneous pain in patients with chronic migraine. Pain 101(1–2):25–32
123. de Tommaso M, Losito L, Difruscolo O, Libro G, Guido M, Livrea P (2005) Changes in cortical processing of pain in chronic migraine. Headache 45(9):1208–1218
124. Lorenzo C, Coppola G, Currà A, Grieco G, Santorelli F, Lepre C et al (2012) Cortical response to somatosensory stimulation in medication overuse headache patients is influenced by angiotensin converting enzyme (ACE) I/D genetic polymorphism. Cephalalgia 32(16): 1189–1197
125. Siniatchkin M, Gerber WD, Kropp P, Vein A (1998) Contingent negative variation in patients with chronic daily headache. Cephalalgia 18(8):565–569; discussion 531
126. Ferraro D, Vollono C, Miliucci R, Virdis D, De A, Pazzaglia C et al (2012) Habituation to pain in "medication overuse headache": a CO2 laser-evoked potential study. Headache 52(5): 792–807
127. Aurora S, Barrodale P, Tipton R, Khodavirdi A (2007) Brainstem dysfunction in chronic migraine as evidenced by neurophysiological and positron emission tomography studies. Headache 47(7):996–1003
128. De Marinis M, Pujia A, Colaizzo E, Accornero N (2007) The blink reflex in "chronic migraine". Clin Neurophysiol 118(2):457–463
129. Sandrini G, Friberg L, Coppola G, Jänig W, Jensen R, Kruit M et al (2011) Neurophysiological tests and neuroimaging procedures in non-acute headache (2nd edition). Eur J Neurol 18(3): 373–381
130. Sacks O, Siegel R (2012) Migraine aura and hallucinatory constants. In: Sacks O (ed) Migraine. Picador, London, pp 273–297

Chapter 9
Neurophysiology of Other Primary Headaches

Anna Ambrosini and Gianluca Coppola

Abbreviations

CH	Cluster headache
DNICs	Diffuse noxious inhibitory controls
EMG	Electromyography
HNCS	Heterotopic noxious conditioning stimulation
LEPs	Laser evoked potentials
nBR	Nociception-specific blink reflex
PPT	Pressure pain thresholds
SUNCT	Short-lasting unilateral neuralgiform headache attacks with conjunctival injection and tearing
TACs	Trigeminal autonomic cephalalgias
TTH	Tension-type headache
TTS	Total tenderness scores
VEPs	Visual evoked potentials

A. Ambrosini (✉)
Department of Headache Medicine, IRCCS Neuromed, Pozzilli (IS), Italy
e-mail: anna.ambrosini@neuromed.it

G. Coppola
Department of Neurophysiology of Vision and Neurophtalmology,
G.B. Bietti Foundation IRCCS, Rome, Italy
e-mail: gianluca.coppola@gmail.com

M. Ashina, P. Geppetti (eds.), *Pathophysiology of Headaches: From Molecule to Man*, Headache, DOI 10.1007/978-3-319-15621-7_9

9.1 Introduction

Primary or idiopathic headache syndromes are disorders where a temporary or permanent dysfunction of the central nervous system is present, but apparent organic lesions cannot be found. They include migraine, tension-type headache (TTH), and the trigeminal autonomic cephalalgias (TACs), among which the most common is "cluster headache" (CH). Primary headaches are highly prevalent and the most common neurological disorders among the general population. They are receiving growing attention over the years on the one hand because they affect people's quality of life and on the other hand because of their significant economic impact. The International Classification of Headache Disorders (ICHD) in its previous [1, 2] and current version [3] are the most sensitive diagnostic tool for headaches, and application of the proposed criteria greatly improved research on headaches allowing for a better comparison of clinical data between headache centers.

The large and still growing scientific knowledge on the primary headaches pathophysiological mechanisms, though scientists have not completely disentangled them, has contributed to raise interest on these neurological conditions. In fact, great advances were made during the last decades through the new research techniques. Clinical neurophysiology methods, in particular, have allowed in vivo measurements of the headache patients' cortical and peripheral responses to various sensory stimuli.

In this chapter we will review the neurophysiological studies in primary headaches other than migraine (treated elsewhere).

9.2 Tension-Type Headache

The most relevant neurophysiological abnormality in migraine is the interictal deficit of cortical habituation to repetitive stimulation. Habituation is defined as "a response decrement as a result of repeated stimulation" [4] and is a common feature of responses to any kind of sensory stimulation. In most studies this phenomenon is not present, and sometimes replaced by potentiation, in episodic migraine patients between attacks, but normalizes just before and during the headache phase. It is likely to be due to interictal cortical hyper-responsivity, which could be a possible expression of thalamocortical dysrhythmia [5]. The phenomenon of habituation has been investigated in tension-type headache as well, but there are only a few reports about it.

In episodic TTH sufferers, the habituation of the latency of P300 (a long-latency cognitive cortical evoked potential) was normal, while P300 amplitude also showed some degree of habituation, although not of statistical significance [6]. No habituation deficits were observed exploring visual evoked or event-related potentials other than P300 in episodic [7] or chronic TTH patients [7, 8]. Mismatch negativity, which is likely to reflect the automatic central processing of a novel stimulus, and

P300 habituation were significantly lower in TTH children than in healthy subjects in one study, where P300 habituation also positively correlated with behavioral symptomatology [9].

Patients affected by chronic TTH showed a normal habituation in scalp potentials evoked by CO_2 laser stimulation (LEPs) of the hand and facial skin [10].

By investigating habituation of sympathetic skin responses (SSR), a tool used to evaluate autonomic dysfunction, Ozkul and Ay found that in both episodic migraine without aura and TTH patients, there was a lack of habituation compared to normal controls [11].

The mild electrophysiological similarities in the cortical habituation behavior between the episodic forms of migraine and TTH suggest that some subgroups of TTH patients might be at the end of the migraine spectrum.

Sensitization is defined as facilitation occurring at the beginning of the stimulus presentation. The few studies in which the dynamic behavior of responses was analyzed using successive blocks of responses did not find any clear evidence for sensitization, expressed as increased amplitude of the first block, neither in episodic TTH for visual evoked potentials (VEPs) [7], visual P300 [6], laser evoked potentials (LEPs) [9], and sympathetic skin responses [11] nor in chronic TTH for visual P300 [8] and LEPs [10].

Some indirect evidence for sensitization was found in TTH, chiefly in its chronic form, with nociceptive specific reflexes and laser evoked cortical potentials, which enable to explore the cortical responses to peripheral nociceptive stimuli.

In chronic TTH patients, the amplitude, area, and latency [12–15] of the blink reflex R2 component were not different respect to values in healthy subjects, but it had a slower recovery cycle that was interpreted as a possible reduced excitability of the brainstem interneurons [14]. When a nociception-specific electrode was used, lower values of the normalized root mean square and area under the curve of the blink with control subjects were found respect to controls [16]. According to the authors, it may reflect consistent increases in eye muscle activity on the painful stimulation side.

One of the most investigated electrophysiological tests in TTH patients is the exteroceptive suppression of the temporalis muscle contraction that is the reflex inhibition of contraction of jaw-closing muscles by electrical stimulation of the infraorbital and mental nerves. It is obtained by surface EMG recordings of contracted muscles, where two different periods of suppression can be identified (SP1 and SP2). The SP2, mediated by a polysynaptic chain of interneurons likely to belong to the bulbar reticular formation, correlates to the level of excitability of these brainstem interneurons [17]. In episodic TTH patients, it was found normal, and in chronic TTH subjects, it was shortened in some studies, but not in others [18], possibly due to methodological differences.

Trigeminocervical reflex obtained from the sternocleidomastoid muscle after electrical stimulation of the supraorbital or infraorbital nerve had reduced latencies in chronic tension-type headache patients, similarly to migraineurs [19–21]. These findings further sustain the presence of a possible dysfunction in brainstem interneuronal activity controlling the pericranial muscles.

The presence of central sensitization in chronic TTH is strongly suggested also by the results of studies on pain sensitivity in pericranial or lower limb tissues. Testing the nociceptive lower limb flexion reflex, significantly lower subjective pain thresholds and reflex threshold in chronic TTH than in controls were found [22], associated with a paradoxical facilitation of the reflex response during the cold pressor test suggesting deficient descending inhibition, an abnormality also found by others [23]. Pressure pain thresholds (PPTs) were found normal in episodic and "mixed" TTH in every studies [24–28] except one [29], but decreased in chronic TTH in almost all studies [28, 30–33] especially on the anterior part of the temporalis muscle [25, 26, 30–32, 34] and in the upper part of the trapezius muscle [35]. Only one study did not confirm these results [26]. In a follow-up study, where PPTs were tested in episodic TTH patients at baseline and retested during the following twelve years, the baseline PPTs were normal but decreased at the follow-up in patients who develop the chronic form, suggesting that it was the headache frequency to induce an increased pain sensitivity and not the opposite [36].

Another indirect measure of sensitization is the temporal summation that is the increase in pain perception to repeated noxious stimulation, obtained by an algometer and heterotopic noxious conditioning stimulation (HNCS). Chronic TTH sufferers had more pain from repeated algometer pressures, both at finger and shoulder, and it was less inhibited by conditioned HNCS compared with controls [37]. Lower pain thresholds in the muscle and skin of the cephalic region but not of the extracephalic region with higher rating to suprathreshold single and repetitive electrical stimulation were reported in patients with chronic TTH than in healthy subjects [38].

When investigating laser evoked potentials in chronic TTH patients, the heat pain threshold was not different respect to controls, at the level of both the hand and pericranial skin, but the total tenderness scores (TTS) at pericranial sites were higher in patients than in controls, which was associated to a greater amplitude of the N2a–P2 LEP complex elicited by stimulation of the pericranial zone [39].

9.3 Cluster Headache and Other Trigeminal Autonomic Cephalalgias (TACs)

Electrophysiological methods were used to investigate cognitive and nociceptive processes in trigeminal autonomic cephalalgias, particularly in cluster headache (CH).

Two visual event-related potential studies in cluster headache either during the bout or outside and in chronic paroxysmal hemicrania showed a normal cognitive habituation [8, 40]. However, intensity dependence of auditory potentials, which is supposed to be an indirect expression of deficient habituation in migraineurs [41], was found markedly increased also in cluster headache patients both during and outside the bout [42].

Formisano et al. were the first to found abnormal habituation of the blink reflex in a small number of CH patients during the attack, but in this study there was not a comparison group of control subjects [43]. Habituation of both the R2 and the R3 blink reflex components are impaired in CH patients on the affected side compared to healthy controls, and this abnormality was even more pronounced than that found in episodic migraine [44]. These results were replicated by using the nociception-specific concentric stimulating electrode: R2 reflex area and habituation were reduced on the affected CH side, and the degree of habituation deficit correlated to number of days elapsed from the beginning of the bout and the daily attack frequency [45]. Contrasting findings were obtained in another study, where the authors failed to detect altered habituation of the nBR R2 in episodic and chronic CH within or outside a bout, but the majority of CH patients investigated were taking one or several prophylactic medications at the time of recordings, which may biased the results [46].

Classical blink reflex studies did not disclose any sign of sensitization in CH [47, 48]. In episodic CH patients within a bout, a significantly faster R2 blink reflex recovery curve on the symptomatic side was found after paired supraorbital stimuli, likely to indirectly reflect sensitization within the spinal trigeminal nucleus. Furthermore, when the supraorbital stimulus was preconditioned by a peripheral stimulation of the index finger, the R2 recovery curve was faster on both affected and unaffected sides in CH patients than in controls. Naloxone injection transiently reverted this bilateral R2 sensitization, suggesting that the faster R2 recovery may reflect hypoactivity of reticular nuclei, due to reduced descending opiatergic inhibition [49], a mechanism that was recently supported by functional neuroimaging studies [50, 51].

Cluster headache patients had lower thresholds for pressure pain [52], electric pain, and nociceptive flexion reflex [53] on the affected than on the unaffected side both in the episodic (in and outside of a bout) and in the chronic type [54]. A phase shift of the normal circadian rhythmic variations in nociceptive flexion reflex threshold in episodic bouts of CH with respect to the remission period and absence of circadian rhythmicity of the nociceptive flexion reflex threshold in chronic CH patients have already been described [54]. The functional activity of the descending diffuse noxious inhibitory controls (DNIC) (or conditioned pain modulation system) was also investigated in a group of episodic CH patients during active and remission phases compared to healthy controls, by measuring the influence of a cold pressor test on the nociceptive withdrawal reflex [55]. Cluster headache patients had a significant facilitation in temporal processing of pain at spinal level during the active phase of the disease, and a facilitation in pain processing reverted during the remission phase of the disease. The cold pressor test activating the DNIC did not induce any significant inhibitory effect on the neurophysiological responses during the active phase of the disease, but was able to induce a clear inhibition during the remission phase. It was thus hypothesized that cluster headache sufferers have a dysfunction of the supraspinal control of pain, which changes according to the clinical activity of the disease and leads to facilitation of pain processing.

Procacci et al. (1989) found cutaneous and deep hyperalgesia to mechanical and electrical stimuli with earlier appearance of pain after an ischemic test in the upper limbs on the affected side of the body in episodic CH patients [56], but when using quantitative sensory testing, perception of warmth, cold, and pressure, the sensation of pain was reduced on the affected side as compared with the contralateral asymptomatic side in episodic and chronic CH patients [57, 58].

Unfortunately, the literature concerning neurophysiological tests in trigeminal autonomic cephalalgias other than cluster headache is really poor. In one study, pain pressure threshold, subjective pain perception after sural nerve stimulation, and nociceptive flexion reflex threshold were tested in patients with chronic paroxysmal hemicrania and with hemicrania continua, and they appeared reduced mostly on the affected side, compared to healthy subjects [59]. Corneal reflex thresholds were significantly reduced on both sides only in chronic paroxysmal hemicrania patients, though there were no abnormalities in the blink reflex. In patients affected by the idiopathic form of another rare type of TACs, "short-lasting unilateral neuralgiform headache attacks with conjunctival injection and tearing" (SUNCT) trigeminal reflexes and laser evoked potentials did not show any abnormalities [60].

9.4 Discussion

Neurophysiological studies have disclosed some abnormalities of the spinal, brainstem, and cortical responsivity to external innocuous or noxious stimuli not only in migraine but also, at a lesser extent, in other primary headaches. These abnormalities can be summarized as follows:

- In subgroups of tension-type headache sufferers, some evidence of deficient habituation chiefly with cognitive potentials (mismatch negativity and P300) and sympathetic skin responses have been found. Indirect evidence for sensitization has been disclosed in chronic TTH patients with nociceptive specific reflexes and grand-averaged evoked potentials. These studies suggest for the subjects chronically affected the presence of generalized increased sensitivity to pain and a dysfunction in supraspinal conditioned pain modulation, which may contribute to the development and/or maintenance of central sensitization in this disorder.
- Habituation deficit of the blink reflex was found in episodic cluster headache patients and more pronounced than in interictal migraineurs, suggesting that additional dysfunctional neurobiological factors are implicated in CH, though CH and migraine probably share some pathophysiological mechanisms, as suggested also by the marked increase of intensity dependence of auditory potentials found in CH [42]. A sensitization of pain processing was observed only during the bout, but not outside. Several causes could be at the basis of this observation: a dysfunctioning descending aminergic, especially dopaminergic, control [61, 62], a malfunctioning hypothalamo-trigeminal control [63], and an altered descending opiatergic pain control system [50, 51]. Future electrophysiological works should aim to understand the role of the descending monoamine and opioid systems in the mechanism of sensitization and lateralization of pain and to unravel the mechanisms of CH periodicity and of its chronification.

References

1. Headache Classification Subcommittee of the International Headache Society (2004) The international classification of headache disorders: 2nd edition. Cephalalgia 1:9–160
2. Olesen J, Bousser MG, Diener HC, Dodick D, First M, Goadsby PJ, Göbel H, Lainez MJA, Lance JW, Lipton RB, Nappi G, Sakai F, Schoenen J, Silberstein SD, Steiner TJ (2006) New appendix criteria open for a broader concept of chronic migraine. Cephalalgia 26(6):742–746
3. Headache Classification Committee of the International Headache Society (IHS) (2013) The international classification of headache disorders, 3rd edition (beta version). Cephalalgia 33(9):629–808
4. Harris J (1943) Habituatory response decrement in the intact organism. Psychol Bull 40: 385–422
5. de Tommaso M, Ambrosini A, Brighina F, Coppola G, Perrotta A, Pierelli F, Sandrini G, Valeriani M, Marinazzo D, Stramaglia S, Schoenen J (2014) Altered processing of sensory stimuli in patients with migraine. Nat Rev Neurol 10(3):144–155
6. Demirci S, Savas S (2002) The auditory event related potentials in episodic and chronic pain sufferers. Eur J Pain 6(3):239–244
7. Wang W, Wang GP, Ding XL, Wang YH (1999) Personality and response to repeated visual stimulation in migraine and tension-type headaches. Cephalalgia 19(8):718–724
8. Evers S, Bauer B, Suhr B, Husstedt I, Grotemeyer K (1997) Cognitive processing in primary headache: a study on event-related potentials. Neurology 48(1):108–113
9. Valeriani M, Galli F, Tarantino S, Graceffa D, Pignata E, Miliucci R, Biondi G, Tozzi A, Vigevano F, Guidetti V (2009) Correlation between abnormal brain excitability and emotional symptomatology in paediatric migraine. Cephalalgia 29(2):204–213
10. Valeriani M, de Tommaso M, Restuccia D, Le Pera D, Guido M, Iannetti GD, Libro G, Truini A, Di Trapani G, Puca F, Tonali P, Cruccu G (2003) Reduced habituation to experimental pain in migraine patients: a CO(2) laser evoked potential study. Pain 105(1–2):57–64
11. Ozkul Y, Ay H (2007) Habituation of sympathetic skin response in migraine and tension type headache. Auton Neurosci 134(1–2):81–84
12. Sand T, Zwart J (1994) The blink reflex in chronic tension-type headache, migraine, and cervicogenic headache. Cephalalgia 14(6):447–450
13. Avramidis T, Podikoglou D, Anastasopoulos I, Koutroumanidis M, Papadimitriou A (1998) Blink reflex in migraine and tension-type headache. Headache 38(9):691–696
14. Aktekin B, Yaltkaya K, Ozkaynak S, Oguz Y (2001) Recovery cycle of the blink reflex and exteroceptive suppression of temporalis muscle activity in migraine and tension-type headache. Headache 41(2):142–149
15. Sand T, Møll-Nilsen B, Zwart J (2006) Blink reflex R2 amplitudes in cervicogenic headache, chronic tension-type headache and migraine. Cephalalgia 26(10):1186–1191
16. Peddireddy A, Wang K, Svensson P, Arendt-Nielsen L (2009) Blink reflexes in chronic tension-type headache patients and healthy controls. Clin Neurophysiol 120(9):1711–1716
17. Schoenen J (1993) Wolff Award 1992. Exteroceptive suppression of temporalis muscle activity in patients with chronic headache and in normal volunteers: methodology, clinical and pathophysiological relevance. Headache 33(1):3–17
18. Schoenen J, Bendtsen L (2006) Neurophysiology of tension-type headache. In: Olesen J, Goadsby P, Ramadan N, Tfelt-Hansen P, Welch K (eds) The headaches, 3rd edn. Lippincott Williams & Wilkins, Philadelphia, pp 644–647
19. Nardone R, Tezzon F (2003) Short latency trigemino-sternocleidomastoid response in patients with migraine. J Neurol 250(6):725–732
20. Nardone R, Tezzon F (2003) The trigemino-cervical reflex in tension-type headache. Eur J Neurol 10(3):307–312
21. Milanov I, Bogdanova D (2003) Trigemino-cervical reflex in patients with headache. Cephalalgia 23(1):35–38

22. Sandrini G, Rossi P, Milanov I, Serrao M, Cecchini AP, Nappi G (2006) Abnormal modulatory influence of diffuse noxious inhibitory controls in migraine and chronic tension-type headache patients. Cephalalgia 26(7):782–789
23. Pielsticker A, Haag G, Zaudig M, Lautenbacher S (2005) Impairment of pain inhibition in chronic tension-type headache. Pain 118(1–2):215–223
24. Göbel H, Weigle L, Kropp P, Soyka D (1992) Pain sensitivity and pain reactivity of pericranial muscles in migraine and tension-type headache. Cephalalgia 12(3):142–151
25. Bovim G (1992) Cervicogenic headache, migraine, and tension-type headache. Pressure-pain threshold measurements. Pain 51(2):169–173
26. Jensen R, Rasmussen B, Pedersen B, Olesen J (1993) Muscle tenderness and pressure pain thresholds in headache. A population study. Pain 52(2):193–199
27. Jensen R (1996) Mechanisms of spontaneous tension-type headaches: an analysis of tenderness, pain thresholds and EMG. Pain 64(2):251–256
28. Engstrøm M, Hagen K, Bjørk M, Stovner L, Stjern M, Sand T (2014) Sleep quality, arousal and pain thresholds in tension-type headache: a blinded controlled polysomnographic study. Cephalalgia 34(6):455–463
29. Mørk H, Ashina M, Bendtsen L, Olesen J, Jensen R (2003) Induction of prolonged tenderness in patients with tension-type headache by means of a new experimental model of myofascial pain. Eur J Neurol 10(3):249–256
30. Schoenen J, Bottin D, Hardy F, Gerard P (1991) Cephalic and extracephalic pressure pain thresholds in chronic tension-type headache. Pain 47(2):145–149
31. Bendtsen L, Jensen R, Olesen J (1996) Decreased pain detection and tolerance thresholds in chronic tension-type headache. Arch Neurol 53(4):373–376
32. Ashina S, Babenko L, Jensen R, Ashina M, Magerl W, Bendtsen L (2005) Increased muscular and cutaneous pain sensitivity in cephalic region in patients with chronic tension-type headache. Eur J Neurol 12(7):543–549
33. Fernández-de-Las-Peñas C, Cuadrado M, Arendt-Nielsen L, Ge H, Pareja J (2007) Increased pericranial tenderness, decreased pressure pain threshold, and headache clinical parameters in chronic tension-type headache patients. Clin J Pain 23(4):346–352
34. Metsahonkala L, Anttila P, Laimi K, Aromaa M, Helenius H, Mikkelsson M, Jäppilä E, Viander S, Sillanpää M, Salminen J (2006) Extracephalic tenderness and pressure pain threshold in children with headache. Eur J Pain 10(7):581–585
35. Fernández-de-las-Peñas C, Madeleine P, Caminero A, Cuadrado M, Arendt-Nielsen L, Pareja J (2010) Generalized neck-shoulder hyperalgesia in chronic tension-type headache and unilateral migraine assessed by pressure pain sensitivity topographical maps of the trapezius muscle. Cephalalgia 30(1):77–86
36. Buchgreitz L, Lyngberg AC, Bendtsen L, Jensen R (2008) Increased pain sensitivity is not a risk factor but a consequence of frequent headache: a population-based follow-up study. Pain 137(3):623–630
37. Cathcart S, Winefield A, Lushington K, Rolan P (2010) Noxious inhibition of temporal summation is impaired in chronic tension-type headache. Headache 50(3):403–412
38. Ashina S, Bendtsen L, Ashina M, Magerl W, Jensen R (2006) Generalized hyperalgesia in patients with chronic tension-type headache. Cephalalgia 26(8):940–948
39. de Tommaso M, Libro G, Guido M, Sciruicchio V, Losito L, Puca F (2003) Heat pain thresholds and cerebral event-related potentials following painful CO2 laser stimulation in chronic tension-type headache. Pain 104(1–2):111–119
40. Evers S, Bauer B, Suhr B, Voss H, Frese A, Husstedt I (1999) Cognitive processing is involved in cluster headache but not in chronic paroxysmal hemicrania. Neurology 53(2):357–363
41. Ambrosini A, Rossi P, De Pasqua V, Pierelli F, Schoenen J (2003) Lack of habituation causes high intensity dependence of auditory evoked cortical potentials in migraine. Brain 126 (Pt 9):2009–2015
42. Afra J, Ertsey C, Bozsik G, Jelencsik I (2005) Cluster headache patients show marked intensity dependence of cortical auditory evoked potentials during and outside the bout. Cephalalgia 25(1):36–40

43. Formisano R, Cerbo R, Ricci M, Agostino R, Cesarino F, Cruccu G, Agnoli A (1987) Blink reflex in cluster headache. Cephalalgia 7(Suppl 6):353–354
44. Perrotta A, Serrao M, Sandrini G, Bogdanova D, Tassorelli C, Bartolo M, Coppola G, Pierelli F, Nappi G (2008) Reduced habituation of trigeminal reflexes in patients with episodic cluster headache during cluster period. Cephalalgia 28(9):950–959
45. Coppola G, Di Lorenzo C, Bracaglia M, Di Lenola D, Parisi V, Perrotta A, Serrao M, Pierelli F (2014) Lateralized nociceptive blink reflex habituation deficit in episodic cluster headache: correlations with clinical features. Cephalalgia. [E-pub ahead of print] doi:10.1177/0333102414550418
46. Holle D, Zillessen S, Gaul C, Naegel S, Kaube H, Diener H, Katsarava Z, Obermann M (2012) Habituation of the nociceptive blink reflex in episodic and chronic cluster headache. Cephalalgia 32(13):998–1004
47. Pavesi G, Granella F, Brambilla S, Medici D, Mancia D, Manzoni G (1987) Blink reflex in cluster headache: evidence of a trigeminal system dysfunction. Cephalalgia 6:100–102
48. Raudino F (1990) The blink reflex in cluster headache. Headache 30(9):584–585
49. Lozza A, Schoenen J, Delwaide PJ (1997) Inhibition of the blink reflex R2 component after supraorbital and index finger stimulations is reduced in cluster headache: an indication for both segmental and suprasegmental dysfunction? Pain 71(1):81–88
50. Sprenger T, Willoch F, Miederer M, Schindler F, Valet M, Berthele A, Spilker ME, Förderreuther S, Straube A, Stangier I, Wester HJ, Tölle TR (2006) Opioidergic changes in the pineal gland and hypothalamus in cluster headache: a ligand PET study. Neurology 66(7):1108–1110
51. Magis D, Bruno M, Fumal A, Gérardy P, Hustinx R, Laureys S, Schoenen J (2011) Central modulation in cluster headache patients treated with occipital nerve stimulation: an FDG-PET study. BMC Neurol 11:25
52. Bono G, Antonaci F, Sandrini G, Pucci E, Rossi F, Nappi G (1996) Pain pressure threshold in cluster headache patients. Cephalalgia 16(1):62–66
53. Sandrini G, Antonaci F, Lanfranchi S, Milanov I, Danilov A, Nappi G (2000) Asymmetrical reduction of the nociceptive flexion reflex threshold in cluster headache. Cephalalgia 20(7):647–652
54. Nappi G, Sandrini G, Alfonsi E, Cecchini A, Micieli G, Moglia A (2002) Impaired circadian rhythmicity of nociceptive reflex threshold in cluster headache. Headache 42(2):125–131
55. Perrotta A, Serrao M, Ambrosini A, Bolla M, Coppola G, Sandrini G, Pierelli F (2013) Facilitated temporal processing of pain and defective supraspinal control of pain in cluster headache. Pain 154(8):1325–1332
56. Procacci P, Zoppi M, Maresca M, Zamponi A, Fanciullacci M, Sicuteri F (1989) Lateralisation of pain in cluster headache. Pain 38(3):275–278
57. Ladda J, Straube A, Förderreuther S, Krause P, Eggert T (2006) Quantitative sensory testing in cluster headache: increased sensory thresholds. Cephalalgia 26(9):1043–1050
58. Ellrich J, Ristic D, Yekta S (2006) Impaired thermal perception in cluster headache. J Neurol 253(10):1292–1299
59. Antonaci F, Sandrini G, Danilov A, Sand T (1994) Neurophysiological studies in chronic paroxysmal hemicrania and hemicrania continua. Headache 34(8):479–483
60. Truini A, Barbanti P, Galeotti F, Leandri M, Cruccu G (2006) Trigeminal sensory pathway function in patients with SUNCT. Clin Neurophysiol 117(8):1821–1825
61. Palmieri A (2006) Chronic cluster headache responsive to pramipexole. Cephalalgia 26(6):761–762
62. Di Lorenzo C, Coppola G, Pierelli F (2013) A case of cluster headache treated with rotigotine: clinical and neurophysiological correlates. Cephalalgia 33(15):1272–1276
63. Malick A, Strassman R, Burstein R (2000) Trigeminohypothalamic and reticulohypothalamic tract neurons in the upper cervical spinal cord and caudal medulla of the rat. J Neurophysiol 84(4):2078–2112

Chapter 10
Biochemistry of Primary Headaches

Paola Sarchielli, Stefano Caproni, Cinzia Costa, Delia Szok, and Janos Tajti

Although the diagnosis of primary headaches remains generally clinical, researchers have made great efforts in the investigation of neurotransmitter pathways, neuropeptides, hormones, and the vascular and trigeminal systems. The increasing evidence regarding the systems involved in migraine, in addition to tension-type and cluster headaches, has permitted better comprehension of the cerebral and extraneurological mechanisms underlying these types of headaches, leading, in some cases, to the identification of a treatment target. This chapter deals with the biochemical alterations that play a pivotal role in the pathophysiology of primary headaches.

10.1 Markers of Trigeminovascular System Activation

10.1.1 Calcitonin Gene-Related Peptide

Calcitonin gene-related principle peptide (CGRP) is a 37-amino-acid neuropeptide that is widely distributed in the peripheral and central nervous systems. Following noxious stimulation, CGRP is released on the periphery, where it induces

P. Sarchielli (✉) • S. Caproni • C. Costa
Department of Medicine, Neurologic Clinic, University of Perugia,
Ospedale S. Maria della Misericordia-S. Andrea delle Fratte, Perugia 06132, Italy
e-mail: paola.sarchielli@unipg.it; sefano.caproni@gmail.com; cinzia.costa@unipg.it

J. Tajti (✉) • D. Szok
Department of Neurology, University of Szeged, Albert Szent-Györgyi Clinical Center,
Semmelweis u. 6, Szeged H-6725, Hungary
e-mail: tajti.janos@med.u-szeged.hu; szok.delia@med.u-szeged.hu

M. Ashina, P. Geppetti (eds.), *Pathophysiology of Headaches: From Molecule to Man*, Headache, DOI 10.1007/978-3-319-15621-7_10

neurogenic inflammation and sensitization of sensory nociceptive fibers, including those distributing to meningeal vessels, and promotes excitatory neurotransmission at the central concentration (dorsal horn of the spinal cord and trigeminal nucleus caudalis), mediating the sensation of pain to the entire body. Head CGRP-containing nerves supply blood vessels to various parts of the body.

Approximately 50 % of trigeminal neurons express CGRP and CGRP receptor. The binding of CGRP to its functional receptor can lead to the activation of multiple pathways that modulate gene expression and ion channel activity with a positive feedback loop, which may in part explain why the peripheral injection of CGRP leads to a delayed migraine-like headache several hours after injection [1, 2].

CGRP may play a role in neuronal–glial interactions in the trigeminal ganglia. The mechanism proposed involves the release of CGRP during the neuronal activation of the trigeminal ganglia. This CGRP then stimulates the satellite glial cells, which release proinflammatory cytokines, thereby further modulating the neuronal response [3].

CGRP involvement in migraine pathophysiology is supported by the ability of this neuropeptide infused intravenously to induce attacks, and by the effectiveness of the CGRP antagonists, olcegepant and telcagepant, as acute antimigraine agents [4]. The latter positive findings are further sustained by other approaches aimed at stopping CGRP activity using antibodies against CGRP and the CGRP receptor.

It has been suggested that CGRP might be a marker of trigeminovascular activation [5]. In migraine patients CGRP concentrations were measured both in peripheral blood and in the extrajugular vein (EJV). While in the former studies an increase in CGRP in the EJV blood of migraine patients assessed during attacks was found compared with controls, this finding was not confirmed in most recent research [5, 6] (Table 10.1).

These discrepant results are difficult to reconcile. According to Tfelt-Hansen and Le, the bulk of EJV blood flow comes from the extracranial tissue of the head and face and only 22 % comes from the internal cerebral circulation in humans. Moreover, the middle meningeal veins contribute only to a small fraction of blood in the EJV. The CGRP in EJV blood, therefore, is most likely derived from part of the vasculature without the blood-brain barrier [7].

The effect of sumatriptan on CGRP is most likely caused by an antimigraine effect, which is not mediated by a decrease in CGRP. In fact, in human volunteers the subcutaneous administration of sumatriptan did not induce variations in EJV concentrations of CGRP [8].

Recent research focusing on the interictal plasma concentrations of chronic migraine (CM) patients showed higher values in these patients in the absence of a migraine attack and with no symptomatic medication. Variables such as age, analgesic overuse, depression, fibromyalgia, vascular risk factors, history of triptan consumption, or type of preventive treatment did not influence CGRP concentrations [9].

Further evidence of the involvement of CGRP in migraine is derived from studies using saliva as a specimen that is easy to obtain, providing information on the pathological states and representing a clinical model for studying neuronal mechanisms involved in migraine. Salivary CGRP concentrations have been reported to be elevated

Table 10.1 Biochemical alterations in migraine with and without aura

		Migraine with aura		Migraine without aura	
Substances	Reference	Ictal	Interictal	Ictal	Interictal
CGRP	[9]	↑ (plasma)	–	↑ (plasma)	–
	[11]	–	–	Increased, then normalized (plasma)	–
	[12]	–	–	Not increased (plasma)	Not increased (plasma)
NKA	[25]	↑ (plasma)	NS	↑ (plasma)	NS
PACAP-38	[44]	–	–	↑ (plasma)	↓ (plasma)
VIP	[48]	–	↑ (plasma)	–	↑ (plasma)
5-HT	[54]	↑ (plasma)	↓ (plasma)	↑ (plasma)	↓ (plasma)
DA	[60]	–	↑ NS (platelet)	–	↑ (platelet)
NE	[71]	–	–	–	↓ (plasma)
Octopamine	[67]	–	NS	–	↑ (plasma)
Synephrine	[67]	–	NS	–	↑ (plasma)
Endocannabinoids	[76]	–	–	–	↑ (platelet)
Glutamate	[82]	↑ (CSF) ↓ (plasma)	–	↑ (CSF) ↓ (plasma)	–
Nitrite (NO)	[88]	–	↑ (plasma)	–	↑ (plasma)
BDNF	[100]	–	↓ (platelet)	–	↓ (platelet)
NGF	[100]	–	↓ (platelet)	–	↓ (platelet)
BDNF	[101]	–	–	↑ (plasma)	–
Adiponectin	[113]	–	↑ (plasma)	–	↑ (plasma)
NPY	[123]	↑ (plasma)	↓ (plasma)	↑ (plasma)	↓ (plasma)
SP	[131]	–	–	↑ (saliva)	–
EDN-1	[140]	–	–	↓ (plasma)	–

CGRP calcitonin gene-related peptide, *NKA* neurokinin A, *PACAP* pituitary adenylate cyclase-activating peptide, *VIP* vasoactive intestinal peptide, *5-HT* 5-hydroxytriptamine, *DA* dopamine, *NE* norepinephrine, *NO* nitric oxide, *BDNF* brain-derived neurotrophic factor, *NGF* nerve growth factor, *NPY* neuropeptide Y, *SP* substance P, *EDN-1* endothelin-1, *NS* not significant, *CSF* cerebrospinal fluid, ↑ increased, ↓ decreased, – no data

in migraineurs during a spontaneous migraine attack, with a return of concentrations to nearly interictal values in response to rizatriptan and onabotulinumtoxinA (onabotA) [10, 11].

CGRP resulted in increases in the EJV blood of cluster headache (CH) patients in the active periods, with normalization after either subcutaneous sumatriptan 6 mg or O_2 inhalation [12–14].

Another study including patients with chronic tension-type headache (CTTH) showed no changes in CGRP concentrations measured in the cubital vein at 10, 20, and 60 min after nitroglycerin infusion, further supporting the specificity of CGRP for primary headaches involving trigeminovascular activation [15]. Similarly, no variations in CGRP concentrations were found in the EJV and antecubital vein samples from patients with cervicogenic headache assessed both in the presence and absence of headache [16].

10.1.2 Neurokinin A

Experimental evidence suggests that NKA, substance P (SP), and CGRP might act as neurotransmitters at the first central synapses of the trigeminal nociceptive pathway to transmit the sensory stimuli to the higher brain centers and could play the role of cotransmitters or comodulators [17]. Immunocytochemical studies have revealed that a contingent supplying cerebral blood is immunoreactive to neurokinin A (NKA) and is derived from the trigeminal ganglion [18].

A significant increase in NKA and in CGRP concentrations was demonstrated in the plasma of young migraine patients during attacks, suggesting, although indirectly, that CGRP and NKA might be involved in the pathogenesis of migraine attacks [19]. Furthermore, an increase in NKA and also CGRP, nitrite, and cyclic guanosine monophosphate (cGMP) concentrations were found in the internal jugular vein (IJV) blood of migraine patients during attacks. They reached their highest values during the first hour, then tended to decrease progressively and returned, after the end of the attacks, to values similar to or below those detected at the time of catheter insertion. Prostaglandin E2 (PGE2) and 6-keto prostaglandin F1α (6-k-PGF1α), in addition to cyclic adenosine monophosphate (cAMP) concentrations significantly increased during the first hour, but reached a peak during the second hour and remained within the same range until the fourth and sixth hours. These findings suggest early activation of the L-arginine/nitric oxide (NO) pathway, which accompanies the release of vasoactive peptides, including NKA, from trigeminal endings and a late rise in the synthesis of prostanoids with algogenic and vasoactive properties, which may intervene in maintaining the headache phase [20].

10.1.3 Pituitary Adenylate Cyclase-Activating Peptide

Pituitary adenylate cyclase-activating peptide (PACAP) is a multifunctional vasodilatory peptide that has recently been implicated in migraine pathogenesis. It belongs to the vasoactive intestinal polypeptide (VIP)-glucagon growth hormone-releasing factor secretin superfamily [21]. PACAP binds to three different G-protein-coupled receptors. The PACAP receptor can be associated with subunits of the CGRP receptor, including receptor activity modifying protein 1 (RAMP1); thus, a possible synergistic effect of CGRP and PACAP receptor cAMP signaling has been suggested, but this effect needs to be tested in future research [22].

In the trigeminovascular system, PACAP is expressed in the trigeminal nucleus caudalis (TNC) and trigeminal ganglia. PACAP and its receptors are also expressed in sphenopalatine ganglia (SPG) neurons, which control dural vessel tone and cranial blood flow. Like CGRP, PACAP binds to a variety of the central nervous system (CNS) sites, such as the dorsal horn of the spinal cord, brainstem, thalamus, and hypothalamus [23–26]. PACAP plays an excitatory role in pain transmission, suggesting that it might play a role in central sensitization [21]. Interestingly, PACAP and CGRP share the ability to induce light aversion in mice and just like CGRP, PACAP has been linked to anxiety-like behavior and chronic stress response [27]. This may be relevant for migraine, which recognizes both anxiety and stress as potential triggers of attacks.

Several lines of evidence relate PACAP to migraine pathophysiology. In human studies, PACAP-38-induced dilatation of human meningeal arteries, but was less potent than VIP. It seems that the ability of PACAP-38 to induce a migraine is not because of its weak vasodilatory effects, but rather because of its other central actions [28]. Moreover, PACAP-38 was able to induce dural mast cell degranulation, more potently than VIP and PACAP-27 [29].

This suggests that it might play role in triggering neurogenic inflammation in the dural vessels, which may be relevant to migraines [30].

In rats, at least, PACAP has also been linked to nitric oxide. This was suggested by the increase in PACAP concentrations within the TNC after nitroglycerin injection, which is more relevant in wild-type than in knockout mice and its increase in both plasma and TNC after electrical stimulation [31, 32]. These findings therefore support the coupling of PACAP and NO in the trigeminovascular system, which has also been observed for CGRP and NO [33].

The role of PACAP in migraine pathophysiology was confirmed by recent findings obtained from migraine patients. Like CGRP, administration of PACAP to migraineurs induced a delayed migraine-like headache after an initial nonspecific headache in contrast with controls, who experienced only the initial headache phase. Similarly, PACAP induced an early vasodilation of the middle cerebral artery and superficial temporal artery in both healthy controls and migraineurs, but only the latter experienced a delayed migraine-like headache [34]. Very recent clinical research in migraineurs demonstrated that PACAP plasma concentrations were elevated during attacks compared with interictal concentrations [35]. Interestingly, in this study interictal PACAP concentrations were lower than those of healthy subjects, whereas ictal PACAP concentrations of migraineurs were within the range of those of healthy subjects. Based on mechanisms partially shared with CGRP, which can be relevant for migraine attacks, PACAP is considered a potential target for antimigraine agents. PACAP and VIP have common receptors, VPAC1 and VPAC2, while the PAC1 receptor is specific to PACAP [36].

Only one PAC1 receptor agonist, maxadilan, is presently available, and has been isolated from the salivary glands of the sand fly *Lutzomyia longipalpis*. Maxadilan is a 61-amino acid peptide. The deletion of the amino acids between 25 and 41 generated a specific PAC1 receptor antagonist, termed M65 [37]. A recent study demonstrated that maxadilan had no effect on CGRP release and M65 did not block the PACAP-38-induced CGRP release in the trigeminal system [38].

10.1.4 Vasoactive Intestinal Peptide

Vasoactive intestinal peptide (VIP) is one of the parasympathetic signaling transmitters contributing to cranial parasympathetic outflow mediated through the SPG. The activation of parasympathetic cranial outflow during migraine and CH attacks due to the trigeminal–autonomic reflex may be mediated by the above neurotransmitter and receptors [12]. VIP is structurally and functionally related to PACAP-38, but differs from PACAP in migraine-inducing properties. The first experience in this regard did not show the ability of VIP to induce migraine-like attacks. It evoked only a very mild headache in healthy volunteers, despite eliciting intracranial vascular dilation.

In a recent study, only 18 % of patients reported migraine-like attacks after VIP infusion against 73 % of patients reporting migraine after PACAP-38 infusion. Both peptides induced marked dilatation of the extracranial but not intracranial arteries. VIP-induced dilatation was normalized after 2 h, whereas vasodilatation induced by PACAP-38 was longer lasting (>2 h) [39]. On this basis, VIP concentrations have been considered a marker of parasympathetic activation in migraine.

Recent evidence suggests that increased interictal VIP concentrations measured in peripheral blood might be a biomarker helping with CM diagnosis, but it was not able to clearly differentiate CM from episodic migraine (EM). Patients with EM showed, in fact, significantly higher concentrations of VIP compared with controls, but lower than those with CM [40].

Concentrations of CGRP and VIP were also indicated to be potential predictors of the efficacy of onabotA in CM. As in the previous study, VIP and CGRP plasma concentrations were significantly increased in CM patients compared with controls. Concentrations of CGRP and, to a lesser degree, VIP, were significantly increased in responders vs nonresponders. The probability of being a responder to onabotA was 28 times higher in patients with a CGRP concentration above the threshold of 72 pg/mL. Although the sensitivity in calculating the threshold for VIP was poor, the probability of CM patients with low CGRP concentrations responding to onabotA was significantly higher than in those patients with high VIP concentrations [41].

10.2 Monoaminergic Systems

10.2.1 Serotonin (5-hydroxytriptamine)

More than 50 years ago Sicuteri et al. [42] hypothesized the involvement of 5-hydroxytriptamine (5-HT) in migraine based on the demonstration of significantly higher concentrations in the urine of 5-hydroxyindoleacetic acid, the major stable metabolite of 5-HT, during migraine attacks. Since then, several studies have tried to verify whether or not biochemical anomalies of 5-HT occur in migraine, some of them using platelets as a peripheral model of serotonergic neurons. In agreement with Sicuteri's results, during a migraine attack, the first studies showed a significantly increased 5-HT turnover in plasma [43] and saliva [44]. Furthermore, increased concentrations of 5-hydroxyindoleacetic acid have been reported in the cerebrospinal fluid (CSF) of migraineurs during attacks [45].

Conversely, interictal concentrations of plasma serotonin have been shown to be low in migraineurs [42]. Furthermore, concentrations of platelet 5-HT appear to fluctuate differently in female migraineurs than in healthy women during the different phases of the menses and, more importantly, the concentrations of indole decrease significantly in the luteal phase in menstrual migraine before the painful attacks [46]. An increase in serotonin transporter activity was found with medication-overuse headache (MOH), which was normalized after withdrawal in parallel to the improvement of headache frequency. However, the study was not able to differentiate whether the increase in serotonin uptake was caused by a regular intake of analgesics or triptans or was a consequence of frequent headache attacks [47]. In addition, a recent interictal neuroimaging study showed increased availability of 5-HT transporters in the brainstem of migraineurs, suggesting an imbalance in the serotonergic system [48].

All the above findings hypothesize that migraine might be a syndrome of a chronically low central serotonin system with consequent 5-HT receptor hypersensitivity, and that migraine attacks might be triggered by a sudden increase in 5-HT release.

10.2.2 Dopamine and Norepinephrine

An alteration in dopaminergic and noradrenergic systems has been suggested in migraine. Both neurotransmitters are derived from the metabolism of tyrosine. In particular, tyrosine hydroxylase generates 3,4-dihydroxyphenylalanine (DOPA). Dopamine (DA) and norepinephrine (NE) are synthesized by DOPA decarboxylase and dopamine b-hydroxylase enzymes (Db-h) respectively.

Increased concentrations of DA were found in platelets of migraine without aura (MwoA) patients, in comparison with healthy control subjects. Even higher concentrations were detected in patients with CH. The increased platelet DA concentrations have been interpreted as an abnormal biochemical phenotypic trait of primary neurovascular headaches [49]. The higher concentrations of DA can be the consequence of a reduction in Db-h activity. Reduced Db-H enzyme activity and reduced NE concentrations have been reported in MwoA patients [50].

More recently, particular polymorphisms in the gene encoding for Db-h protein have been associated with migraine [51]. All these findings support the notion that complex abnormalities in the metabolism of tyrosine could result in the possible derangement of dopaminergic and noradrenergic pathways in migraine and also in CH.

10.2.3 Elusive Amines

The involvement of the elusive amines, tyramine and octopamine, in migraine pathophysiology was proposed several years ago, but only in the last few years has there been a renewed interest in their role in migraine. These amines, together with

DA and NE, are products of the alternative metabolic pathways of tyrosine. The renewed interest in elusive amines has also been derived from the discovery of a new class of G-protein coupled receptors, the trace amines receptors (TAARs), and there is a high affinity for these amines in rodents and humans [52]. TAARs have been identified in various tissues, including specific brain areas such as the rhinencephalon, limbic system, amygdala, hypothalamus, extrapyramidal system, and locus coeruleus. Some of these areas are relevant parts of the pain matrix modulating pain threshold. Their functions are mainly regulated by synapses using DA and NE as neurotransmitters.

Plasma concentrations of all trace amines have been determined in plasma and platelets of patients with primary neurovascular headaches. In particular, significantly higher concentrations of octopamine, synephrine, and tyramine were found in the plasma and platelets of CH patients, in both the remission and active phases, compared with control subjects or migraine patients. Additionally, intraplatelet concentrations were higher in CH patients than in control subjects.

In migraine patients assessed during headache-free periods, plasma concentrations of octopamine and synephrine were also higher than in controls, but in migraine with aura (MwA) patients, the difference did not reach the concentration of statistical significance [53]. Based on these results, an abnormality in the metabolism of tyrosine toward the increased synthesis of products of the decarboxylase pathway in migraine, resulting in increased synthesis of octopamine, synephrine, and tyramine in association with a decrease in NE, has been hypothesized.

10.3 Endocannabinoid System

Several lines of evidence support the central role of the endocannabinoid system in pain modulation [54]. It has been demonstrated that cannabinoid (CB) type-1 receptor (CB1R)-dependent retrograde mechanisms are involved in the release of neurotransmitters controlling nociceptive inputs and that the concentrations of these lipids are high in sensory terminals, skin, and dorsal root ganglia known to be involved in the transmission and modulation of pain signals. In the trigeminal system, activation CB1R has been shown to inhibit trigeminal neuron activity, which is pivotal in the pathogenic events of migraine [55].

Endocannabinoids are also involved in the descending modulation mediated by brainstem ventrolateral periaqueductal gray (PAG) of basal trigeminovascular neuronal tone and Aδ-fiber dural-nociceptive responses via CB1. Interaction between serotonergic and endocannabinoid systems in the processing of somatosensory nociceptive information suggests that at least part of the therapeutic action of triptans might occur via endocannabinoid-containing neurons in the ventrolateral PAG [56]. The endocannabinoid system is also involved in neurobiological mechanisms underlying drug addiction as it influences glutamatergic cortico-striatal transmission and the activity of the mesolimbic reward system [57].

Reduced concentrations of AEA were shown in the CSF of CM patients with and without medication overuse, reflecting impairment of the endocannabinoid system in these patients, which may contribute to chronic head pain and seems to be related to increased CGRP and NO production [58]. The activity of active endocannabinoids, anandamide membrane transporter (AMT) and fatty acid amide hydrolase (FAAH), and the concentration of cannabinoid receptors were also measured in peripheral platelets from patients with episodic MwoA, episodic tension-type headache (ETTH), and healthy controls. A significant increase in the activity of AMT and in FAAH activity was observed in female but not male patients with MwoA. No differences in the cannabinoid receptors emerged among the study groups. These findings suggest an increase in anandamide (*N*-arachidonoylethanolamine – AEA) degradation by platelets in women with migraine, which may be responsible for a reduction in AEA concentration in the blood, and which can lower the pain threshold and may explain the prevalence of migraine in women [59].

The activity of AMT and of FAAH was assessed in platelets of CM patients without symptomatic drug overuse, MOH patients, episodic migraine (EM) patients, and control subjects. AMT and FAAH appeared to be significantly reduced in both CM and MOH patients, compared with either controls or EM patients. This reduction was observed in both male and female patients in the CM and MOH groups. These changes observed in the endogenous cannabinoid degrading system have been hypothesized to reflect adaptive behavior to chronic headache and/or drug overuse [60].

In a more recent study a significant reduction in FAAH activity was demonstrated in MOH patients before withdrawal of treatment compared with controls. This was coupled with a significant improvement (reduction) in the facilitation of spinal cord pain processing (increased temporal summation threshold and related reduction in pain sensation) at both 10 and 60 days after treatment withdrawal. These findings could be a consequence of mechanisms aimed at reducing the endocannabinoid degradation and increased endocannabinoid system activity, resulting in an antinociceptive effect in patients undergoing withdrawal of treatment [61]. This supports the potential role of the CB1 receptor as a possible therapeutic target in CM [58].

A further confirmation of the dysfunction of the endocannabinoid system associated with a deficiency in the serotonergic system in CM and MOH is derived from a study investigating 2-acylglycerol (2-AG) and AEA concentrations in the platelets of these patients. These concentrations were significantly lower in MOH patients and CM patients than in the control subjects, with no significant differences between the two patient groups. Endocannabinoid and serotonergic systems appear to be both mutually related and defective in CM and MOH patients, and can contribute to headache chronification [62].

10.4 Glutamate

Glutamate, the major excitatory neurotransmitter in the CNS, is involved in several aspects of migraine pathophysiology. Preclinical data suggest its involvement in cortical spreading depression, trigeminovascular activation, and central sensitization. Based on these findings, glutamate concentrations have been determined in migraine patients both in peripheral blood and in CSF to assess possible dysfunction of the glutamatergic system in these patients [63]. An alteration in glutamate concentrations has been reported in migraine patients, with higher concentrations of CSF and plasma glutamate in episodic migraine patients compared with controls [64]. Furthermore, higher glutamate concentrations in CSF have been reported in chronic daily headache (CDH) patients related to the increase in the end-products of nitric oxide [65].

Platelet glutamate uptake and release have also been evaluated in patients with MwA and MwoA. Both glutamate release from stimulated platelets and plasma glutamate concentrations appeared to be increased in migraine patients, more markedly in MwA patients. Platelet glutamate uptake, assessed as 3H-glutamate intake, also appeared to be increased in MwA, whereas it was reduced in MwoA patients compared with controls. These findings suggest the involvement of different pathophysiological mechanisms in the two migraine types, with a more pronounced upregulation of glutamatergic metabolism in MwA [66].

Glutamate concentrations were also assessed using MR spectroscopy in vivo in women with migraine, and showed higher glutamate to glutamine ratios in occipital regions in women with migraine during the interictal state. These findings support the hypothesis that this increased ratio could arise from neuronal–glial coupling of glutamatergic metabolism differences compared with controls or an increased neuron/astrocyte ratio in the occipital cortex in migraine patients [67].

Glutamate and glutamine are strongly compartmentalized (in neurons for glutamate and in astrocytes for glutamine). The visual cortex is the brain region with a higher neuron/astrocyte ratio (the highest neuronal density and the relatively lowest density of astrocytes. Elevations in extracellular glutamate or potassium above certain thresholds are likely candidates to be the final common steps in the multiple distinct processes that can lead to cortically spreading depression. Astrocytes play a key role in this phenomenon, by acting as a sink for extracellular glutamate and potassium, in addition to generally acting as a buffer for the ionic and neurochemical changes that initiate and propagate cortical spreading depression.

In parallel to serotonin synthesis, the major route of tryptophan catabolism is the kynurenine pathway, which produces neuroactive metabolites. Among these substances, kynurenic acid has potential neuroprotective action blocking glutamate release and glutamatergic neurotransmission. Thus, kynurenines may affect several pathogenic events involving glutamate pathogenesis in migraine directly, by acting on glutamate receptors and exerting other neuromodulatory effects, and indirectly via an altered serotonin metabolism [68].

10.5 Nitric Oxide

Nitric oxide (NO) is a labile molecule with only a few seconds half-life, which is synthesized by endothelium, neuronal cells, and immune cells by specific NO synthases (NOS; endothelia, neuronal and inducible NOS). It is rapidly oxidized by tissue oxygen to the stable end products, nitrate and nitrite; the total concentration of both is the best index of overall NO production in the circulation. The most important effect of NO is the activation of the soluble guanylate cyclase. This enzyme induces the synthesis of cGMP and NO-cGMP pathways in smooth vascular muscles and is responsible for vascular dilatation and relaxation. Through these effects, NO is involved in the regulation of cerebral vessel tone in the cerebral blood flow.

Nitrite concentrations in plasma were investigated in patients with migraine and CH patients compared with a group of healthy nonheadache controls. Significantly higher nitrite concentrations were found in migraine patients, with and without aura, and in cluster headache patients, during the remission and cluster phases, than in controls, suggesting the involvement of a basal dysfunctioning in the L-arginine-NO pathway in the peripheral mechanisms, predisposing patients with neurovascular headaches to individual attacks [69].

In another study, basal plasma concentrations of NO metabolite nitrites (measured spectrophotometrically after the conversion of nitrates to NO_2-), similar in patients and controls, were found to be significantly increased after glyceryl trinitrate (GTN) infusion at pain peak in patients and after 45 min in controls, but not after 120 min, with no differences between groups. These data do not support the presence of basal hyperactivity of the L-arginine-NO pathway in CH patients. The authors hypothesized that increased NO production might be of relevance in the mechanisms leading to CH attacks, but other factors are likely to render CH patients hyperresponsive to NO, ultimately causing the occurrence of pain and associated features [70].

It has been demonstrated that increased NO may interact with reactive oxygen substances (ROS). Enhanced endothelial NO and superoxide anion release may be involved in the induction of migraine through cerebral blood flow changes [71].

Dysfunction of the L-arginine/NO pathway activation was also investigated in migraine and tension type headache. It was demonstrated that the activation of the L-arginine/NO pathway in migraine patients, especially those with aura, increased basal and collagen-stimulated production of NO, and cGMP in the platelet cytosol of migraine patients, accompanied by a decrease in collagen-induced aggregation, especially in patients with aura, compared with controls and this production was further increased during attacks [72]. The increase in NO production in association with a decrease in platelet aggregation to collagen was also confirmed in female migraineurs compared with female controls with no headaches. This decrease was most evident at mid-cycle in nonmenstrual migraine patients and in the luteal phase in menstrual migraine patients, and was more accentuated during migraine attacks in both subgroups. The activation of the L-arginine/NO pathway was more accentuated in the luteal phase in menstrual migraine patients, and this could be responsible for the increased susceptibility to migraine attacks during perimenstrual and menstrual periods in these patients [73].

In patients with chronic daily headache (CDH) evolving from a previous episodic migraine and in patients with CTTH a significant reduction in platelet collagen aggregation was coupled with increased NO and cGMP production and a significant increase in cytosolic Ca(2+) concentration, which was more evident in patients with analgesic abuse compared with healthy control individuals. This was accompanied by a reduced platelet content and collagen-induced secretion of serotonin, and in patients with tension-type headache by increased glutamate content [74].

Based on the above findings, antagonizing NO production or the blockade of steps in the NO-cGMP pathway, or scavenging of NO have been proposed to be potential targets in the treatment of primary headaches for new drugs in treating migraine and other headaches. Nonselective NOS inhibitors are likely to have side effects, whereas the selective compounds, nNOS and i-NOS inhibitors, are now under investigation. Antagonizing the rate-limiting cofactor tetrahydrobiopterin could be another very likely new treatment. It is more unlikely that the antagonism of cGMP or its formation will also be feasible, but augmenting its breakdown via phosphodiesterase activation is a possibility, in addition to other ways of inhibiting the NO-cGMP pathway.

10.6 Neurotrophins

The involvement of nerve growth factor (NGF) in peripheral nociception clearly emerged from the evidence in experimental pain models of upregulation and increased delivery of this neurotrophin. Overexpression of the high affinity tyrosine kinase A (TrkA) receptors by nociceptive terminals was also observed and accompanied by enhanced SP and CGRP release [75]. The role of NGF as a peripheral pain mediator is further suggested by the effectiveness of NGF neutralizing molecules as analgesic agents in many models of persistent pain; these molecules have also been evaluated in exploratory clinical trials. A further mechanism mediated by NGF in experimental models of hyperalgesia is the enhancement of N-methyl-D-ASPARTATE (NMDA)-evoked responses through the induction of brain-derived neurotrophic factor (BDNF) synthesis by TrkA-positive sensory neurons and its interaction with Trk-B receptors.

Experimental data also suggest that NGF might activate mast cells through the collaborative interaction with lysophosphatidyl serine expressed on membranes of activated platelets. This could be of relevance for the neurogenic inflammation mechanisms, assumed to constitute an experimental model of acute migraine attack, where the occurrence of activated platelets has been demonstrated.

All these findings strongly suggest the potential implication of both NGF and BDNF maintaining chronic pain states in humans, including headache.

Higher concentrations of BDNF were also demonstrated on the periphery in patients affected by neurovascular primary headaches. In one report, patients with all primary headaches showed significantly decreased platelet concentrations of BDNF, but a selective reduction of platelet NGF was found only in migraineurs and not in CH patients. No changes were observed in the plasma concentrations of either of the neurotrophins in the two patient groups compared with controls. These findings further suggested the potential involvement of BDNF and NGF in the pathophysiology of both headache disorders, and raised the possibility that it might be helpful in differentiating migraine biologically from CH [76].

Tumor necrosis factor (TNF)-alpha, sTNF-R1, sTNF-R2 receptors were also determined in a further study aimed at verifying changes in BDNF serum concentrations in migraine patients assessed during migraine attacks. BDNF was increased during a migraine attack, whereas there was no significant difference in the serum concentrations of TNF-alpha, sTNF-R1, and sTNF-R2 between attacks and headache-free periods. This report therefore reinforces the hypothesis of the involvement of BDNF in migraine pathophysiology [77].

Further studies have been performed to assess NGF and BDNF in the CSF of patients with chronic headache. Significantly higher concentrations of NGF were demonstrated in the CSF of patients with CDH, with no differences between patients with and those without analgesic abuse [78]. Similar findings were found in a subsequent study revealing higher concentrations of both neurotrophins in the CSF of both CM and fibromyalgia patients. Values of BDNF and those of NGF correlated positively with those of glutamate [79].

In the spinal cord after the stimulation of primary sensory neurons various substrates are released from the central terminals, such as ATP or BDNF. ATP is able to influence the activity of the microglia, which causes the release of BDNF from the microglia through the activation of the P2X4 receptors. BDNF activates the TrkB receptors, which results in the downregulation of the K^+-Cl^--contransporter of the second-order neurons, which convey information to the thalamus. This process is associated with the development of allodynia.

Experimental findings demonstrated that somatostatin (SOM) is a regulatory peptide in both the central and peripheral nervous systems. In humans, SOM is used to treat opioid-resistant pain. In animal models, SOM and its stable analog octreotide (OCT) have analgesic effects and SOM released from sensory neurone peripheral endings exerts anti-inflammatory actions. Activity-induced release of endogenous SOM can be modulated by the trophic factor glial-cell-line-derived neurotrophic factor (GDNF).

Table 10.2 Biochemical alterations in different migraine types

Substances	Reference	Migraine type	Sample	Alteration
Orexin-A	[109]	Chronic migraine	CSF	Increased
Adiponectin	[112]	Migraine	Plasma	Increased (ictal)
				Decreased (interictal)
Leptin	[121]	Episodic migraine	Plasma	Decreased (interictal)
GABA	[126]	Chronic migraine with depression	CSF	Decreased
Substance P	[132]	Chronic migraine	Saliva	Increased
			Plasma	Increased
Endothelin-1	[139]	Migraine with and without aura	Plasma	Increased (6 h after the onset of the attack)

GABA gamma-aminobutyric acid

Glial-cell-line-derived neurotrophic factor and SOM were measured in the CSF of CM and fibromyalgia patients both with and without analgesic abuse and control subjects. Significantly lower concentrations of GDNF and SOM were found in the CSF of both CM and fibromyalgia patients compared with controls, with no significant differences between those with analgesic overuse and those without. The abuse of simple or combination analgesics does not seem to influence the biochemical changes investigated, which appear to be more strictly related to the chronic pain state [80] (Table 10.2).

10.7 Neurometabolic Systems

10.7.1 Hypothalamic Orexinergic System

The hypothalamus has been involved in several headache disorders, including migraine. Its role in migraine has initially been suggested owing to the observations of premonitory symptoms in migraineurs, such as changes in thirst, food cravings, and mood and sleep disturbances. More recently, functional imaging showed hypothalamic activation during acute migraine attacks [81].

It has also been demonstrated that several hypothalamic peptides, proteins, and neurotransmitters involved in feeding are also involved in migraine pathophysiology. Among them orexin, adiponectin, and leptin should be mentioned.

Specifically, orexin (OX) A- and OXB-containing neurons are primarily located in the lateral hypothalamus. They project to the cortex, thalamus, hypothalamus, brainstem (including the locus coeruleus and the raphe nucleus), in addition to the gastrointestinal tract. Orexins interact with 2G-protein coupled receptors, OXR1 and OXR2, which have been shown to contribute to regulate food intake. Further evidence suggests that orexin might also modulate adipose tissue metabolism by inhibiting lipolysis.

The hypothalamic orexinergic system may act as a key regulator of involvement in the modulation of trigeminovascular activation and processing and may therefore be involved in the pathophysiology of a variety of primary headaches, including cluster headaches and chronic migraine [82].

Based on the above evidence, OXA and corticotropin-releasing factor (CRF) were determined in the CSF of CM patients without analgesic abuse and in that of MOH patients. Significantly higher concentrations of both OXA and CRF were found in the CSF of MOH and to a lesser extent in patients with CM without analgesic abuse compared with controls. A significant positive correlation was also found between CSF OXA and CRF values, monthly drug intake group, and scores of a self-completion ten-item instrument to measure dependence upon a variety of substances, the Leeds Dependence Questionnaire (LDQ) in the MOH patient group. The significantly higher OXA concentrations found in CM, mainly in MOH, can be interpreted as a compensatory response to chronic head pain or, alternatively, as an expression of hypothalamic response to stress due to chronic pain. A potential role for orexin-A in driving drug-seeking in MOH patients through the activation of stress pathways in the brain can also be hypothesized [83].

Changes in the concentrations of hypothalamic neuropeptides in migraineurs under preventive treatment with amitriptyline and flunarizine were assessed in another study. Thirty-nine migraine patients with a body mass index <25 kg/m^2 and with no endocrinological or metabolic diseases were assigned to receive amitriptyline, or flunarizine, for 3 months. Orexin-A and orexin-B concentrations were significantly reduced in both groups. Conversely, plasma neuropeptide-Y concentrations were markedly increased, with the highest concentrations at the second and third months, in both patient groups. Correlation of orexin-A concentrations with weight gain suggest their involvement in body weight increase occurring in migraineurs during amitriptyline or flunarizine prophylactic treatment [84].

10.7.2 Adipokines

Adiponectin (ADP) is a protein primarily secreted from adipocytes, and its receptors are expressed in the brain (particularly in the proopiomelanocortin [POMC] and neuropeptide Y [NPY] neurons of the hypothalamus), the blood vessel endothelium, in addition to liver and muscle.

As far as migraine is concerned, a pattern similar to that observed with serotonin has been hypothesized in migraineurs, with low concentrations of ADP interictally and increased concentrations during acute attacks. ADP concentrations appeared to be raised in migraine and this increase was independent of psychiatric comorbidities, migraine severity and allodynia [85].

Conversely, in a pilot study of female episodic migraineurs, the high molecular weight (HMW]):low molecular weight (LMW) ADP ratio concentration was associated with migraine severity and was predictive of acute treatment response to sumatriptan. In responder patients, the HMW:LMW ratio concentration was in fact greater at pain onset compared with nonresponders. Responders also showed a decrease in the HMW:LMW ratio at 60 and 120 min after treatment compared with onset. These changes in responders remained significant after adjusting for covariates, including measured body mass index (m-BMI). Although nonresponders showed no significant changes in unadjusted T-ADP or ADP oligomer or ratio concentrations, the HMW:LMW ratio appeared to be increased in nonresponders after adjustments [86]. ADP and the HMW:LMW ratio of ADP were therefore suggested to be potential novel biomarkers and drug targets for episodic migraine.

In the recent Atherosclerosis Risk in Communities Study, the prevalence of migraine was significantly associated with total adiponectin only in older men, but not in older women [87].

10.7.3 Leptin

Leptin is an adipocytokine involved in appetite suppression and the modulation of inflammatory processes. It is primarily produced by adipocytes such as adiponectin and leptin, but can also be produced by several other tissues, including the brain. Interestingly, leptin receptors are abundantly expressed in the arcuate nucleus (ARC) and dorsomedial nucleus (DM) hypothalamus. Leptin has also been shown to modulate inflammation.

In the first study leptin concentrations in migraineurs were determined in serum pre- and post-treatment in two small groups of patients treated with amitriptyline or flunarizine. Body mass index (BMI) and serum leptin concentrations were found to be increased at 4 and 12 weeks post-treatment with both preventive drugs compared with baseline concentrations. It would not be possible to discriminate from the above results if the changes in leptin concentrations were entirely due to weight gain or a therapeutic response [88].

In more recent research, Guldiken et al. [89] assessed interictal serum leptin concentrations in migraineurs compared with age- and gender-matched controls. Lower leptin concentrations and lower fat mass were found in episodic migraineurs. However, there was no significant difference in leptin concentrations between the groups after adjusting for fat mass.

No conclusive results can be drawn from the above research on leptin. Further research is therefore needed evaluating serum leptin concentrations in migraineurs that should take into account disease duration, sex hormones, and the phase of the menstrual cycle in women.

10.7.4 Neuropeptide Y

Neuropeptide Y is widely distributed throughout sympathetic nerve endings where it is costored and cosecreted with noradrenaline. It is considered a marker of noradrenergic function. NPY participates in the autonomic control of cerebral circulation and can be involved in disorders characterized by neurogenically mediated changes in the cerebral blood flow, such as migraine, cluster headache, and stroke.

Studies on this marker are dated. Significantly lower plasma concentrations of NPY in young MwA patients and, to a lesser extent, those with MwoA were found in the interictal period, compared with a control group. Plasma NPY concentrations tended to increase significantly during migraine attacks, particularly in patients with MwA. No significant variations were observed between headache-free periods and attacks in tension-type headache patients.

Reduced NPY concentrations in the interictal period have been interpreted as evidence of the derangement of the sympathetic function in the course of migraine (especially MwA), whereas the increase in NPY concentrations during migraine attacks was suggested to be an expression of sympathetic activation, even though this system is less functionally efficient [90].

These results denied an increased sympathetic tone in migraine patients with either migraine or subarachnoid hemorrhage (SAH), and suggested that the higher CSF NPY might originate centrally. Findings in subarachnoid hemorrhage patients argue in favor of a decreased sympathetic tone, which could represent a homeostatic response counterbalancing vasoconstriction mediated by other mechanisms [91].

10.7.5 Gamma-Amino Butyric Acid

Gamma-amino butyric acid (GABA) is the major inhibitory neurotransmitter in the CNS where it plays an important role in pain modulation. The most effective drugs for migraine prevention, valproate and topiramate, exert a potent GABAergic agonism. Direct evidence of the role of GABA in migraine is scarce. A dated research reported that this neurotransmitter could be detected during attacks only in migraineurs, but not in TTH patients, suggesting an increase in this inhibitory neurotransmitter in the brain of migraine patients ictally aimed at antagonizing pain [92].

The GABA concentrations were also measured in the CSF of CM patients, and only those with comorbid depression showed significantly lower concentrations of this inhibitory neurotransmitter with no difference when comparing patients with controls supporting the possibility that a GABA deficiency might underlie mechanisms of depression in CM and that preventive therapies modulating GABA neurotransmission might be useful in this condition [93].

More recently, GABA concentrations were measured using magnetic resonance spectroscopy (MRS), in migraine patients. They did not differ significantly in migraineurs and controls during attacks and interictally. GABA concentrations, however, did vary significantly as a function of severe headache attacks and of attack-related disability, further suggesting the role of GABA in suppressing headache attacks [94].

Despite the potential role of the GABAergic system in migraine, no association was found between common variants of the GABA A receptor and migraine, denying their major contribution to migraine susceptibility [95].

10.8 Substance P

The neuropeptide substance P (SP) is widely distributed in both the central and the peripheral nervous system. In the trigeminal system, SP, present in sensory afferent neurons, mediates vasodilation and nociceptive information.

Experimental findings demonstrated that stimulation of the trigeminal ganglion in cats and humans elicits release of SP and CGRP. Immunoreactivity of SP was found to be increased in the saliva of migraine and cluster headache patients during attacks. This increase was accompanied by a rise in CGRP in both migraine and cluster headache patients, and also VIP, but only in cluster headache patients [96]. A further study reported significantly higher concentrations of SP, together with those of CGRP and NGF in the plasma and saliva of patients with CM compared with control subjects. Plasma concentrations of SP and CGRP correlated significantly with their concentrations in saliva. There was a significant positive correlation between NGF and both neuropeptide concentrations in plasma, and between the neuropeptide concentrations in both plasma and saliva. NGF and both neuropeptides were highly associated with pain intensity [97].

A central sensitization has been advocated to explain chronic daily headache due to sustained trigeminovascular system activation. Glutamate-NO-cGMP-mediated mechanisms and increased release of sensory neuropeptides have been shown to play an important role in the maintenance of chronic head pain. In particular, among sensory neuropeptides from the trigeminal system, SP, CGRP, and, to a lesser extent, neurokinin A, appeared to be significantly increased in the CSF of CDH patients.

Substance P and somatostatin-like immunoreactivity in addition to enkephalinase activity have also been evaluated in a dated study in the CSF during spontaneous and histamine-induced attacks in cluster headache patients in the active phase. During the histamine-provoked attacks, CSF and plasma somatostatin and enkephalinase activity were unchanged, while plasma SP decreased significantly. During spontaneous attacks, a significant lowering of SP without changes in somatostatin was found in plasma compared with controls. Notably, both during and between attacks in the cluster phase, plasma enkephalinase activity was increased in comparison with the values in controls and this could explain the reduced concentrations of SP detected [98].

In a further study, SP and 5-HT concentrations were determined in platelets of migraine and TTH patients. SP concentration was significantly higher, whereas 5-HT concentration was significantly lower in both patient groups compared with controls. Furthermore, there was significant negative correlation between the concentrations of platelet SP and those of platelet 5-HT. These findings may reflect similar changes in monoaminergic pathways involved in trigeminal pain processing in both types of primary headache [99].

10.9 Endothelins

Endothelin-1 (EDN-1), encoded by the EDN1 gene on chromosome 6p24, is one of the most potent vasoconstrictors in humans, but also exerts neuronal effects. EDN-1 is in fact able to induce cortical spreading depression (CSD) in animal models and its involvement in migraine is demonstrated by increased concentrations during attacks [100]. In preclinical studies expression of ET(A) and ET(B) receptor mRNA was detected in human cerebral arteries with different effects: endothelin A receptor agonism mediates strong vasoconstriction, whereas agonism to ET (B) receptor determines concentration-dependent vasodilation. In one study, EDN-1 ictal values were markedly elevated at the beginning of the migraine crisis (<2 h) and declined to interictal or even lower concentrations later (4–6 h) in the course of an attack. The local vasoconstriction at the beginning of a migraine attack was interpreted to be

EDN-1-mediated secondary to serotonin activation [101]. Vasopressin, which is known to induce EDN-1 synthesis in endothelial cells, seems to play a role as a mediator of elevated plasma EDN-1 in migraine, as suggested by the finding of increased vasopressin concentrations 3 h after the attack onset, preceding an EDN-1 increase at 6 h [102]. In contrast to the above results, only one study reported decreased concentrations of plasma EDN-1 during migraine attacks compared with interictal conditions, which returned to basal values after pain relief [103]. In a more recent study, EDN-1 appeared to correlate with the duration of migraine, systolic and diastolic blood pressure, carotid artery intima media thickness, and impaired endothelial vasoreactivity. An increase in EDN-1 was indicated to be an expression of endothelial injury in migraineurs, independent of attacks and associated with vascular risk factors [104].

10.10 Hormonal Alterations in Primary Headaches

In primary headache disorders, contributions are made to the premorbid state by genetic and epigenetic factors. Hormonal fluctuations alter the triggering of the migraine attack. Estrogen has mainly an excitatory effect, while progesterone has an inhibitory effect in the CNS. The progesterone concentration is decreased in the premenstrual phase and its withdrawal could cause cortical hyperexcitability. CSD is a propagating transient negative direct potential shift, which occurs in the aura phase of MwA. CSD is influenced by estrogen [105]. Estrogen receptors, which are expressed by the microglia, become activated during CSD, with the consequence of the release of cytokines [106]. In the migraine-related structures (e.g., hypothalamus, periaqueductal gray), estrogen has been revealed. Neuronal hyperexcitability during the menstrual cycle is modulated by the brain-derived neurotrophic factor, which is induced by estrogen and is decreased by progesterone. It is supposed that degranulation of the dural mast cells could activate the perivascular nerve fibers. The mast cells synthesize gonadotropin-releasing hormone, which can modulate the neuronal activity [105].

10.10.1 Migraine

The prevalence of migraine changes with age. Migraine occurs in 3–10 % of prepubertal children (puberty begins at 8–14 years in girls and 9–15 years in boys), and the rates are similar among boys and girls. In the postpubertal period, the hypothalamus resets its neurohormonal systems [107]. Migraine affects close to three times as many adult women (15–17 %) as adult men (6 %), and the prevalence of migraine decreases in postmenopausal women. Alterations in the hormonal milieu of the reproductive cycles of women are associated with the frequency of migraine headache. The patterns of migraine correlate with menarche, pregnancy, lactation, and the menopause.

10.10.2 Tension-Type Headache

A literature review has indicated that TTH is triggered by the menses. Population studies have revealed that 12.7 % of women have menstrual TTH, but pregnancy, the clinical features of TTH may improve. TTH remains unchanged or worsens in 70 % of postmenopausal women. A connection between the hormonal fluctuation and the pathomechanism of TTH is currently lacking [108].

10.10.3 Cluster Headache

No strong link has been detected between CH and hormones, nor is there any association between CH and menarche, menstruation, hormonal contraceptive use or menopause. In spite of this, data exist that point to the remission of CH during pregnancy, with a relapse several days after delivery and a worsening in the perimenopausal period. The results of clinical studies revealed no relationship between menstruation and CH or between the use of hormonal contraceptives and CH [108].

10.11 Immunological Findings in Primary Headaches

10.11.1 Migraine

Among the mechanisms thought to intervene in migraine pathophysiology the involvement of allergic factors has been hypothesized for many years [109]. High comorbidity between migraine and atopic diseases such as eczema and asthma has been shown in dated studies, whereas the role of dietary factors as triggers of migraine via immunological mechanisms has been criticized over the years [110].

Evidence for the involvement of the immune system in migraine comes from findings of changes of serum concentrations in complement and immunoglobulins, histamine, and immune cells in migraine patients, without clarification of the mechanisms involved [111].

Research into cytokines in patients with migraine has demonstrated fluctuations of cytokine concentrations in migraine. The involvement of proinflammatory cytokines in migraine attacks is emphasized in a study showing a trend toward an increase in plasma concentrations of TNF-α in children with migraine compared with controls. In the same study, TNF-α and IL-1 α, in addition to sTNF-RI concentrations, were significantly higher in MwA than in MwoA [107]. Prophylactic drugs seemed to reduce all proinflammatory cytokine concentrations with no significant differences from each other [112].

More recent research concerning the distribution of undetectable and detectable anti-inflammatory cytokines in children and adolescents with migraine or TTH did not demonstrate differences among the two groups and age-matched control subjects [113].

Chronic migraine with MOH is also characterized by alterations of some immune parameters compared with episodic migraine patients indicating an inflammatory state, as in other chronic pain conditions, especially with comorbid depression [114]. They include changes in white blood cell and total lymphocyte count and number and percentages of some lymphocyte subsets, such as CD4, CD19, CD8, and CD3 [115, 116]. These changes are accompanied by a reduction in beta-endorphin lymphocyte concentrations, according to previous findings in migraine and tension-type headache.

The assessment of cytokine concentrations in the cerebrospinal fluid of headache patients assessed during attacks showed significant group differences in IL-1ra, transforming growth factor-beta (TGF)-β1, and monocyte chemoattractant protein (MCP)-1 in episodic TTH and migraine compared with controls, and a significant difference in MCP-1 between cervicogenic headache and MwA. Intrathecal MCP-1 appeared to correlate with IL-1ra, IL-10, and TGF- β 1 in episodic TTH, and MCP-1 with IL-10 in MwA patients. Cytokine changes are modest compared with those found in other serious neurological conditions, and have been interpreted as being a mild response to pain [117].

More information on the involvement of immune cells in migraine is derived from research into meningeal sterile inflammation. Sterile inflammation is believed to play a role in the persistent activation of meningeal trigeminal nociceptors and related blood vessels underlying migraine attack precipitation and maintenance.

Mast cells surrounding dural nociceptors are activated in the course of sterile meningeal inflammation. Activated mast cells degranulate and release several proinflammatory mediators such as histamine and proinflammatory cytokines, which contribute to sustained trigeminal nociceptor activation. Among proinflammatory mediators, tumor necrosis factor (TNF)-α, in particular, is thought to intervene in the sensitization of meningeal nociceptors and the induction of intracranial throbbing pain through a complex meningeal immunovascular mechanism implicating activation of the TNF receptor (TNFR) 1 and 2 on both vascular and immune cells, in addition to downstream activation of meningeal NOS, COX-1, and COX-2, and phosphorylation of MAP kinase p38 in vascular cells. IL-1β, but not IL-6, also promotes the activation and increased mechanosensitivity of intracranial meningeal nociceptors.

Other mediators potentially implicated in promoting the persistent activation of meningeal nociceptors after CSD might be vasoconstricting mediators released during the prolonged reduction in CBF that follows CSD [118]. One potential factor in this regard is the arachidonic acid-derived 20-hydroxyeicosatetraenoic acid (20-HETE), because of its ability to activate the ion-channel TRPV1 on nociceptors.

The involvement of TNF-α in trigeminovascular activation is supported by its transient increase together with that of soluble intercellular adhesion molecule (ICAM)-1 in the internal jugular blood of MwoA patients assessed ictally. The increase in TNF-α can be induced by sensory neuropeptides, mainly CGRP released from activated trigeminal endings. sICAM-1 concentrations then progressively decrease and the very late activation antigen (VLA)-1 expression by lymphocytes is downregulated during migraine attacks. This could antagonize their transvascular migration in the brain, to further support the hypothesis of "sterile" inflammation in the dura mater [119].

A further finding in the jugular venous blood of MwoA patients is the transient increase in the concentrations of the neutrophil chemotactic chemokine IL-8, but not the monocyte chemotactic chemokine MCP-1 or the lymphocyte chemotactic chemokine RANTES during attacks, consistent with the increase in CGRP. Experimental evidence demonstrated a CGRP-induced activation of IL-8 gene expression, but not RANTES and MCP-1, via transcriptional factor AP-2 transduction in response to cyclic adenosine monophosphate in sensory neurons involved in inflammatory pain models. Although IL-8 is transiently increased during migraine attacks, it is unlikely to be due to an accumulation of leukocytes secondary to neurogenic inflammation, as it is for other neuroinflammatory conditions affecting CNS [120].

Several preclinical and clinical data suggest the putative role of matrix metalloproteinases in migraine. These proteolytic enzymes involved in the remodeling of the majority of protein components of the extracellular matrix may alter the composition and function of the blood–brain barrier and promote a local inflammatory process with accumulation of inflammatory mediators capable of discharging and sensitizing meningeal nociceptors. In experimental models, they are upregulated by CSD [121].

10.11.2 Tension-Type Headache

Few studies have been performed in TTH patients, suggesting dysfunction of the immune system in patients, mainly in the chronic form (CTTH).

Decreased serum concentrations of IL-2 were found in patients with CTTH and in those with CM compared with controls. Decreased serum IL-2 concentrations have been interpreted to be a reflection of the reduction in 5-HT or catecholamine concentrations in CNS [122]. Conversely, concentrations of interleukin (IL)-1β were significantly elevated in participants diagnosed with CTTH compared with healthy controls, while IL-18 concentrations were found to be significantly elevated in men with CTTH [123].

In a recent study, children with CTTH showed lower IgA concentrations, but not lower cortisol concentrations. A significant negative association between the number of years with headache and IgA concentration was found [124].

10.11.3 Cluster Headache

Some evidence suggests the involvement of an inflammatory process in CH. Reduced activity of the NK cells and an increase in lymphokine-activated killer (LAK) generation has been shown in patients with CH compared with controls [125, 126].

Impairment of the cytolytic and proliferative responsiveness of peripheral blood mononuclear cells to IL-2 was also found [127].

Further findings in cluster headache patients concern the increase in the number of monocyte and NK cell populations (despite the reduced activity of these cells), alterations in NK+, CD3+, and CD4+ concentrations found in the cluster period, probably pain- or stress-related, reduced activity of Gi proteins, and a marked downregulation of Gi α mRNA [128, 129]. An increase in IL-1 β concentrations in CH patients between attacks has been shown compared with controls. IL-1 β was further increased during the ictal phase of CH compared with patients between attacks and normal individuals [130]. These data were contradicted by a more recent study on systemic changes in the IL-1 or IL-6 systems in CH patients [130]. In the same study, elevated soluble IL-2 receptors indicative of T cell activation were also found in cluster headache patients assessed during attacks. It has been hypothesized that IL-2 might be involved in the activation of the hypothalamus by stimulating CRF release.

10.11.4 Other Primary Headaches

Elevated TNF alpha concentrations were found in the CSF, but not in the plasma, of almost all new daily persistent headache (NDPH) patients assessed; similar concentrations were found in CM patients. An inflammatory status in the CNS has been hypothesized in both conditions and is considered one of the causes of refractoriness to treatment [131].

10.12 Magnesium in Primary Headaches

Magnesium (Mg) is an essential element that controls several cellular functions. Its physiological roles include control of neuronal activity, cardiac excitability, muscular contraction, neuromuscular transmission, vasomotor tone, blood pressure, and peripheral and cerebral blood flow. A deficiency of Mg in migraineurs was proposed for the first time by Durlach [132], who found that migraineurs may excrete excessive amounts of Mg owing to stress. More recently, a Mg load test study [133] revealed that migraine patients had greater retention of Mg than healthy controls, suggesting its systemic deficiency. Lower concentrations of Mg were found in CSF and were detected by ^{31}P-magnetic resonance spectroscopy in the brain of migraine patients assessed interictally [134]. Low free Mg in the brain was also associated with deficient energy metabolism in patients with migraine and cluster headaches [135].

Low Mg concentrations have been associated with vasoconstriction, platelet aggregation, neurotransmitter release, including SP. Based on the preclinical and clinical evidence, Mg was assessed in the serum, but contrasting results were obtained in this regard. In particular, they did not appear to be significantly different among episodic (both with and without aura) and chronic migraine patients [136], and in hemiplegic migraine [137], or conversely to be reduced interictally and more consistently ictally [138]. Interestingly, an inverse correlation between increased P100 amplitude and lowered serum Mg concentrations was found in a study involving children suffering from migraine with and without aura assessed in a headache-free period. Both groups expressed neuronal hyperexcitability of the visual pathways related to a lowered threshold for migraine attacks, which can be normalized in most patients by treatment with oral magnesium pidolate [139].

Studies on Mg in erythrocytes and lymphocytes are more consistent, all showing on average a significant reduction in migraine patients compared with controls or TTH patients with no significant variations between ictal and interictal concentrations [140, 141]. It has been hypothesized that this reduction might be due to abnormal regulation of intracellular Mg that can reflect at the peripheral concentration changes observed in the brain of a migraineur. Red blood Mg deficiency may also explain premenstrual syndrome including migraine.

Serum ionized magnesium (IMg2+) was also measured by an ion-selective electrode for Mg2+ in patients with various headache syndromes. Low serum IMg2+ and a high ICa2+/IMg2+ ratio were found in 42 % of migraine patients assessed with attacks, but in only 23 % of patients with other severe, continuous headaches. Conversely, total serum Mg was normal in all patients [142]. In another study, 30.8 % of chronic migraine patients had low serum ionized but not total Mg concentrations, and 61.5 % had high ionized Ca/Mg ratios. Proportions for chronic tension-type headache patients were 4.5 and 36.4 % respectively. A high incidence of IMg2+ deficiency and elevated ICa2+/IMg2+ ratio has also been demonstrated during menstrual attacks (45 %), confirming previous suggestions of the involvement of Mg deficiency in its development [143].

Taking into account the above results, some Mg formulations have been used for prophylactic treatment (magnesium pidolate, trimagnesium dicitrate, magnesium oxide) and for the treatment of acute headaches (magnesium sulfate).

Two double-blind, randomized, placebo-controlled trials have shown the therapeutic efficacy of Mg supplementation in migraine patients, one involving women with menstrual migraine treated with two cycles of Mg or placebo taken daily from ovulation to the first day of flow [144, 145]. A third trial using a different Mg salt showed no effect of oral Mg on migraine [146].

In a randomized, double-blind, placebo-controlled study involving children and adolescents (aged 3–17 years), the administration of magnesium oxide induced a statistically significant trend toward reduction in headache frequency [147].

Further trials have also shown that intravenous magnesium sulfate is effective in the treatment of acute migraine [148]. Intravenous magnesium sulfate with prochlorperazine was also able to abort a prolonged migrainous aura [149].

In addition, Mg was found to be effective for the treatment of CH. After i.v. administration of magnesium sulfate, 41 % of patients reported meaningful relief defined as a complete cessation of attacks or relief for more than 3 days [150].

Based on the promising findings described above, Mg in the sulfate salt formulation can be recommended for migraine and possibly CH attacks. This may be particularly relevant for migraine patients) considering that up to 50 % of patients during attacks have low ionized Mg concentrations. Furthermore, Mg in chelated formulations may be recommended for the prophylactic treatment of patients with MwA and MwoA. Mg may also be indicated for patients with other headaches, such as TTH, with pericranial muscle tenderness, if peripheral Mg deficiency is demonstrated [151].

10.13 Conclusion

In this chapter we reviewed the most important aspects of the biochemistry of primary headaches. The most consistent results have been obtained for migraine (Tables 10.1 and 10.2). Further investigations are needed to broaden the horizons of this topic. The integration of the different fields of the pathophysiology of headaches could, in future, allow for a significant improvement in the knowledge and the management of such a relevant and common ailment.

References

1. Eftekhari S, Salvatore CA, Calamari A et al (2010) Differential distribution of calcitonin gene-related peptide and its receptor components in the human trigeminal ganglion. Neuroscience 169:683–696
2. Walker CS, Conner AC, Poyner DR et al (2010) Regulation of signal transduction by calcitonin gene-related peptide receptors. Trends Pharmacol Sci 31:476–483
3. Thalakoti S, Patil VV, Damodaram S et al (2007) Neuron-glia signaling in trigeminal ganglion: implications for migraine pathology. Headache 47:1008–1023
4. Durham PL, Vause CV (2010) Calcitonin gene-related peptide (CGRP) receptor antagonists in the treatment of migraine. CNS Drugs 24:539–548
5. Goadsby PJ, Edvinsson L (1993) The trigeminovascular system and migraine: studies characterizing cerebrovascular and neuropeptide changes seen in humans and cats. Ann Neurol 33:48–56
6. Sarchielli P, Alberti A, Codini M et al (2000) Nitric oxide, prostaglandin and trigeminal vasoactive peptides in internal jugular vein blood during spontaneous migraine attacks. Cephalalgia 20:903–918
7. Tvedskov JF, Lipka K, Ashina M et al (2005) No increase of calcitonin gene-related peptide in jugular blood during migraine. Ann Neurol 58:561–563

8. Tfelt-Hansen P, Le H (2009) Calcitonin gene-related peptide in blood: is it increased in the external jugular vein during migraine and cluster headache? A review. J Headache Pain 10:137–143
9. Hansen JM, Petersen J, Wienecke T et al (2009) Sumatriptan does not change calcitonin gene-related peptide in the cephalic and extracephalic circulation in healthy volunteers. J Headache Pain 10:85–91
10. Cernuda-Morollón E, Larrosa D, Ramón C et al (2013) Interictal increase of CGRP levels in peripheral blood as a biomarker for chronic migraine. Neurology 81:1191–1196
11. Cady RK, Vause CV, Ho TW et al (2009) Elevated saliva calcitonin gene-related peptide levels during acute migraine predict therapeutic response to rizatriptan. Headache 49:1258–1266
12. Cady R, Turner I, Dexter K et al (2014) An exploratory study of salivary calcitonin gene-related peptide levels relative to acute interventions and preventative treatment with onabotulinumtoxinA in chronic migraine. Headache 54:269–277
13. Goadsby PJ, Edvinsson L (1994) Human in vivo evidence for trigeminovascular activation in cluster headache. Neuropeptide changes and effects of acute attack therapies. Brain 117:427–434
14. Fanciullacci M, Alessandri M, Figini M et al (1995) Increase in plasma calcitonin gene-related peptide from the extracerebral circulation during nitro-glycerin-induced cluster headache attack. Pain 60:119–123
15. Fanciullacci M, Alessandri M, Sicuteri R et al (1997) Responsiveness of the trigeminovascular system to nitroglycerin in cluster headache patients. Brain 20:283–288
16. Ashina M, Bendtsen L, Jensen R et al (2001) Calcitonin gene-related peptide during nitric-oxide headache in patients with chronic tension-type headache. Eur J Neurol 8:173–178
17. Uddman R, Edvinsson L (1989) Neuropeptides in the cerebral circulation. Cerebrovasc Brain Metab Rev 1:230–252
18. Frese A, Schilgen M, Edvinsson L et al (2005) Calcitonin gene-related peptide in cervicogenic headache. Cephalalgia 25:700–703
19. Samsam M, Coveñas R, Ahangari R et al (2000) Simultaneous depletion of neurokinin A, substance P and calcitonin gene-related peptide from the caudal trigeminal nucleus of the rat during electrical stimulation of the trigeminal ganglion. Pain 84:389–395
20. Gallai V, Sarchielli P, Floridi A et al (1995) Vasoactive peptide levels in the plasma of young migraine patients with and without aura assessed both interictally and ictally. Cephalalgia 15:384–390
21. Vaudry D, Gonzalez BJ, Basille M et al (2000) Pituitary adenylate cyclase-activating polypeptide and its receptors: from structure to functions. Pharmacol Rev 52:269–324
22. Schytz HW, Olesen J, Ashina M (2010) The PACAP receptor: a novel target for migraine treatment. Neurotherapeutics 7:191–196
23. Csati A, Tajti J, Kuris A et al (2012) Distribution of vasoactive intestinal peptide, pituitary adenylate cyclase-activating peptide, nitric oxide synthase, and their receptors in human and rat sphenopalatine ganglion. Neuroscience 202:158–168
24. Masuo Y, Ohtaki T, Masuda Y et al (1992) Binding sites for pituitary adenylate cyclase activating polypeptide (PACAP): comparison with vasoactive intestinal polypeptide (VIP) binding site localization in rat brain sections. Brain Res 575:113–123
25. Narita M, Dun SL, Dun NJ et al (1996) Hyperalgesia induced by pituitary adenylate cyclase-activating polypeptide in the mouse spinal cord. Eur J Pharmacol 311:121–126
26. Tajti J, Uddman R, Edvinsson L (2001) Neuropeptide localization in the "migraine generator" region of the human brainstem. Cephalalgia 21:96–101
27. Sandor K, Kormos V, Botz B et al (2010) Impaired nocifensive behaviours and mechanical hyperalgesia, but enhanced thermal allodynia in pituitary adenylate cyclase-activating polypeptide deficient mice. Neuropeptides 44:363–371
28. Markovics A, Kormos V, Gaszner B et al (2012) Pituitary adenylate cyclase-activating polypeptide plays a key role in nitroglycerol-induced trigeminovascular activation in mice. Neurobiol Dis 45:633–644
29. Robert C, Bourgeais L, Arreto CD et al (2013) Paraventricular hypothalamic regulation of trigeminovascular mechanisms involved in headaches. J Neurosci 33:8827–8840

30. Vécsei L, Tuka B, Tajti J (2014) Role of PACAP in migraine headaches. Brain 137:650–651
31. Chan KY, Baun M, de Vries R et al (2011) Pharmacological characterization of VIP and PACAP receptors in the human meningeal and coronary artery. Cephalalgia 31:181–189
32. Hashimoto H, Shintani N, Baba A (2006) New insights into the central PACAPergic system from the phenotypes in PACAP- and PACAP receptor-knockout mice. Ann N Y Acad Sci 1070:75–89
33. Baun M, Pedersen MH, Olesen J et al (2012) Dural mast cell degranulation is a putative mechanism for headache induced by PACAP-38. Cephalalgia 32:337–345
34. Schytz HW, Birk S, Wienecke T et al (2009) PACAP38 induces migraine-like attacks in patients with migraine without aura. Brain 132:16–25
35. Tuka B, Helyes Z, Markovics A et al (2013) Alterations in PACAP-38-like immunoreactivity in the plasma during ictal and interictal periods of migraine patients. Cephalalgia 33:1085–1095
36. Dickson L, Finlayson K (2013) VPAC and PAC receptors: from ligands to function. Pharmacol Ther 121:294–316
37. Lerner EA, Iuga AO, Reddy VB (2007) Maxadilan, a PAC1 receptor agonist from sand flies. Peptides 28:1651–1654
38. Jansen-Olesen I, Baun M, Amrutkar DV et al (2014) PACAP-38 but not VIP induces release of CGRP from trigeminal nucleus caudalis via a receptor distinct from the PAC1 receptor. Neuropeptides 48:53–64
39. Amin FM, Hougaard A, Schytz HW et al (2014) Investigation of the pathophysiological mechanisms of migraine attacks induced by pituitary adenylate cyclase-activating polypeptide-38. Brain 137:779–794
40. Morollón E, Martínez-Camblor P, Alvarez R et al (2014) Increased VIP levels in peripheral blood outside migraine attacks as a potential biomarker of cranial parasympathetic activation in chronic migraine. Cephalalgia
41. Cernuda-Morollón E, Martínez-Camblor P, Ramón C et al (2014) CGRP and VIP levels as predictors of efficacy of Onabotulinumtoxin type A in chronic migraine. Headache 54:987–995
42. Sicuteri F, Testi A, Anselmi B (1961) Biochemical investigations in headache: Increase of hydroyindoleacetic acid excretion during migraine attack. Int Arch Allergy 19:265–271
43. Ferrari MD, Odink J, Tapparelli C et al (1989) Serotonin metabolism in migraine. Neurology 39:1239–1242
44. Marukawa H, Shimomura T, Takahashi K (1996) Salivary substance P, 5-hydroxytryptamine, and gamma-aminobutyric acid levels in migraine and tension-type headache. Headache 36:100–104
45. Kovacs K, Bors L, Tothfalusi L et al (1989) Cerebrospinal fluid (CSF) investigations in migraine. Cephalalgia 9:53–57
46. Fioroni L, D'Andrea G, Alecci M et al (1996) Platelet serotonin pathway in menstrual migraine. Cephalalgia 16:427–430
47. Ayzenberg I, Obermann M, Leineweber K et al (2008) Increased activity of serotonin uptake in platelets in medication overuse headache following regular intake of analgesics and triptans. J Headache Pain 9:109–112
48. Schuh-Hofer S, Richter M, Geworski L et al (2007) Increased serotonin transporter availability in the brainstem of migraineurs. J Neurol 254:789–796
49. D'Andrea G, Granella F, Perini F et al (2006) Platelet levels of dopamine are increased in migraine and cluster headache. Headache 46:585–591
50. Gallai V, Gaiti A, Sarchielli P et al (1992) Evidence for an altered dopamine b-hydroxylase activity in migraine and tension type headache. Acta Neurol Scand 86:403–446
51. Fernandez F, Lea RA, Colson NJ et al (2006) Association between a 19 bp deletion polymorphism at dopamine betahydroxylase (DBH) locus and migraine with aura. J Neurol Sci 251:118–123
52. Borowsky B, Adham N, Jones KA et al (2001) Trace amines: identification of a family of mammalian G protein-coupled receptors. Proc Natl Acad Sci U S A 98:933–941

53. D'Andrea G, Terrazzino S, Leon A et al (2004) Elevated levels of circulating trace amines in primary headaches. Neurology 62:1701–1705
54. Wilson RI, Nicoll RA (2002) Endocannabinoid signaling in the brain. Science 296:678–682
55. Greco R, Gasperi V, Maccarrone M, Tassorelli C (2010) The endocannabinoid system and migraine. Exp Neurol 224:85–91
56. Akerman S, Holland PR, Lasalandra MP et al (2013) Endocannabinoids in the brainstem modulate dural trigeminovascular nociceptive traffic via CB1 and "triptan" receptors: implications in migraine. J Neurosci 33:14869–14877
57. Maldonado R, Valverde O, Berrendero F (2006) Involvement of the endocannabinoid system in drug addiction. Trends Neurosci 29:225–232
58. Cupini LM, Bari M, Battista N et al (2006) Biochemical changes in endocannabinoid system are expressed in platelets of female but not male migraineurs. Cephalalgia 26:277–281
59. Cupini LM, Costa C, Sarchielli P et al (2008) Degradation of endocannabinoids in chronic migraine and medication overuse headache. Neurobiol Dis 30:186–189
60. Perrotta A, Arce-Leal N, Tassorelli C et al (2012) Acute reduction of anandamide-hydrolase (FAAH) activity is coupled with a reduction of nociceptive pathways facilitation in medication-overuse headache subjects after withdrawal treatment. Headache 52:1350–1361
61. Sarchielli P, Pini LA, Coppola F et al (2007) Endocannabinoids in chronic migraine: CSF findings suggest a system failure. Neuropsychopharmacology 32:1384–1390
62. Rossi C, Pini LA, Cupini ML et al (2008) Endocannabinoids in platelets of chronic migraine patients and medication-overuse headache patients: relation with serotonin levels. Eur J Clin Pharmacol 64:1–8
63. Ramadan NM (2003) The link between glutamate and migraine. CNS Spectr 8:446–449
64. Zukerman E, Minatti-Hannuch SN, Mazzacoratti MGN et al (1993) Cerebrospinal fluid neurotransmitter amino acids in migraine. Cephalalgia 13(Suppl 13):92
65. Gallai V, Alberti A, Gallai B et al (2003) Glutamate and nitric oxide pathway in chronic daily headache: evidence from cerebrospinal fluid. Cephalalgia 23:166–174
66. Vaccaro M, Riva C, Tremolizzo L et al (2007) Platelet glutamate uptake and release in migraine with and without aura. Cephalalgia 27:35–40
67. González de la Aleja J, Ramos A, Mato-Abad V et al (2013) Higher glutamate to glutamine ratios in occipital regions in women with migraine during the interictal state. Headache 53:365–375
68. Párdutz A, Fejes A, Bohár Z et al (2012) Kynurenines and headache. J Neural Transm 119: 285–296
69. D'Amico D, Ferraris A, Leone M et al (2002) Increased plasma nitrites in migraine and cluster headache patients in interictal period: basal hyperactivity of L-arginine-NO pathway? Cephalalgia 22:33–36
70. Costa A, Ravaglia S, Sances G et al (2003) Nitric oxide pathway and response to nitroglycerin in cluster headache patients: plasma nitrite and citrulline levels. Cephalalgia 23: 407–413
71. Silva FA, Rueda-Clausen CF, Silva SY et al (2007) Endothelial function in patients with migraine during the interictal period. Headache 47:45–51
72. Gallai V, Floridi A, Mazzotta G et al (1996) L-arginine/nitric oxide pathway activation in platelets of migraine patients with and without aura. Acta Neurol Scand 94:151–160
73. Sarchielli P, Tognoloni M, Russo S et al (1996) Variations in the platelet arginine/nitric oxide pathway during the ovarian cycle in females affected by menstrual migraine. Cephalalgia 16:468–475
74. Sarchielli P, Alberti A, Russo S et al (1999) Nitric oxide pathway, Ca2+, and serotonin content in platelets from patients suffering from chronic daily headache. Cephalalgia 19:810–816
75. Mendell LM, Albers KM, Davis BM (1999) Neurotrophins, nociceptors, and pain. Microsc Res Tech 45:252–261
76. Blandini F, Rinaldi L, Tassorelli C et al (2006) Peripheral levels of BDNF and NGF in primary headaches. Cephalalgia 26:136–142
77. Tanure MT, Gomez RS, Hurtado RC et al (2010) Increased serum levels of brain-derived neurotropic factor during migraine attacks: a pilot study. J Headache Pain 11:427–430

78. Sarchielli P, Alberti A, Floridi A et al (2001) Levels of nerve growth factor in cerebrospinal fluid of chronic daily headache patients. Neurology 57:132–134
79. Sarchielli P, Mancini ML, Floridi A et al (2007) Increased levels of neurotrophins are not specific for chronic migraine: evidence from primary fibromyalgia syndrome. J Pain 8:737–745
80. Sarchielli P, Alberti A, Candeliere A et al (2006) Glial cell line-derived neurotrophic factor and somatostatin levels in cerebrospinal fluid of patients affected by chronic migraine and fibromyalgia. Cephalalgia 26:409–415
81. Denuelle M, Fabre N, Payoux P et al (2007) Hypothalamic activation in spontaneous migraine attacks. Headache 47:1418–1426
82. Holland P, Goadsby PJ (2007) The hypothalamic orexinergic system: pain and primary headaches. Headache 47:951–962
83. Sarchielli P, Rainero I, Coppola F et al (2008) Involvement of corticotrophin-releasing factor and orexin-A in chronic migraine and medication-overuse headache: findings from cerebrospinal fluid. Cephalalgia 28:714–722
84. Caproni S, Corbelli I, Pini LA et al (2011) Migraine preventive drug-induced weight gain may be mediated by effects on hypothalamic peptides: the results of a pilot study. Cephalalgia 31:543–549
85. Duarte H, Teixeira AL, Rocha NP, Domingues RB (2014) Increased serum levels of adiponectin in migraine. J Neurol Sci 342:186–188
86. Peterlin BL, Tietjen GE, Gower BA et al (2013) Ictal adiponectin levels in episodic migraineurs: a randomized pilot trial. Headache 53:474–490
87. Dearborn JL, Schneider AL, Gottesman RF, Kurth T et al (2014) Adiponectin and leptin levels in migraineurs in the Atherosclerosis Risk in Communities Study. Neurology 83:2211–2218
88. Berilgen MS, Bulut S, Gonen M et al (2005) Comparison of the effects of amitriptyline and flunarizine on weight gain and serum leptin, C peptide and insulin levels when used as migraine preventive treatment. Cephalalgia 25:1048–1053
89. Guldiken B, Guldiken S, Demir M et al (2008) Low leptin levels in migraine: a case control study. Headache 40:1103–1107
90. Gallai V, Sarchielli P, Trequattrini A et al (1994) Neuropeptide Y in juvenile migraine and tension-type headache. Headache 34:35–40
91. Valenzuela RF, Donoso MV, Mellado PA et al (2000) Migraine, but not subarachnoid hemorrhage, is associated with differentially increased NPY-like immunoreactivity in the CSF. J Neurol Sci 173:140–146
92. Welch KMA, Chabi E, Nell JH et al (1975) Cerebrospinal fluid gamma aminobutyric acid levels and migraine. Br Med J 3:516–517
93. Vieira DS, Naffah-Mazacoratti MG, Zukerman E et al (2006) Cerebrospinal fluid GABA levels in chronic migraine with and without depression. Brain Res 1090:197–201
94. Bigal ME, Hetherington H, Pan J et al (2008) Occipital levels of GABA are related to severe headaches in migraine. Neurology 70:2078–2080
95. Chen T, Murrell M, Fowdar J et al (2012) Investigation of the role of the GABRG2 gene variant in migraine. J Neurol Sci 318:112–114
96. Nicolodi M, Del Bianco E (1990) Sensory neuropeptides (substance P, calcitonin gene-related peptide) and vasoactive intestinal polypeptide in human saliva: their pattern in migraine and cluster headache. Cephalalgia 10:39–50
97. Jang MU, Park JW, Kho HS et al (2011) Plasma and saliva levels of nerve growth factor and neuropeptides in chronic migraine patients. Oral Dis 17:187–193
98. Sicuteri F, Fanciullacci M, Geppetti P et al (1985) Substance P mechanism in cluster headache: evaluation in plasma and cerebrospinal fluid. Cephalalgia 5:143–149
99. Nakano T, Shimomura T, Takahashi K et al (1993) Platelet substance P and 5-hydroxytryptamine in migraine and tension-type headache. Headache 33:528–532
100. Gallai V, Sarchielli P, Firenze C et al (1994) Endothelin 1 in migraine and tension-type headache. Acta Neurol Scand 89:47–55
101. Kallela M, Färkkilä M, Saijonmaa O et al (1998) Endothelin in migraine patients. Cephalalgia 18:329–332

102. Hasselblatt M, Köhler J, Volles E et al (1999) Simultaneous monitoring of endothelin-1 and vasopressin plasma levels in migraine. Neuroreport 10:423–425
103. Nattero G, Mengozzi G, Inconis T et al (1996) Nitric oxide, endothelin-1, and transcranial Doppler in migraine. Findings in interictal conditions and during migraine attack. Headache 36:307–311
104. Hamed SA, Hamed EA, Ezz Eldin AM et al (2010) Vascular risk factors, endothelial function, and carotid thickness in patients with migraine: relationship to atherosclerosis. J Stroke Cerebrovasc Dis 19:92–103
105. Borsook D, Erpelding N, Lebel A et al (2014) Sex and the migraine brain. Neurobiol Dis 68:200–214
106. Mor G, Nilsen J, Horvath T et al (1999) Estrogen and microglia: a regulatory system that affects the brain. J Neurobiol 40:484–496
107. Alstadhaug KB (2009) Migraine and the hypothalamus. Cephalalgia 29:809–817
108. Lieba-Samal D, Wöber C (2011) Sex hormones and primary headaches other than migraine. Curr Pain Headache Rep 15:407–414
109. Kemper RH, Meijler WJ, Korf J et al (2001) Migraine and function of the immune system: a meta-analysis of clinical literature published between 1966 and 1999. Cephalalgia 21:549–557
110. Mortimer MJ, Kay J, Gawkrodger DJ et al (1993) The prevalence of headache and migraine in atopic children: an epidemiological study in general practice. Headache 33:427–431
111. Gazerani P, Pourpak Z, Ahmadiani A et al (2003) A correlation between migraine, histamine and immunoglobulin e. Scand J Immunol 57:286–290
112. Boćkowski L, Sobaniec W, Zelazowska-Rutkowska B (2009) Proinflammatory plasma cytokines in children with migraine. Pediatr Neurol 41:17–21
113. Boćkowski L, Smigielska-Kuzia J, Sobaniec W et al (2010) Anti-inflammatory plasma cytokines in children and adolescents with migraine headaches. Pharmacol Rep 62:287–291
114. Forcelini CM, Dantas DCM, Luz C et al (2011) Analysis of leukocytes in medication-overuse headache, chronic migraine and episodic migraine. Headache 51:1228–1238
115. Grazzi L, Corsini E, Ciusani E et al (2014) Evaluation of immune parameters in chronic migraine with medication overuse. Neurol Sci 35(Suppl 1):171–173
116. Bø SH, Davidsen EM, Gulbrandsen P et al (2009) Cerebrospinal fluid cytokine levels in migraine, tension-type headache and cervicogenic headache. Cephalalgia 29:365–372
117. Levy D (2012) Endogenous mechanisms underlying the activation and sensitization of meningeal nociceptors: the role of immuno-vascular interactions and cortical spreading depression. Curr Pain Headache Rep 16(3):270–277
118. Sarchielli P, Alberti A, Baldi A et al (2006) Proinflammatory cytokines, adhesion molecules, and lymphocyte integrin expression in the internal jugular blood of migraine patients without aura assessed ictally. Headache 46:200–207
119. Sarchielli P, Alberti A, Vaianella L et al (2004) Chemokine levels in the jugular venous blood of migraine without aura patients during attacks. Headache 44:961–968
120. Gursoy-Ozdemir Y, Qiu J, Matsuoka N et al (2004) Cortical spreading depression activates and upregulates MMP-9. J Clin Invest 113:1447–1455
121. Shimomura T, Araga S, Esumi E et al (1991) Decreased serum interleukin-2 level in patients with chronic headache. Headache 31:310–313
122. Della Vedova C, Cathcart S, Dohnalek A et al (2013) Peripheral interleukin-1ß levels are elevated in chronic tension-type headache patients. Pain Res Manag 18:301–306
123. Fernández-de-Las-Peñas C, Fernández-Mayoralas DM, Arroyo-Morales M et al (2011) Lower immunglobulin A levels but not lower cortisol or α-amylase activity in children with chronic tension-type headache. Cephalalgia 31:481–487
124. Martelletti P, Stirparo G, De Stefano L et al (1987) Reduced activity of the NK cells from patients with cluster headache and the "in vitro" response to beta-interferon. Headache 27:548–551
125. Giacovazzo M, Stirparo G, DeStefano L et al (1989) Lymphokine-activated killer (LAK) cell phenomenon in cluster headache. "In vitro" activation by recombinant interleukin-2. Headache 29:177–179
126. Stirparo G, Martelletti P, Giacovazzo M et al (1992) Impairment of cytolytic and proliferative responsiveness of peripheral blood mononuclear cells from cluster headache patients to IL-2. Pharmacol Res 26(Suppl 2):194–195

127. Bussone G, Salmaggi A, Leone M et al (1992) Immunological alterations in cluster headache during remission and cluster period. Comparison with low back pain patients. Cephalalgia 12:250–253
128. Galeotti N, Ghelardini C, Zoppi M et al (2001) Hypofunctionality of Gi proteins as aetio-pathogenic mechanism for migraine and cluster headache. Cephalalgia 21:38–45
129. Martelletti P, Granata M, Giacovazzo M (1993) Serum interleukin-1 beta is increased in cluster headache. Cephalalgia 13:343–345
130. Empl M, Förderreuther S, Schwarz M, Müller N, Straube A (2003) Soluble interleukin-2 receptors increase during the active periods in cluster headache. Headache 43(1):63–68
131. Rozen T, Swidan SZ (2007) Elevation of CSF tumor necrosis factor alpha levels in new daily persistent headache and treatment refractory chronic migraine. Headache 47:1050–1055
132. Durlach J (1976) Neurological manifestations of magnesium imbalance. In: Vinken PJ, Bruyn GW (eds) Handbook of clinical neurology. North-Holland Publishing Co, Amsterdam, pp 545–579
133. Trauninger A, Pfund Z, Koszegi T et al (2002) Oral magnesium load test in patients with migraine. Headache 42:114–119
134. Ramadan NM, Halvorson H, Vande-Linde A et al (1989) Low brain magnesium in migraine. Headache 29:590–593
135. Lodi R, Iotti S, Cortelli P et al (2001) Deficient energy metabolism is associated with low free magnesium in the brains of patients with migraine and cluster headache. Brain Res Bull 54:437–441
136. Schoenen J, Sianard-Gainko J, Lenaerts M (1991) Blood magnesium levels in migraine. Cephalalgia 11:97–99
137. Smeets MC, Vernooy CB, Souverijn JH et al (1994) Intracellular and plasma magnesium in familial hemiplegic migraine and migraine with and without aura. Cephalalgia 14:29–32
138. Sarchielli P, Coata G, Firenze C et al (1992) Serum and salivary magnesium levels in migraine and tension-type headache. Results in a group of adult patients. Cephalalgia 12:21–27
139. Aloisi P, Marrelli A, Porto C et al (1997) Visual evoked potentials and serum magnesium levels in juvenile migraine patients. Headache 37:383–385
140. Gallai V, Sarchielli P, Morucci P et al (1993) Red blood cell magnesium levels in migraine patients. Cephalalgia 13:94–98
141. Gallai V, Sarchielli P, Morucci P et al (1994) Magnesium content of mononuclear blood cells in migraine patients. Headache 34:160–165
142. Mauskop A, Altura BT, Cracco RQ et al (1993) Deficiency in serum ionized magnesium but not total magnesium in patients with migraines. Possible role of ICa2+/IMg2+ ratio. Headache 33:135–138
143. Mauskop A, Altura BT, Altura BM (2002) Serum ionized magnesium levels and serum ionized calcium/ionized magnesium ratios in women with menstrual migraine. Headache 42:242–248
144. Facchinetti F, Sances G, Borella P et al (1991) Magnesium prophylaxis of menstrual migraine: effects on intracellular magnesium. Headache 31:298–301
145. Peikert A, Wilimzig C, Kohne-Volland R (1996) Prophylaxis of migraine with oral magnesium: results from a prospective, multicenter, placebo-controlled and double-blind randomized study. Cephalalgia 16:257–263
146. Pfaffenrath V, Wessely P, Meyer C et al (1996) Magnesium in the prophylaxis of migraine-A double-blind, placebo-controlled study. Cephalalgia 16:436–440
147. Wang F, Van Den Eeden SK, Ackerson LM (2003) Oral magnesium oxide prophylaxis of frequent migraine headache in children: a randomized, double-blind, placebo-controlled trial. Headache 43:601–610
148. Bigal ME, Bordini CA, Tepper SJ et al (2002) Intravenous magnesium sulphate in the acute treatment of migraine without aura and migraine with aura. A randomized, double-blind, placebo-controlled study. Cephalalgia 22:345–353
149. Rozen TD (2003) Aborting a prolonged migrainous aura with intravenous prochlorperazine and magnesium sulfate. Headache 43:901–903
150. Mauskop A, Altura BT, Cracco RQ, Altura BM (1995) Intravenous magnesium sulfate relieves cluster headaches in patients with low serum ionized magnesium levels. Headache 35:597–600
151. Mauskop A, Varughese J (2012) Why all migraine patients should be treated with magnesium. J Neural Transm 119:575–579

Chapter 11
Pathophysiology of Migraine: Current Status and Future Directions

Jakob Møller Hansen and Dan Levy

11.1 Introduction

Around 10 % of the global adult population has active migraine [1]. The public health burden of migraine is high because migraine attacks are associated with temporary disability and substantial impairment in activities [2]. As such, migraine is ranked as one of the most disabling conditions [3, 4]. The widespread disability produced by migraine [5] is therefore an important target for treatment.

The hallmark of migraine is the head pain, but a plethora of other clinical symptoms is needed for a headache to be qualified as a migraine according to the current diagnostic criteria; see Table 11.1.

There has been tremendous progress in our acceptance, understanding and treatment possibilities of migraine, but to optimize migraine management, it is important that we continue to improve our understanding of the basic migraine mechanisms. An understanding of migraine pathophysiology must encompass the varied clinical symptoms and relate these findings to anatomy and physiology.

J.M. Hansen, MD, PhD (✉)
Department of Neurology, Faculty of Health Sciences, Glostrup Hospital,
Danish Headache Center, University of Copenhagen, Glostrup, Copenhagen, Denmark
e-mail: jmh@dadlnet.dk

D. Levy, PhD (✉)
Department of Anesthesia Critical Care and Pain Medicine, Beth Israel Deaconess
Medical Center, Harvard Medical School, Boston, MA, USA
e-mail: dlevy1@bidmc.harvard.edu

M. Ashina, P. Geppetti (eds.), *Pathophysiology of Headaches: From Molecule to Man*, Headache, DOI 10.1007/978-3-319-15621-7_11

Table 11.1 Diagnostic criteria for migraine, according to the current classification of headache disorders of the International Headache Society [14], reproduced with permission

Diagnose	Migraine without aura (MO)	Migraine with aura (MA)
Diagnostic criteria	(A) At least 5 attacks fulfilling criteria B–D (B) Headache attacks lasting 4–72 h (untreated or unsuccessfully treated) (C) Headache has at least two of the following four characteristics: Unilateral location, Pulsating quality, Moderate or severe pain intensity, Aggravation by or causing avoidance of routine physical activity (e.g. walking or climbing stairs) (D) During headache at least one of the following: Nausea and/or vomiting Photophobia and phonophobia (E) Not better accounted for by another ICHD-3 diagnosis	(A) At least 2 attacks fulfilling criteria B–D (B) One or more of the following fully reversible aura symptoms: 1. Visual 2. Sensory 3. Speech and/or language 4. Motor 5. Brainstem 6. Retinal (C) At least two of the following four characteristics: 1. At least one aura symptom spreads gradually over ≥5 min, and/or two or more symptoms occur in succession 2. Each individual aura symptom lasts 5–60 min 3. At least one aura symptom is unilateral 4. The aura is accompanied, or followed within 60 min, by headache (D) Not better accounted for by another ICHD-3 diagnosis, and transient ischemic attacks have been excluded

11.2 Genetic Predisposition in Migraine

Due to both clinical and genetic heterogeneity in migraine patients, as well as a long list of possible environmental causes and triggering factors, it has been difficult to identify genes and their importance for the common types of migraine [6].

A genetic basis for migraine with aura (MA) was reported [7, 8] with loci for visual aura mapped to chromosome nine [9] at 9q21-q22 [10]. Genome-wide association studies in the general population have suggested susceptibility loci for migraine [11] that were later confirmed [12] and also for migraine without aura [13].

Important insights into the genetic and molecular pathophysiology of migraine have come from studies of rare monogenic subtypes of migraine, where mutations in a single gene are linked to a migraine phenotype. These include familial hemiplegic migraine (FHM) and migraine in familial advanced sleep phase syndrome (FASPS).

FHM is a dominantly inherited subtype of migraine with aura, phenotypically characterized by fully reversible half-sided weakness and other aura symptoms preceding or accompanying a migraine headache and at least one affected first-degree relative [14]. Mutations in a number of different genes cause the FHM phenotype

[15–19]. The migraine aura is believed to be mediated by cortical spreading depression (CSD) and is a self-propagating wave of cellular depolarization in the cerebral cortex (see also below) [20, 21].

In vivo studies of mice with *knock-ins* of two different FHM genes showed increased susceptibility to CSD as well as increased velocity of propagation of CSD compared with wild-type animals [22–25]. FHM mutations may thus predispose to CSD, migraine aura and possibly headache [20, 26]. The fact that the FHM-1 knock-in mice also show a relevant pain phenotype [27, 28] suggests that these transgenic mice are an important model to improve our understanding of migraine pathogenesis and may be used as a platform for testing novel anti-migraine drugs.

Another genetic factor that may drive migraine is a mutation in the gene encoding casein kinase Iδ (CKIδ). This mutation underlies familial advanced sleep phase syndrome (FASPS), which is characterized by early sleep times and early-morning awakening, and affects a phosphorylation site within the CKI-binding domain of the human PER2 gene [29].

A recent study reported of two families, each with a distinct missense mutation in the gene encoding CKIδ, in which the mutation co-segregated with both the presence of migraine and advanced sleep phase [30]. The authors engineered a knock-in mouse carrying one of the mutations (CKIδ-T44A allele) and showed in a beautiful translational set-up that these mice were more sensitive to pain after treatment with the migraine trigger nitroglycerin [31], had a reduced threshold for CSD and astrocytes - which could play a role in CSD [32]- from the KI mice exhibited increased calcium signalling. Interestingly, the FHM-1 knock-in migraine mouse model also shows a change in circadian phenotype towards an enhanced circadian resetting [33].

It is important to note that the common types of migraine, MA and MO, are generally not associated with any of the known FHM mutations [34–37]. Furthermore, many of the traits found in these monogenic subtypes of migraine (e.g. hemiplegia during aura, progressive ataxia in FHM and FASP) are not found in common types of migraine. The translation from the promising animal studies of the monogenic migraines into a human clinical setting, and the relation between these rare subtypes and MO and MA, may thus not be as straightforward as hoped.

The search for the elusive migraine genes is ongoing and is expected to provide further suggestions for molecular pathways and possible treatment targets; see also Chap. 4.

11.3 What Causes a Migraine Attack?

11.3.1 Triggering Factors for Migraine

A migraine trigger is any factor that on exposure or withdrawal leads to the development of a migraine attack [38]. Migraine triggers are often stereotypic within each patient, with a large degree of overlap between patients, and constant across geographic, cultural, ethnic or racial boundaries [39–41].

Triggering factors for migraine are found across the migraine spectrum, from migraine without aura to migraine with aura, and in familial hemiplegic migraine [42–47]. Despite a large number of reported trigger factors, the mechanisms by which migraine triggers exert their effect is not clarified. It can be speculated that different migraine triggers activate a wide variety of brain areas, ultimately resulting in release of nociceptive molecules from the parenchyma or meninges that are capable of activating and sensitizing meningeal nociceptors [48] and ultimately the migraine pain pathway [49].

The link between known triggering factors and the genesis of a migraine attack requires, however, a better scrutiny. While retrospective and diary studies often identify numerous factors, clinical studies paint a different picture [40].

One study exposed 27 MA patients to their self-reported triggers in order to assess the causal relation between trigger factors and migraine attacks. The study tried to trigger a migraine aura with factors that the patients themselves had identified as strong triggers. These triggers included different types of photostimulation, strenuous exercise or a combination of these factors. It was found that only 11 % reported attacks of MA following provocation, which led the authors to conclude that experimental provocation using self-reported natural trigger factors causes MA only in a small subgroup of patients with MA [50].

Testing the role of low-intensity factors such as chocolate as a migraine trigger also indicated mixed results with two studies concluded as negative [51, 52], while another was positive [53].

To better evaluate the role of migraine triggers in the onset of new or first attacks of migraine and to determine if avoidance of trigger factors improves migraine status, prospective cohort studies are required.

11.4 Signalling Molecules in Migraine

Migraine is considered a neurovascular headache [54]. It is believed to arise from a primary brain dysfunction, leading to activation and sensitisation of primary afferent neurons that innervate the cranial meninges, i.e. peripheral trigeminovascular neurons and the ensuing release of vasoactive neuropeptides [55, 56].

Release of vasoactive peptides during attacks could be the result of activation of a brainstem reflex [57], where activation of trigeminal nerves and subsequent nociceptive signaling to the central nervous system mediates a parasympathetic reflex arc during migraine, leading to the release of e.g. neuropeptides [58].

11.4.1 Calcitonin Gene-Related Peptide (CGRP)

A prominent example is CGRP, a 37-amino-acid neuropeptide identified in the early 1980s [59, 60]. CGRP is broadly distributed in the nervous system [61, 62] and in the trigeminal pain pathway at peripheral [63–65] and central levels [66–68]. CGRP is released from perivascular trigeminal nerve terminals [69, 70] but does not activate or sensitize the nociceptors in the rat meninges [71] or in the human skin and muscle [72]. Animal models of headache and pain suggest modulatory role of CGRP in nociceptive transmission [71, 73–77].

In 1988, it was reported that CGRP is released into the extracerebral circulation of humans during thermocoagulation of the trigeminal ganglion [78]. Studies in migraine patients showed elevation of CGRP during [79] and outside of migraine attacks [80]. However, a newer study challenged these reports, showing no changes in plasma CGRP during migraine attacks compared to outside of attacks [81]. Regardless, the importance of CGRP in migraine pathogenesis was underlined after large randomized controlled trials confirmed that CGRP antagonists are effective in subsets of patients in treating acute migraine attacks [82–85] and also in migraine prophylaxis [86]. The exact pathways involved in mediating the role of CGRP in migraine attacks and mechanisms by which CGRP antagonists abort or prevent migraine are not fully clarified but may involve both peripheral and central sites of action [68, 87].

The newest chapter in the saga of CGRP-based migraine treatment was the finding that monoclonal antibodies to CGRP are effective in migraine prophylaxis [88, 89]. Interestingly, both the anti-migraine action of the CGRP antibodies and their large molecular weight point to a peripheral site of action for CGRP in migraine.

11.4.2 Vasoactive Intestinal Peptide (VIP) and Pituitary Adenylate Cyclase-Activating Polypeptide (PACAP)

Two other neuropeptides that are also of interest in migraine pathophysiology are VIP and PACAP. These peptides are secreted from perivascular parasympathetic nerve fibres [90]. PACAP is also found in trigeminal nerve fibres surrounding cerebral blood vessels [91]. The release of VIP and PACAP regulates cerebrovascular tone and haemodynamics of the brain [92]. Studies of VIP in double-blinded placebo-controlled crossover studies in both healthy subjects [93] and migraine patients [94] showed a very modest headache/migraine induction rate.

Infusion of PACAP in healthy subjects induced vasodilatation of a similar magnitude to VIP but longer lasting [95]. PACAP induced sustained cephalic vasodilatation as well as migraine attacks in migraine patients without aura [96]. In a randomized head-to-head comparison study of intravenous administration of PACAP and VIP, more patients (73 %) reported migraine-like attacks after PACAP38 than after VIP (18 %) [97].

PACAP-induced migraine was associated with sustained dilatation of extracranial arteries and elevated plasma PACAP before onset of migraine-like attacks. Two receptors, $VPAC_1$ [98] and $VPAC_2$ [99], are activated with equal affinity by PACAP and VIP, but a third receptor, PAC_1, is selectively activated by PACAP [100]. PACAP has a much higher affinity for the PAC1 receptor. Thus, it can be speculated that the PACAP38-evoked migraine involves the activation of the PAC1 receptor, which may be a future anti-migraine drug target [101]. See also Chap. 10.

11.5 The Premonitory Symptoms

Many migraine patients experience a prodromal phase with so-called premonitory symptoms, hours or even days before the aura or headache. The most commonly reported premonitory symptoms are yawning, stiff neck, fatigue, irritability, food cravings, difficulty concentrating, mood change, light hypersensitivity and nausea [102–104].

It is worth noting that some typical premonitory symptoms (heightened sensitivity to light, sound and smell) are also found on the list of migraine triggers. It may well be that migraine patients identify as triggers particular sensory stimuli to which they are already more sensitive because their acute migraine attack has already begun [105].

Cortical hypersensitivity may underlie some of the premonitory symptoms exhibited by migraineurs. For example, in a PET study of patients reporting premonitory symptoms, activation of extrastriate visual cortex in patients with sensitivity to light (i.e. photophobia) was significantly greater in patients with photophobia compared to those without, suggesting that photophobia may be linked to activation of the visual cortex during the premonitory phase of migraine [106].

Several prodromal- and attack-related migraine symptoms have been hypothesized to be caused by alterations in dopaminergic function (e.g. yawning, nausea, vomiting and fatigue), and dopamine antagonist medications have been successfully employed in the acute treatment of migraine [107, 108]. There is also indirect evidence for a role of dopamine in acute migraine attacks based on previous studies [109]. For example, in visually triggered migraine, blood oxygen level-dependent (BOLD) fMRI (functional magnetic resonance imaging) showed that baseline T2-weighted signal intensities increased in the red nucleus and substantia nigra before the onset of visually triggered symptoms, suggesting activation of brainstem dopaminergic structures during migraine attacks [110].

A recent PET study set out to identify other brain areas that may be involved in the premonitory phase of migraine [111] using i.v. infusion of nitroglycerin (NTG) that is able to induce both migraine headache [31] and premonitory symptoms [112].

The authors reported brain activations, in particular of the hypothalamus, in the premonitory phase, notably at a time point when the patients were totally pain-free. The hypothalamus might also be an important brain region during the attack [113, 114]. The hypothalamus may thus be a possible site of drug action for very early treatment of impending migraine attacks. Possible targets might include hypothalamic peptides, especially the orexins [115]. Orexin receptors antagonism attenuates trigeminal nociceptive activity and increases the threshold for cortical spreading depression (CSD), but these actions may not necessarily involve the hypothalamus.

Although targeting the hypothalamic orexinergic system has been suggested as a novel mechanism for the preventive treatment of migraine with and without aura [116], a recent randomized controlled trial testing of the orexin receptor antagonist filorexant for migraine prevention did not find any effect compared to placebo [117].

11.6 The Migraine Aura

About a third of patients with migraine have attacks with aura [118], a usually transient clinical disturbance that a can be attributed to brain dysfunction [20]. In migraine with typical aura, the most prevalent aura symptoms are visual disturbances [119]; see Fig. 11.1.

Following the original description of the cortical spreading depression (CSD) phenomenon in rabbits by Leão [120], it has been hypothesized that CSD and the ensuing cerebral oligemia are the underlying mechanisms of the migraine visual aura [21, 121]. While a CSD has never been recorded during a migraine attack, spreading depolarizations closely resembling CSD have been well documented in humans following ischemic stroke, intracerebral hematoma, subarachnoid haemorrhage and brain trauma [122–124].

A seminal finding however that provided critical support for the role of CSD in mediating the aura was a functional MRI study in a patient showing perturbations in

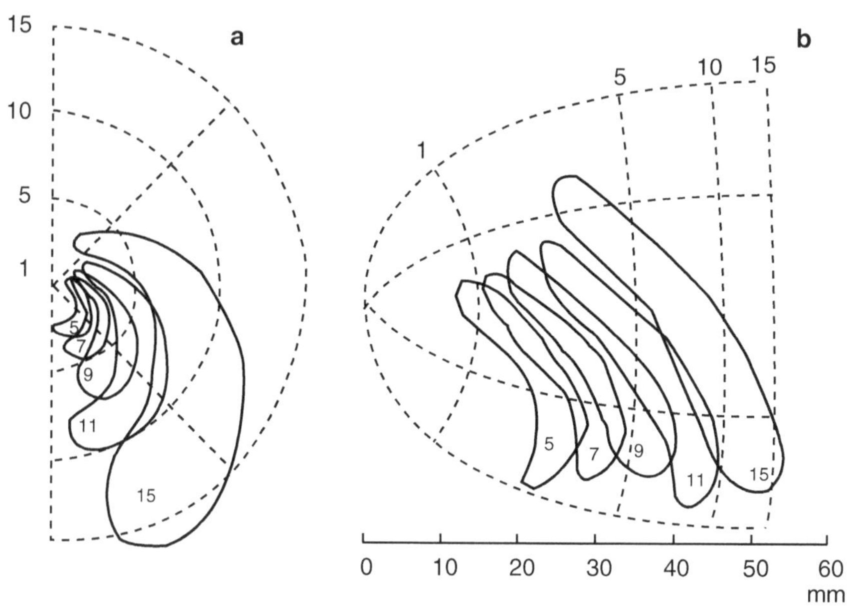

Fig. 11.1 Typical propagation pattern of a visual migraine aura. (**a**) The figure depicts the right visual hemifield and the travelling visual migraine aura, with the numbers indicating the time passed (in minutes) since first occurrence. (**b**) Here, the visual disturbance is projected onto a flat model of the primary visual cortex by reversed retinotopic mapping (Used with permission and adapted from [159])

the occipital cortex during visual aura that are consistent with features of CSD in the rodent cortex [125].

CSD may be one of the cortical events that lead to the genesis of migraine pain.

Animal studies have shown that CSD initiated in the rat visual cortex can lead to long-lasting activation of nociceptors that innervate the dura mater [126]. Interestingly, the activation of meningeal nociceptors following CSD was not mediated by the parasympathetic system, suggesting again that meningeal vasodilatation may not play a role in the emergence of the headache in MA. In the same model, it was also shown that CSD can also activate spinal trigeminal nucleus that receive input from the dura, thereby linking CSD to activation of the trigeminovascular pathway [127]. This supports previous findings in rodent models that CSD activates trigeminovascular afferents and evokes a series of cortical meningeal and brainstem events consistent with the development of headache [26] – possibly through the opening of stressed neuronal pannexin1 channels [128]. In addition to activating peripheral and central trigeminovascular neurons, CSD has been shown to mediate meningeal excitability by modulating corticotrigeminal networks [129].

A study based on quantitative examination of the timing of migraine symptoms relative to aura reported that headache and other migraine symptoms commonly occur simultaneously with aura, at least in a substantial number of patients [130]. This result suggests that in some attacks, the pain and associated symptoms of migraine are caused by parallel mechanisms occurring at the same time as the aura rather than as a direct downstream consequence of the mechanism underlying the aura, such as CSD. Whether CSD and its related migraine aura is a trigger of headache is still debatable, but well-planned prospective recordings of migraine attacks in patients and validated animal studies may help solve this long-standing riddle.

11.7 Migraine Headache: How and Where?

Structures that produce, or are perceived to produce, head pain can be found in the trigeminovascular system; see Fig. 11.2. Within the skull, its periosteal lining, the meninges and large vessels, but not the brain itself, are able to generate nociceptive signals [131].

Peptidergic nerve fibres with cell bodies in the trigeminal ganglion innervate intracranial meningeal blood vessels, the dura mater and venous sinuses (i.e. the trigeminovascular system) [132, 133] which are considered the major pain pathways that mediate migraine pain [134]. The major pain pathway from the vessels and dura mater is the first (ophthalmic) division of the trigeminal nerve [135]. Traction on cerebral vessels was suggested as the cause of headache pain [136] leading to the pathophysiologic concept of migraine as vascular headaches [131]. Clinical studies of various vasodilators [93, 94, 97, 137–139] and functional brain imaging [140, 141] suggest that vascular changes, though present, are unlikely to be the primary cause for head pain in migraine [142, 143]. This view is supported by

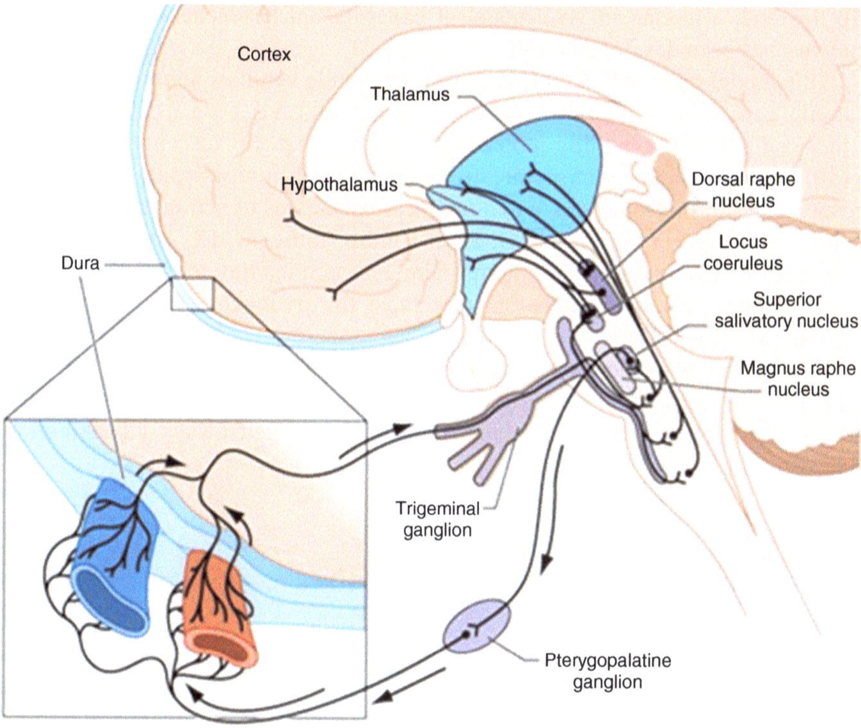

Fig. 11.2 The major structures involved in activation of the trigeminovascular system. Trigeminovascular input from the meningeal-vascular structures passes through the trigeminal ganglion and synapses on second-order neurons in the trigeminocervical complex. These neurons in turn project in the quintothalamic tract and, after decussating in the brainstem, synapse on neurons in the thalamus before projecting to the sensory cortex (Reproduced with permission and adapted from [160])

recent treatment trials where drugs without (prominent) vascular effects are effective in both abortive [144] and prophylactic migraine treatments [86]. A different view, suggesting that migraine pain could arise as a result of a brain disturbance in aminergic sensory modulatory systems, including brainstem, hypothalamic and thalamic structures, has also been suggested [145]. This debate is far from resolution [146–148]; see also Chap. 1.

11.8 Migraine Triggers in a Human Experimental Model

Because migraine attacks are fully reversible and can be aborted by therapy, the headache- or migraine-provoking property of naturally occurring signalling molecules can be tested in a human model. If a naturally occurring substance can provoke migraine in human patients, then it is possible, although not certain, that blocking its signalling will be effective in the treatment of acute migraine attacks.

To this end, a human in vivo model of experimental headache and migraine in humans has been developed [149].

This model has predicted efficacy of nitric oxide synthase inhibition and calcitonin gene-related peptide receptor blockade and has been used to examine other endogenous signalling molecules as well as genetic susceptibility factors [150]. Human models of migraine offer unique possibilities to study mechanisms responsible for migraine and to explore the mechanisms of action of existing and future anti-migraine drugs [151]. Furthermore, these models have played an important role in the translational migraine research leading to the identification of three new principally different targets in the treatment of acute migraine attacks [84, 152, 153].

New additions to the model, such as advanced MR modalities and PET [111, 114, 138, 143, 154, 155], may lead to a better understanding of the complex events that constitutes a migraine attack and better and more targeted ways of intervention. See also Chap. 5.

11.9 Conclusion and Future Perspectives

There has been tremendous progress in our understanding of migraine mechanisms and treatment over the last decades. Migraine, however, remains under-diagnosed and under-treated and its research underfunded. A number of important questions will be in focus in the time to come.

New migraine treatments are needed. As the case of CGRP shows, it is possible to go from animal studies based on anatomy and physiology to human experimental studies and to clinical trials to yield a working drug based on stringent translational thinking (and as always, a certain amount of serendipity). Along this line of thinking, future migraine drugs could act well on PACAP, orexins or other endogenous signalling molecules. The future development of prophylactic treatments for migraine thus will require a better understanding of the endogenous factors that mediate the activation of the trigeminovascular system.

Though still new to the migraine field, the recent successes in migraine prophylaxis with monoclonal antibodies to CGRP [88, 89] suggest that antibody-based treatments could be of potential use in migraine. New trials of other antibodies are currently under way (e.g. ClinicalTrials.gov Identifier: NCT02163993, NCT01688739).

Botulinum neurotoxin has been tested for prophylactic treatment of episodic migraine headaches and has consistently been found to be no better than placebo – both from a clinical and statistical perspective [156]. Interestingly, in chronic daily headaches and chronic migraines, botulinum neurotoxin showed a small to modest benefit [157]. The reasons for different treatment effects between low- and high-frequency migraine are still a matter of debate. To date, botulinum neurotoxin injection is the only FDA-approved treatment for chronic migraine. A recent study examined the hypothesis that botulinum neurotoxin affects specifically meningeal nociceptors and reported that botulinum neurotoxin was able to inhibit mechanical nociception in these neurons [158], thus providing mechanistic insights into how botulinum neurotoxin might exert any anti-migraine effect.

Future studies are needed to dissect the different aspects of the migraine biophenotype. These studies must be based on good human prospective studies in addition to experimental and observational basic science and animal studies to ensure translational progress in the migraine field.

References

1. Jensen R, Stovner LJ (2008) Epidemiology and comorbidity of headache. Lancet Neurol 7(4):354–361
2. Lipton RB, Stewart WF, Diamond S, Diamond ML, Reed M (2001) Prevalence and burden of migraine in the United States: data from the American Migraine Study II. Headache 41(7):646–657
3. Menken M, Munsat TL, Toole JF (2000) The global burden of disease study: implications for neurology. Arch Neurol 57(3):418–420
4. Murray CJ, Vos T, Lozano R, Naghavi M, Flaxman AD, Michaud C et al (2012) Disability-adjusted life years (DALYs) for 291 diseases and injuries in 21 regions, 1990-2010: a systematic analysis for the Global Burden of Disease Study 2010. Lancet 380(9859):2197–2223
5. Lipton RB, Stewart WF, Scher AI (2001) Epidemiology and economic impact of migraine. Curr Med Res Opin 17(Suppl 1):s4–s12
6. Ligthart L, de Vries B, Smith AV, Ikram MA, Amin N, Hottenga JJ et al (2011) Meta-analysis of genome-wide association for migraine in six population-based European cohorts. Eur J Hum Genet 19(8):901–907
7. Anttila V, Stefansson H, Kallela M, Todt U, Terwindt GM, Calafato MS et al (2010) Genome-wide association study of migraine implicates a common susceptibility variant on 8q22.1. Nat Genet 42(10):869–873
8. Lafreniere RG, Cader MZ, Poulin JF, Andres-Enguix I, Simoneau M, Gupta N et al (2010) A dominant-negative mutation in the TRESK potassium channel is linked to familial migraine with aura. Nat Med 16(10):1157–1160
9. Deprez L, Peeters K, Van Paesschen W, Claeys KG, Claes LR, Suls A et al (2007) Familial occipitotemporal lobe epilepsy and migraine with visual aura: linkage to chromosome 9q. Neurology 68(23):1995–2002
10. Tikka-Kleemola P, Artto V, Vepsalainen S, Sobel EM, Raty S, Kaunisto MA et al (2010) A visual migraine aura locus maps to 9q21-q22. Neurology 74(15):1171–1177
11. Chasman DI, Schurks M, Anttila V, de Vries B, Schminke U, Launer LJ et al (2011) Genome-wide association study reveals three susceptibility loci for common migraine in the general population. Nat Genet 43(7):695–698
12. Esserlind AL, Christensen AF, Le H, Kirchmann M, Hauge AW, Toyserkani NM et al (2013) Replication and meta-analysis of common variants identifies a genome-wide significant locus in migraine. Eur J Neurol 20(5):765–772
13. Freilinger T, Anttila V, de Vries B, Malik R, Kallela M, Terwindt GM et al (2012) Genome-wide association analysis identifies susceptibility loci for migraine without aura. Nat Genet 44(7):777–782
14. IHS (2013) The international classification of headache disorders, 3rd edition (beta version). Cephalalgia 33(9):629–808
15. Ophoff RA, Terwindt GM, Vergouwe MN, van Eijk R, Oefner PJ, Hoffman SM et al (1996) Familial hemiplegic migraine and episodic ataxia type-2 are caused by mutations in the Ca2+ channel gene CACNL1A4. Cell 87(3):543–552

16. Dichgans M, Freilinger T, Eckstein G, Babini E, Lorenz-Depiereux B, Biskup S et al (2005) Mutation in the neuronal voltage-gated sodium channel SCN1A in familial hemiplegic migraine. Lancet 366(9483):371–377
17. De Fusco M, Marconi R, Silvestri L, Atorino L, Rampoldi L, Morgante L et al (2003) Haploinsufficiency of ATP1A2 encoding the Na+/K+ pump alpha2 subunit associated with familial hemiplegic migraine type 2. Nat Genet 33(2):192–196
18. Cuenca-Leon E, Corominas R, Montfort M, Artigas J, Roig M, Bayes M et al (2009) Familial hemiplegic migraine: linkage to chromosome 14q32 in a Spanish kindred. Neurogenetics 10(3):191–198
19. Riant F, Roze E, Barbance C, Meneret A, Guyant-Marechal L, Lucas C et al (2012) PRRT2 mutations cause hemiplegic migraine. Neurology 79(21):2122–2124
20. Pietrobon D, Moskowitz MA (2014) Chaos and commotion in the wake of cortical spreading depression and spreading depolarizations. Nat Rev Neurosci 15(6):379–393
21. Lauritzen M (1994) Pathophysiology of the migraine aura. The spreading depression theory. Brain 117(Pt 1):199–210
22. van den Maagdenberg AM, Pietrobon D, Pizzorusso T, Kaja S, Broos LA, Cesetti T et al (2004) A Cacna1a knock-in migraine mouse model with increased susceptibility to cortical spreading depression. Neuron 41(5):701–710
23. Eikermann-Haerter K, Dilekoz E, Kudo C, Savitz SI, Waeber C, Baum MJ et al (2009) Genetic and hormonal factors modulate spreading depression and transient hemiparesis in mouse models of familial hemiplegic migraine type 1. J Clin Invest 119(1):99–109
24. van den Maagdenberg AM, Pizzorusso T, Kaja S, Terpolilli N, Shapovalova M, Hoebeek FE et al (2010) High cortical spreading depression susceptibility and migraine-associated symptoms in Ca(v)2.1 S218L mice. Ann Neurol 67(1):85–98
25. Leo L, Gherardini L, Barone V, De Fusco M, Pietrobon D, Pizzorusso T et al (2011) Increased susceptibility to cortical spreading depression in the mouse model of familial hemiplegic migraine type 2. PLoS Genet 7(6):e1002129
26. Bolay H, Reuter U, Dunn AK, Huang Z, Boas DA, Moskowitz MA (2002) Intrinsic brain activity triggers trigeminal meningeal afferents in a migraine model. Nat Med 8(2): 136–142
27. Chanda ML, Tuttle AH, Baran I, Atlin C, Guindi D, Hathaway G et al (2013) Behavioral evidence for photophobia and stress-related ipsilateral head pain in transgenic Cacna1a mutant mice. Pain 154(8):1254–1262
28. Hullugundi SK, Ansuini A, Ferrari MD, van den Maagdenberg AM, Nistri A (2014) A hyper-excitability phenotype in mouse trigeminal sensory neurons expressing the R192Q Cacna1a missense mutation of familial hemiplegic migraine type-1 (FHM1). Neuroscience 266: 244–254
29. Xu Y, Padiath QS, Shapiro RE, Jones CR, Wu SC, Saigoh N et al (2005) Functional consequences of a CKIdelta mutation causing familial advanced sleep phase syndrome. Nature 434(7033):640–644
30. Brennan KC, Bates EA, Shapiro RE, Zyuzin J, Hallows WC, Huang Y et al (2013) Casein kinase idelta mutations in familial migraine and advanced sleep phase. Sci Transl Med 5(183):183ra56, 1–11
31. Thomsen LL, Kruuse C, Iversen HK, Olesen J (1994) A nitric oxide donor (nitroglycerin) triggers genuine migraine attacks. Eur J Neurol 1(1):73–80
32. Baca S, Barth A, Mody I, Charles A (ed) (2014) Optogenetic elicitation of cortical spreading depression in unanesthetized, head-restrained mice. 4th European Headache and Migraine Trust International Congress: EHMTIC 2014, Copenhagen. 18 Sept 2014
33. van Oosterhout F, Michel S, Deboer T, Houben T, van de Ven RC, Albus H et al (2008) Enhanced circadian phase resetting in R192Q Cav2.1 calcium channel migraine mice. Ann Neurol 64(3):315–324
34. Kirchmann M, Thomsen LL, Olesen J (2006) The CACNA1A and ATP1A2 genes are not involved in dominantly inherited migraine with aura. Am J Med Genet B Neuropsychiatr Genet 141B(3):250–256

35. Netzer C, Todt U, Heinze A, Freudenberg J, Zumbroich V, Becker T et al (2006) Haplotype-based systematic association studies of ATP1A2 in migraine with aura. Am J Med Genet B Neuropsychiatr Genet 141B(3):257–260
36. Jen JC, Kim GW, Dudding KA, Baloh RW (2004) No mutations in CACNA1A and ATP1A2 in probands with common types of migraine. Arch Neurol 61(6):926–928
37. Wieser T, Mueller C, Evers S, Zierz S, Deufel T (2003) Absence of known familial hemiplegic migraine (FHM) mutations in the CACNA1A gene in patients with common migraine: implications for genetic testing. Clin Chem Lab Med 41(3):272–275
38. Martin VT, Behbehani MM (2001) Toward a rational understanding of migraine trigger factors. Med Clin North Am 85(4):911–941
39. Andress-Rothrock D, King W, Rothrock J (2010) An analysis of migraine triggers in a clinic-based population. Headache 50(8):1366–1370
40. Pavlovic JM, Buse DC, Sollars CM, Haut S, Lipton RB (2014) Trigger factors and premonitory features of migraine attacks: summary of studies. Headache 54(10):1670–1679
41. Lipton RB, Pavlovic JM, Haut SR, Grosberg BM, Buse DC (2014) Methodological issues in studying trigger factors and premonitory features of migraine. Headache 54(10):1661–1669
42. Ierusalimschy R, Moreira Filho PF (2002) Precipitating factors of migraine attacks in patients with migraine without aura. Arq Neuropsiquiatr 60(3-A):609–613
43. Hauge A, Kirchmann M, Olesen J (2010) Trigger factors in migraine with aura. Cephalalgia 30(3):346–353
44. Hansen JM, Hauge AW, Ashina M, Olesen J (2011) Trigger factors for familial hemiplegic migraine. Cephalalgia 31(12):1274–1281
45. Kelman L (2007) The triggers or precipitants of the acute migraine attack. Cephalalgia 27(5):394–402
46. Yadav RK, Kalita J, Misra UK (2010) A study of triggers of migraine in India. Pain Med 11(1):44–47
47. Friedman DI, De ver Dye T (2009) Migraine and the environment. Headache 49(6):941–952
48. Levy D (2012) Endogenous mechanisms underlying the activation and sensitization of meningeal nociceptors: the role of immuno-vascular interactions and cortical spreading depression. Curr Pain Headache Rep 16(3):270–277
49. Noseda R, Burstein R (2013) Migraine pathophysiology: anatomy of the trigeminovascular pathway and associated neurological symptoms, CSD, sensitization and modulation of pain. Pain 154 Suppl 1:10.1016/j.pain.2013.07.021
50. Hougaard A, Amin FM, Hauge AW, Ashina M, Olesen J (2013) Provocation of migraine with aura using natural trigger factors. Neurology 80(5):428–431
51. Moffett AM, Swash M, Scott DF (1974) Effect of chocolate in migraine: a double-blind study. J Neurol Neurosurg Psychiatry 37(4):445–448
52. Marcus DA, Scharff L, Turk D, Gourley LM (1997) A double-blind provocative study of chocolate as a trigger of headache. Cephalalgia 17(8):855–862; discussion 00
53. Gibb CM, Davies PT, Glover V, Steiner TJ, Clifford Rose F, Sandler M (1991) Chocolate is a migraine-provoking agent. Cephalalgia 11(2):93–95
54. May A, Goadsby PJ (1999) The trigeminovascular system in humans: pathophysiologic implications for primary headache syndromes of the neural influences on the cerebral circulation. J Cereb Blood Flow Metab 19(2):115–127
55. Edvinsson L, Goadsby PJ (1995) Neuropeptides in the cerebral circulation: relevance to headache. Cephalalgia 15(4):272–276
56. Pietrobon D, Moskowitz MA (2013) Pathophysiology of migraine. Annu Rev Physiol 75:365–391
57. Edvinsson L, Petersen KA (2007) CGRP-receptor antagonism in migraine treatment. CNS Neurol Disord Drug Targets 6(4):240–246
58. Burstein R, Jakubowski M (2005) Unitary hypothesis for multiple triggers of the pain and strain of migraine. J Comp Neurol 493(1):9–14
59. Amara SG, Jonas V, Rosenfeld MG, Ong ES, Evans RM (1982) Alternative RNA processing in calcitonin gene expression generates mRNAs encoding different polypeptide products. Nature 298(5871):240–244

60. Rosenfeld MG, Mermod JJ, Amara SG, Swanson LW, Sawchenko PE, Rivier J et al (1983) Production of a novel neuropeptide encoded by the calcitonin gene via tissue-specific RNA processing. Nature 304(5922):129–135
61. van Rossum D, Hanisch UK, Quirion R (1997) Neuroanatomical localization, pharmacological characterization and functions of CGRP, related peptides and their receptors. Neurosci Biobehav Rev 21(5):649–678
62. Tajti J, Uddman R, Edvinsson L (2001) Neuropeptide localization in the "migraine generator" region of the human brainstem. Cephalalgia 21(2):96–101
63. Eftekhari S, Warfvinge K, Blixt FW, Edvinsson L (2013) Differentiation of nerve fibers storing CGRP and CGRP receptors in the peripheral trigeminovascular system. J Pain 14(11):1289–1303
64. Eftekhari S, Edvinsson L (2011) Calcitonin gene-related peptide (CGRP) and its receptor components in human and rat spinal trigeminal nucleus and spinal cord at C1-level. BMC Neurosci 12:112
65. Hostetler ED, Joshi AD, Sanabria-Bohorquez S, Fan H, Zeng Z, Purcell M et al (2013) In vivo quantification of calcitonin gene-related peptide (CGRP) receptor occupancy by telcagepant in rhesus monkey and human brain using the positron emission tomography (PET) tracer [11C]MK-4232. J Pharmacol Exp Ther 347(2):478–486
66. Cumberbatch MJ, Williamson DJ, Mason GS, Hill RG, Hargreaves RJ (1999) Dural vasodilation causes a sensitization of rat caudal trigeminal neurones in vivo that is blocked by a 5-HT1B/1D agonist. Br J Pharmacol 126(6):1478–1486
67. Storer RJ, Akerman S, Goadsby PJ (2004) Calcitonin gene-related peptide (CGRP) modulates nociceptive trigeminovascular transmission in the cat. Br J Pharmacol 142(7):1171–1181
68. Summ O, Charbit AR, Andreou AP, Goadsby PJ (2010) Modulation of nociceptive transmission with calcitonin gene-related peptide receptor antagonists in the thalamus. Brain 133(Pt 9):2540–2548
69. Zaidi M, Bevis PJ, Abeyasekera G, Girgis SI, Wimalawansa SJ, Morris HR et al (1986) The origin of circulating calcitonin gene-related peptide in the rat. J Endocrinol 110(1):185–190
70. Hoffmann J, Wecker S, Neeb L, Dirnagl U, Reuter U (2012) Primary trigeminal afferents are the main source for stimulus-induced CGRP release into jugular vein blood and CSF. Cephalalgia 32(9):659–667
71. Levy D, Burstein R, Strassman AM (2005) Calcitonin gene-related peptide does not excite or sensitize meningeal nociceptors: implications for the pathophysiology of migraine. Ann Neurol 58(5):698–705
72. Pedersen-Bjergaard U, Nielsen LB, Jensen K, Edvinsson L, Jansen I, Olesen J (1991) Calcitonin gene-related peptide, neurokinin A and substance P: effects on nociception and neurogenic inflammation in human skin and temporal muscle. Peptides 12(2):333–337
73. Sun RQ, Lawand NB, Willis WD (2003) The role of calcitonin gene-related peptide (CGRP) in the generation and maintenance of mechanical allodynia and hyperalgesia in rats after intradermal injection of capsaicin. Pain 104(1–2):201–208
74. Sun RQ, Lawand NB, Lin Q, Willis WD (2004) Role of calcitonin gene-related peptide in the sensitization of dorsal horn neurons to mechanical stimulation after intradermal injection of capsaicin. J Neurophysiol 92(1):320–326
75. Mao J, Coghill RC, Kellstein DE, Frenk H, Mayer DJ (1992) Calcitonin gene-related peptide enhances substance P-induced behaviors via metabolic inhibition: in vivo evidence for a new mechanism of neuromodulation. Brain Res 574(1–2):157–163
76. Oku R, Satoh M, Fujii N, Otaka A, Yajima H, Takagi H (1987) Calcitonin gene-related peptide promotes mechanical nociception by potentiating release of substance P from the spinal dorsal horn in rats. Brain Res 403(2):350–354
77. Russo AF (2014) Calcitonin gene-related peptide (CGRP): a new target for migraine. Annu Rev Pharmacol Toxicol 55:533–552
78. Goadsby PJ, Edvinsson L, Ekman R (1988) Release of vasoactive peptides in the extracerebral circulation of humans and the cat during activation of the trigeminovascular system. Ann Neurol 23(2):193–196

79. Goadsby PJ, Edvinsson L, Ekman R (1990) Vasoactive peptide release in the extracerebral circulation of humans during migraine headache. Ann Neurol 28(2):183–187
80. Ashina M, Bendtsen L, Jensen R, Schifter S, Olesen J (2000) Evidence for increased plasma levels of calcitonin gene-related peptide in migraine outside of attacks. Pain 86(1–2): 133–138
81. Tvedskov JF, Lipka K, Ashina M, Iversen HK, Schifter S, Olesen J (2005) No increase of calcitonin gene-related peptide in jugular blood during migraine. Ann Neurol 58(4): 561–568
82. Ho TW, Mannix LK, Fan X, Assaid C, Furtek C, Jones CJ et al (2008) Randomized controlled trial of an oral CGRP receptor antagonist, MK-0974, in acute treatment of migraine. Neurology 70(16):1304–1312
83. Ho TW, Ferrari MD, Dodick DW, Galet V, Kost J, Fan X et al (2008) Efficacy and tolerability of MK-0974 (telcagepant), a new oral antagonist of calcitonin gene-related peptide receptor, compared with zolmitriptan for acute migraine: a randomised, placebo-controlled, parallel-treatment trial. Lancet 372(9656):2115–2123
84. Olesen J, Diener HC, Husstedt IW, Goadsby PJ, Hall D, Meier U et al (2004) Calcitonin gene-related peptide receptor antagonist BIBN 4096 BS for the acute treatment of migraine. N Engl J Med 350(11):1104–1110
85. Connor KM, Shapiro RE, Diener HC, Lucas S, Kost J, Fan X et al (2009) Randomized, controlled trial of telcagepant for the acute treatment of migraine. Neurology 73(12):970–977
86. Ho TW, Connor KM, Zhang Y, Pearlman E, Koppenhaver J, Fan X et al (2014) Randomized controlled trial of the CGRP receptor antagonist telcagepant for migraine prevention. Neurology 83(11):958–966
87. Sixt ML, Messlinger K, Fischer MJ (2009) Calcitonin gene-related peptide receptor antagonist olcegepant acts in the spinal trigeminal nucleus. Brain 132(11):3134–3141
88. Dodick DW, Goadsby PJ, Silberstein SD, Lipton RB, Olesen J, Ashina M et al (2014) Safety and efficacy of ALD403, an antibody to calcitonin gene-related peptide, for the prevention of frequent episodic migraine: a randomised, double-blind, placebo-controlled, exploratory phase 2 trial. Lancet Neurol 13(11):1100–1107
89. Dodick DW, Goadsby PJ, Spierings EL, Scherer JC, Sweeney SP, Grayzel DS (2014) Safety and efficacy of LY2951742, a monoclonal antibody to calcitonin gene-related peptide, for the prevention of migraine: a phase 2, randomised, double-blind, placebo-controlled study. Lancet Neurol 13(9):885–892
90. Jansen-Olesen I, Gulbenkian S, Engel U, Cunha e Sa M, Edvinsson L (2004) Peptidergic and non-peptidergic innervation and vasomotor responses of human lenticulostriate and posterior cerebral arteries. Peptides 25(12):2105–2114
91. Baeres FM, Moller M (2004) Origin of PACAP-immunoreactive nerve fibers innervating the subarachnoidal blood vessels of the rat brain. J Cereb Blood Flow Metab 24(6):628–635
92. Gulbenkian S, Uddman R, Edvinsson L (2001) Neuronal messengers in the human cerebral circulation. Peptides 22(6):995–1007
93. Hansen JM, Sitarz J, Birk S, Rahmann AM, Oturai PS, Fahrenkrug J et al (2006) Vasoactive intestinal polypeptide evokes only a minimal headache in healthy volunteers. Cephalalgia 26(8):992–1003
94. Rahmann A, Wienecke T, Hansen JM, Fahrenkrug J, Olesen J, Ashina M (2008) Vasoactive intestinal peptide causes marked cephalic vasodilation, but does not induce migraine. Cephalalgia 28(3):226–236
95. Birk S, Sitarz JT, Petersen KA, Oturai PS, Kruuse C, Fahrenkrug J et al (2007) The effect of intravenous PACAP38 on cerebral hemodynamics in healthy volunteers. Regul Pept 140(3):185–191
96. Schytz HW, Birk S, Wienecke T, Kruuse C, Olesen J, Ashina M (2009) PACAP38 induces migraine-like attacks in patients with migraine without aura. Brain 132(Pt 1):16–25
97. Amin FM, Hougaard A, Schytz HW, Asghar MS, Lundholm E, Parvaiz AI et al (2014) Investigation of the pathophysiological mechanisms of migraine attacks induced by pituitary adenylate cyclase-activating polypeptide-38. Brain 137(Pt 3):779–794

98. Hosoya M, Onda H, Ogi K, Masuda Y, Miyamoto Y, Ohtaki T et al (1993) Molecular cloning and functional expression of rat cDNAs encoding the receptor for pituitary adenylate cyclase activating polypeptide (PACAP). Biochem Biophys Res Commun 194(1):133–143
99. Lutz EM, Sheward WJ, West KM, Morrow JA, Fink G, Harmar AJ (1993) The VIP2 receptor: molecular characterisation of a cDNA encoding a novel receptor for vasoactive intestinal peptide. FEBS Lett 334(1):3–8
100. Harmar AJ, Arimura A, Gozes I, Journot L, Laburthe M, Pisegna JR et al (1998) International Union of Pharmacology. XVIII. Nomenclature of receptors for vasoactive intestinal peptide and pituitary adenylate cyclase-activating polypeptide. Pharmacol Rev 50(2):265–270
101. Schytz HW, Olesen J, Ashina M (2010) The PACAP receptor: a novel target for migraine treatment. Neurotherapeutics 7(2):191–196
102. Giffin NJ, Ruggiero L, Lipton RB, Silberstein SD, Tvedskov JF, Olesen J et al (2003) Premonitory symptoms in migraine: an electronic diary study. Neurology 60(6):935–940
103. Kelman L (2004) The premonitory symptoms (prodrome): a tertiary care study of 893 migraineurs. Headache 44(9):865–872
104. Schoonman GG, Evers DJ, Terwindt GM, van Dijk JG, Ferrari MD (2006) The prevalence of premonitory symptoms in migraine: a questionnaire study in 461 patients. Cephalalgia 26(10):1209–1213
105. Charles A (2013) The evolution of a migraine attack – a review of recent evidence. Headache 53(2):413–419
106. Maniyar FH, Sprenger T, Schankin C, Goadsby PJ (2014) Photic hypersensitivity in the premonitory phase of migraine–a positron emission tomography study. Eur J Neurol 21(9):1178–1183
107. Salazar G, Fragoso M, Vergez L, Sergio P, Cuello D (2011) Metoclopramide as an analgesic in severe migraine attacks: an open, single-blind, parallel control study. Recent Pat CNS Drug Discov 6(2):141–145
108. Tfelt-Hansen P, Henry P, Mulder LJ, Scheldewaert RG, Schoenen J, Chazot G (1995) The effectiveness of combined oral lysine acetylsalicylate and metoclopramide compared with oral sumatriptan for migraine. Lancet 346(8980):923–926
109. Bergerot A, Storer RJ, Goadsby PJ (2007) Dopamine inhibits trigeminovascular transmission in the rat. Ann Neurol 61(3):251–262
110. Cao Y, Aurora SK, Nagesh V, Patel SC, Welch KM (2002) Functional MRI-BOLD of brainstem structures during visually triggered migraine. Neurology 59(1):72–78
111. Maniyar FH, Sprenger T, Monteith T, Schankin C, Goadsby PJ (2014) Brain activations in the premonitory phase of nitroglycerin-triggered migraine attacks. Brain 137(Pt 1):232–241
112. Afridi SK, Kaube H, Goadsby PJ (2004) Glyceryl trinitrate triggers premonitory symptoms in migraineurs. Pain 110(3):675–680
113. Borsook D, Burstein R (2012) The enigma of the dorsolateral pons as a migraine generator. Cephalalgia 32(11):803–812
114. Moulton EA, Becerra L, Johnson A, Burstein R, Borsook D (2014) Altered hypothalamic functional connectivity with autonomic circuits and the locus coeruleus in migraine. PLoS One 9(4):e95508
115. Holland P, Goadsby PJ (2007) The hypothalamic orexinergic system: pain and primary headaches. Headache 47(6):951–962
116. Hoffmann J, Supronsinchai W, Akerman S, Andreou AP, Winrow CJ, Renger J et al (2014) Evidence for orexinergic mechanisms in migraine. Neurobiol Dis 74C:137–143
117. Chabi A, Zhang Y, Jackson S, Cady R, Lines C, Herring WJ et al (2014) Randomized controlled trial of the orexin receptor antagonist filorexant for migraine prophylaxis. Cephalalgia 2014 Aug 8. pii: 0333102414544979. doi: 10.1177/0333102414544979 [Epub ahead of print]
118. Russell MB, Rasmussen BK, Thorvaldsen P, Olesen J (1995) Prevalence and sex-ratio of the subtypes of migraine. Int J Epidemiol 24(3):612–618
119. Russell MB, Olesen J (1996) A nosographic analysis of the migraine aura in a general population. Brain 119(Pt 2):355–361
120. Leão AAP (1944) Spreading depression of activity in the cerebral cortex. J Neurophysiol 7(6):359–390

121. Charles A, Brennan K (2009) Cortical spreading depression-new insights and persistent questions. Cephalalgia 29(10):1115–1124
122. Woitzik J, Hecht N, Pinczolits A, Sandow N, Major S, Winkler MK et al (2013) Propagation of cortical spreading depolarization in the human cortex after malignant stroke. Neurology 80(12):1095–1102
123. Drenckhahn C, Winkler MK, Major S, Scheel M, Kang EJ, Pinczolits A et al (2012) Correlates of spreading depolarization in human scalp electroencephalography. Brain 135(Pt 3):853–868
124. Strong AJ, Fabricius M, Boutelle MG, Hibbins SJ, Hopwood SE, Jones R et al (2002) Spreading and synchronous depressions of cortical activity in acutely injured human brain. Stroke 33(12):2738–2743
125. Hadjikhani N, Sanchez Del Rio M, Wu O, Schwartz D, Bakker D, Fischl B et al (2001) Mechanisms of migraine aura revealed by functional MRI in human visual cortex. Proc Natl Acad Sci U S A 98(8):4687–4692
126. Zhang X, Levy D, Noseda R, Kainz V, Jakubowski M, Burstein R (2010) Activation of meningeal nociceptors by cortical spreading depression: implications for migraine with aura. J Neurosci 30(26):8807–8814
127. Zhang X, Levy D, Kainz V, Noseda R, Jakubowski M, Burstein R (2011) Activation of central trigeminovascular neurons by cortical spreading depression. Ann Neurol 69(5):855–865
128. Karatas H, Erdener SE, Gursoy-Ozdemir Y, Lule S, Eren-Kocak E, Sen ZD et al (2013) Spreading depression triggers headache by activating neuronal Panx1 channels. Science 339(6123):1092–1095
129. Noseda R, Constandil L, Bourgeais L, Chalus M, Villanueva L (2010) Changes of meningeal excitability mediated by corticotrigeminal networks: a link for the endogenous modulation of migraine pain. J Neurosci 30(43):14420–14429
130. Hansen JM, Lipton RB, Dodick DW, Silberstein SD, Saper JR, Aurora SK et al (2012) Migraine headache is present in the aura phase: a prospective study. Neurology 79(20):2044–2049
131. Wolff H (1963) Headache and other head pain. Oxford University Press, New York
132. Mayberg M, Langer RS, Zervas NT, Moskowitz MA (1981) Perivascular meningeal projections from cat trigeminal ganglia: possible pathway for vascular headaches in man. Science 213(4504):228–230
133. Liu-Chen LY, Mayberg MR, Moskowitz MA (1983) Immunohistochemical evidence for a substance P-containing trigeminovascular pathway to pial arteries in cats. Brain Res 268(1):162–166
134. Olesen J, Burstein R, Ashina M, Tfelt-Hansen P (2009) Origin of pain in migraine: evidence for peripheral sensitisation. Lancet Neurol 8(7):679–690
135. Cushing H (1904) The sensory distribution of the fifth cranial nerve. Bull Johns Hopk Hosp XV:213–232
136. Ray B, Wolff H (1940) Experimental studies on headache. Pain sensitive structures of the head and their significance in headache. Arch Surg 41:813–856
137. Lassen LH, Jacobsen VB, Haderslev PA, Sperling B, Iversen HK, Olesen J et al (2008) Involvement of calcitonin gene-related peptide in migraine: regional cerebral blood flow and blood flow velocity in migraine patients. J Headache Pain 9(3):151–157
138. Schoonman GG, van der Grond J, Kortmann C, van der Geest RJ, Terwindt GM, Ferrari MD (2008) Migraine headache is not associated with cerebral or meningeal vasodilatation–a 3T magnetic resonance angiography study. Brain 131(Pt 8):2192–2200
139. Asghar MS, Hansen AE, Amin FM, van der Geest RJ, van der Koning P, Larsson HBW et al (2011) Evidence for a vascular factor in migraine. Ann Neurol 69(4):635–645
140. Weiller C, May A, Limmroth V, Juptner M, Kaube H, Schayck RV et al (1995) Brain stem activation in spontaneous human migraine attacks. Nat Med 1(7):658–660
141. Afridi SK, Giffin NJ, Kaube H, Friston KJ, Ward NS, Frackowiak RS et al (2005) A positron emission tomographic study in spontaneous migraine. Arch Neurol 62(8):1270–1275
142. Ahn AH (2010) On the temporal relationship between throbbing migraine pain and arterial pulse. Headache 50(9):1507–1510

143. Amin FM, Asghar MS, Hougaard A, Hansen AE, Larsen VA, de Koning PJ et al (2013) Magnetic resonance angiography of intracranial and extracranial arteries in patients with spontaneous migraine without aura: a cross-sectional study. Lancet Neurol 12(5):454–461
144. Farkkila M, Diener HC, Geraud G, Lainez M, Schoenen J, Harner N et al (2012) Efficacy and tolerability of lasmiditan, an oral 5-HT(1F) receptor agonist, for the acute treatment of migraine: a phase 2 randomised, placebo-controlled, parallel-group, dose-ranging study. Lancet Neurol 11(5):405–413
145. Goadsby PJ, Charbit AR, Andreou AP, Akerman S, Holland PR (2009) Neurobiology of migraine. Neuroscience 161(2):327–341
146. Goadsby PJ, Akerman S (2012) The trigeminovascular system does not require a peripheral sensory input to be activated–migraine is a central disorder. Focus on 'Effect of cortical spreading depression on basal and evoked traffic in the trigeminovascular sensory system'. Cephalalgia 32(1):3–5
147. Burstein R, Strassman A, Moskowitz M (2012) Can cortical spreading depression activate central trigeminovascular neurons without peripheral input? Pitfalls of a new concept. Cephalalgia 32(6):509–511
148. Levy D (2010) Migraine pain and nociceptor activation–where do we stand? Headache 50(5):909–916
149. Iversen HK, Olesen J, Tfelt-Hansen P (1989) Intravenous nitroglycerin as an experimental model of vascular headache. Basic characteristics. Pain 38(1):17–24
150. Ashina M, Hansen JM (2010) Pharmacological migraine provocation: a human model of migraine. Handb Clin Neurol 97:773–779
151. Olesen J, Tfelt-Hansen P, Ashina M (2009) Finding new drug targets for the treatment of migraine attacks. Cephalalgia 29(9):909–920
152. Lassen LH, Ashina M, Christiansen I, Ulrich V, Olesen J (1997) Nitric oxide synthase inhibition in migraine. Lancet 349(9049):401–402
153. Read SJ, Hirst WD, Upton N, Parsons AA (2001) Cortical spreading depression produces increased cGMP levels in cortex and brain stem that is inhibited by tonabersat (SB-220453) but not sumatriptan. Brain Res 891(1–2):69–77
154. Schwedt TJ, Larson-Prior L, Coalson RS, Nolan T, Mar S, Ances BM et al (2013) Allodynia and descending pain modulation in migraine: a resting state functional connectivity analysis. Pain Med 15(1):154–165
155. Mainero C, Boshyan J, Hadjikhani N (2011) Altered functional magnetic resonance imaging resting-state connectivity in periaqueductal gray networks in migraine. Ann Neurol 70(5):838–845
156. Shuhendler AJ, Lee S, Siu M, Ondovcik S, Lam K, Alabdullatif A et al (2009) Efficacy of botulinum toxin type A for the prophylaxis of episodic migraine headaches: a meta-analysis of randomized, double-blind, placebo-controlled trials. Pharmacotherapy 29(7):784–791
157. Jackson JL, Kuriyama A, Hayashino Y (2012) Botulinum toxin A for prophylactic treatment of migraine and tension headaches in adults: a meta-analysis. JAMA 307(16):1736–1745
158. Burstein R, Zhang X, Levy D, Aoki KR, Brin MF (2014) Selective inhibition of meningeal nociceptors by botulinum neurotoxin type A: therapeutic implications for migraine and other pains. Cephalalgia 34(11):853–869
159. Dahlem MA, Hadjikhani N (2009) Migraine aura: retracting particle-like waves in weakly susceptible cortex. PLoS One 4(4):e5007
160. Fauci A, Braunwald E, Kasper DL et al (2008) Harrison's principles of internal medicine, 17th edn. McGraw-Hill, New York

Chapter 12
Pathophysiology of TTH: Current Status and Future Directions

Sait Ashina and Lars Bendtsen

12.1 Introduction

Tension-type headache (TTH) is a highly prevalent primary headache with enormous socioeconomic costs [1, 2]. Frequent or chronic TTH has a major impact on both the individual and population levels. Our understanding of mechanisms of TTH has improved due to recent advances in basic, clinical, and epidemiologic research. Abnormalities in peripheral and central nociceptive systems in combination with environmental, emotional, and genetic factors are involved in the pathophysiology of TTH. Risk factors for poor prognosis and/or chronification of TTH have been identified in longitudinal epidemiological studies.

12.2 Psychological Factors

Emotional factors such as stress and mental tension have been shown to be risk factors for the development of TTH. A positive correlation between headache and stress was demonstrated in patients with TTH [3, 4]. Stress also induces more

S. Ashina, MD (✉)
Headache Program, Department of Neurology, Mount Sinai Beth Israel,
Icahn School of Medicine at Mount Sinai, 10 Union Square East, Suite 2Q/R,
New York, NY 10003, USA

Department of Neurology, Danish Headache Center, Glostrup Hospital,
University of Copenhagen, Glostrup, Copenhagen, Denmark
e-mail: sashina@chpnet.org

L. Bendtsen, MD, PhD
Department of Neurology, Danish Headache Center, Glostrup Hospital,
University of Copenhagen, Glostrup, Copenhagen, Denmark

M. Ashina, P. Geppetti (eds.), *Pathophysiology of Headaches: From Molecule to Man*, Headache, DOI 10.1007/978-3-319-15621-7_12

headache in patients with CTTH than in healthy controls possibly by inducing hyperalgesia in the setting of already sensitized nociceptive pathways [5, 6]. Interestingly, stress did not enhance abnormal temporal summation of pain, a clinical correlate of wind-up of nociception in humans, or modulate diffuse noxious inhibitory control [7]. Clinical depression is a common psychiatric condition in frequent TTH and may possibly be involved in the pathophysiological mechanisms of TTH, but the relationship between these two conditions can be bidirectional [8]. Janke et al. [9] demonstrated that depression increased vulnerability to TTH in patients with frequent headaches during and following the laboratory stress test. Moreover, depression was associated with increased pericranial muscle tenderness. Thus, depression may contribute to central sensitization or increased excitability of central pain pathways in patients with CTTH [9, 10]. Furthermore, maladaptive coping strategies such as catastrophizing and avoidance seem to be common in TTH [11]. The neurobiological mechanisms through which emotional factors contribute to TTH are not fully clarified. Further studies are warranted and needed to explore the possible relation of corticolimbic circuits to central sensitization in TTH.

12.3 Genetic Factors

Genetic factors seem to play a role in the pathophysiology of TTH, but the inheritance in this type of headache is likely polygenic due to high prevalence and variability in frequency. In CTTH, population-relative risk in first- and second-degree relatives was increased threefold compared to controls [12, 13]. The mode of inheritance in CTTH was investigated by complex segregation analysis [13]. The analysis demonstrated the multifactorial inheritance in CTTH supporting multifactorial model. The relative importance of genetic and environmental influence for the development of TTH was studied in population of twins [14]. The environmental influence was found to be of major importance for episodic TTH (ETTH), and a genetic factor was minor. Currently, we believe that the majority of the population, perhaps all, has the potential to develop TTH if exposed to sufficiently strong environmental risk factors.

12.4 Peripheral Factors

12.4.1 Muscle Tenderness

Peripheral factors involving muscles and peripheral nociceptors have long been considered of importance in the pathophysiology of TTH. Increased tenderness in pericranial muscles during attacks and headache-free periods is a frequent finding and abnormality in patients with ETTH and CTTH [15–21]. Tenderness has been demonstrated to be uniformly increased throughout the pericranial region and is positively associated with both the intensity and the frequency of TTH [17, 19, 20]. Tenderness and hardness have been found to be increased both on days with and without headache, indicating that hardness is not the consequence of actual headache [10, 22–24]. A recent series of pilot studies reported an increased number of active trigger points both in patients with frequent ETTH and in patients with CTTH [25–27].

12.4.2 Muscle Strain

Sustained pericranial muscle contraction has long been suggested to be an etiologic factor in TTH. Sustained experimental tooth clenching was reported to induce more headaches in TTH patients than healthy controls [28], and these patients are more likely to develop shoulder and neck pain in response to static exercise than controls [29]. Several laboratory-based electromyographic (EMG) studies have reported normal or only slightly increased muscle activity in TTH [30]. However, EMG activity has been reported to be increased in myofascial trigger points [31], and it is possible that continuous activity in a few motor units over a long time could be sufficient for excitation or sensitization of peripheral nociceptors.

12.4.3 Muscle Blood Flow

A microdialysis study in TTH demonstrated an increase in muscle blood flow during exercise that was lower in patients than in controls, but lactate levels in a tender site in the trapezius muscle did not differ between patients and controls [32]. It was suggested that the altered blood flow was caused by altered sympathetic outflow to blood vessels in striated muscle secondary to central sensitization of nociceptive pathways.

12.4.4 Inflammation

The increased myofascial pain sensitivity in TTH could be secondary to the release of inflammatory mediators resulting in excitation and sensitization of peripheral sensory afferents. Infusion of a combination of endogenous substances into the trapezius muscle resulted in more pain in patients with frequent ETTH compared to controls [33]. The “inflammatory” hypothesis has been challenged by Ashina et al. [34] who demonstrated that the in vivo interstitial concentrations of adenosine 5-triphosphate, glutamate, glucose, pyruvate, urea, and prostaglandin E_2 in tender muscles during rest and static exercise did not differ between patients with CTTH and controls.

12.4.5 Summary of Peripheral Factors

Overall, previous studies indicate that muscle pain in TTH is not caused by generalized excessive muscle contraction and muscle ischemia. It cannot be excluded that a locally increased muscle tone without EMG activity may result in micro-trauma of muscle fibers and tendon insertions or that excessive activity in a few motor units may excite or sensitize peripheral nociceptors. Pericranial myofascial pain sensitivity is increased in patients with TTH, and peripheral activation or sensitization of myofascial nociceptors could play a role in the increased pain sensitivity.

12.5 Central Factors

12.5.1 Pain Sensitivity and Central Sensitization

The increased myofascial pain sensitivity in TTH could also be due to central factors such as sensitization of second-order neurons at the level of the spinal dorsal horn/trigeminal nucleus, sensitization of supraspinal neurons, and decreased antinociceptive activity from supraspinal structures. Pain detection thresholds have been reported normal in patients with ETTH in studies performed before the separation between the infrequent and frequent form of TTH in the second edition of the International Classification of Headache Disorders (ICHD-2) [18, 35–38]. Pain detection thresholds may be decreased in patients with frequent ETTH [33, 39], and both pain detection and tolerance thresholds were found to be decreased in patients with CTTH in studies performed with sufficient or large sample size [17, 39–44]. The difference in pain sensitivity is even more pronounced in cephalic and extracephalic regions in CTTH when recording pain sensitivity to clinically relevant stimuli, i.e., suprathreshold stimuli [40].

Hypersensitivity to various sensory stimulus modalities has been demonstrated in CTTH, including pressure [17, 41, 44], thermal [41], electrical [40, 42, 45–47], and inflammatory (intramuscular infusions of painful substances) stimuli [33, 39, 48, 49]. In addition, sensitivity to various stimulus modalities in various tissues, i.e., muscle, skin, tendons and peripheral nerves, is increased both at cephalic and extracephalic sites, both during and outside of a headache phase [17, 39–42, 44, 49].

Increased temporal summation to pressure pain and trend toward increased temporal summation to electrical pain in patients with CTTH compared to controls has been reported [7, 40]. These studies point toward the generalized increase in pain sensitivity in patients with frequent TTH. Widespread pain sensitivity in frequent TTH and CTTH supports the role of central sensitization. A population-based study demonstrated a close relationship between altered pain perception and chronification of headache confirming findings from experimental studies [50]. Furthermore, it was demonstrated that the increase in TTH prevalence over a 12-year period was related to increased pain sensitivity [50]. Increased pain sensitivity is a consequence of frequent TTH, not a risk factor, and the results support that central sensitization plays an important role for the chronification of TTH [51]. In addition, stimulus-response function for pressure versus pain in pericranial muscles is not only quantitatively but also qualitatively altered in patients with CTTH. The above hypothesis was further supported by a study demonstrating decrease in the volume of gray matter brain structures involved in pain processing in patients with CTTH [52]. This decrease was positively correlated with duration of headache and was most likely a consequence of central sensitization generated by prolonged input from pericranial myofascial structures [52]. However, generalized hypersensitivity and central sensitization can only partially account for the increased pericranial tenderness in patients with CTTH because of incomplete correlation between general pain hypersensitivity and pericranial tenderness [17, 19].

12.5.2 Central Pain Modulation

Decreased antinociceptive activity from supraspinal structures, i.e., deficient descending inhibition, may also contribute to the increased pain sensitivity in chronic TTH [7, 42, 46, 47, 53, 54]. A high-density brain electroencephalogram mapping study found impaired inhibition of nociceptive input in CTTH [53]. Impaired descending inhibition could be the primary abnormality, or it could contribute to or be a consequence of central sensitization [23]. Further studies are needed to clarify this issue. Both central sensitization and deficient descending inhibition may contribute to the development of CTHH from its episodic form.

12.6 Risk Factors for Chronification

Poor outcome or chronification of TTH has been studied in longitudinal epidemiological studies [55, 56]. Lyngberg et al. [55] defined poor outcome of tension-type headache as 180 TTH days or more per year at follow-up either due to increased frequency from ETTH at baseline into CTTH or unremitting CTTH. Predictors of poor outcome in multivariate model were found to be baseline CTTH, active migraine, sleeping problems, and demographic factor such as not being married. Increasing age (per 10-year increase), poor self-rated health, smoking, no physical activity, and frequent use of pain relievers tended to be associated with poor outcome ($p<0.20$). In another study, Ashina et al. [56] found that unilateral headache, nausea, and individual headache attack duration of 72 h or more were associated with poor outcome or chronification of pure TTH, but the associations did not reach statistical significance likely due to small sample size.

12.7 Biochemical Factors

12.7.1 Serotonin

The role of serotonin (5-hydroxytryptamine, 5-HT), an important transmitter in the antinociceptive pathways descending from the dorsal raphe nucleus in the brain stem to the spinal dorsal horn, in TTH has been investigated by studying 5-HT in peripheral blood [23]. However, the results of the studies are inconsistent due to clinical and methodological differences [23]. In summary, patients with ETTH have a decreased platelet 5-HT uptake and increased plasma 5-HT, whereas patients with CTTH have normal platelet 5-HT uptake, decreased platelet 5-HT, and normal plasma 5-HT [23]. In addition, there is a negative correlation between plasma 5-HT and headache frequency that has been demonstrated in CTTH [57]. Thus, patients with CTTH may have an impaired ability to increase plasma 5-HT and thus possibly synaptic 5-HT levels in response to increased nociceptive inputs from the peripheral nociceptors. A serotonergic dysfunction may contribute to central sensitization of nociceptive pathways and transformation of ETTH to CTTH, but further studies are needed to confirm this hypothesis. Moreover, we do not know whether the peripheral changes in 5-HT actually reflect similar mechanisms in central neurons. Studies involving newer methods including functional brain imaging are warranted.

12.7.2 Neuropeptides

There has been increasing interest in the role of neuropeptides in primary headaches. Bach et al. [58] reported normal calcitonin gene-related peptide (CGRP) levels in the cerebrospinal fluid (CSF) in patients with CTTH. Ashina et al. [59, 60] measured plasma levels of CGRP, substance P, NPY, and VIP in the cranial and peripheral circulation of patients and control subjects. It was shown that plasma levels of all four neuropeptides were normal in the cranial and peripheral circulation of patients. Moreover, plasma levels of neuropeptides were largely unrelated to the headache state, and no relationship between CGRP levels and muscular factors was demonstrated [59, 60]. However, it is important to state that findings of normal levels of neuropeptides in the cranial and peripheral circulation cannot exclude that abnormalities of these neuropeptides at the neuronal or peripheral or muscular levels may play a role in the pathophysiology of CTTH. Further studies with new sensitive methods of analysis are necessary to clarify the role of neuropeptides in the pathophysiology of CTTH.

12.7.3 Glutamate, Enkephalins, and Endorphins

Plasma levels of the excitatory amino acid glutamate were shown to be normal in patients with TTH [61]. In addition, in vivo concentrations of glutamate in a tender point of the trapezius muscle in patients with CTTH did not differ from control subjects in the resting state and in response to static exercise [34]. However, Sarchielli et al. [62] demonstrated an increase in platelet glutamate content in CTTH patients and suggested that glutamate in the central nervous system may be involved in the induction and maintenance of headache. Interestingly, in a randomized, double-blind, placebo-controlled, crossover trial, *N*-methyl D-aspartate (NMDA) antagonist memantine was reported to reduce pain intensity in CTTH patients to a limited extent [63].

The level of met-enkephalin in the CSF was demonstrated to be increased in patients with CTTH [64]. However, no correlation was found among CSF met-enkephalin-immunoreactivity and pericranial tenderness, nociceptive flexion-reflex threshold, or thermal pain threshold.

In contrast, there was no difference in the β-endorphin level in the CSF between CTTH patients and control subjects [58]. In patients with ETTH, β-endorphin levels in peripheral blood mononuclear cells were lower than in controls [65]. Moreover, a positive correlation was found between pressure-pain threshold values and β-endorphin levels in patients and control subjects [65]. These findings suggest that patients with TTH may have dysfunction of the endogenous antinociceptive systems.

12.8 Mechanism-Based Treatment

The hypothesis of central sensitization in TTH is supported by pharmacological studies. Amitriptyline, a tricyclic antidepressant medication, is the drug of choice for the prophylaxis of TTH. Amitriptyline was shown to reduce both headache and pericranial myofascial tenderness in patients with CTTH [66]. Bendtsen et al. suggested that the reduction of myofascial tenderness during treatment with amitriptyline may be caused by a segmental reduction of central sensitization in combination with an enhanced efficacy of noradrenergic or serotonergic descending inhibition [66]. Mirtazapine, a noradrenergic and specific serotonergic antidepressant that acts by antagonizing the adrenergic alpha2-autoreceptors and alpha2-heteroreceptors as well as by blocking 5-HT2 and 5-HT3 receptors, is considered a second drug of choice in the prophylaxis of TTH [67]. Venlafaxine, another second drug of choice for the prophylaxis of TTH, is a potent inhibitor of serotonin and norepinephrine reuptake and weak inhibitor of dopamine reuptake in the CNS [67]. Findings of lower serotonin (5-HT) reuptake inhibition and higher analgesic efficacy of amitriptyline than citalopram, a selective serotonin reuptake inhibitor, suggest that 5-HT reuptake inhibition is not the major mechanism of action of amitriptyline in TTH [68]. The analgesic and prophylactic effect of anti-depressants in CTTH is not solely due to 5-HT reuptake inhibition, and other mechanisms such as norepinephrine reuptake inhibition, NMDA receptor antagonism, and blockade of muscarinic receptors and ion channels are possibly involved and should be addressed in future studies.

Nitric oxide (NO) has also been a subject of research in TTH. Ashina et al. demonstrated that infusion of the NO donor, glycerol trinitrate, provokes TTH-like headache attacks in patients with chronic TTH [69]. In addition, a NO synthase (NOS) inhibitor significantly reduced headache [70] as well as pericranial myofascial tenderness and hardness [71] in patients with CTTH. In addition,

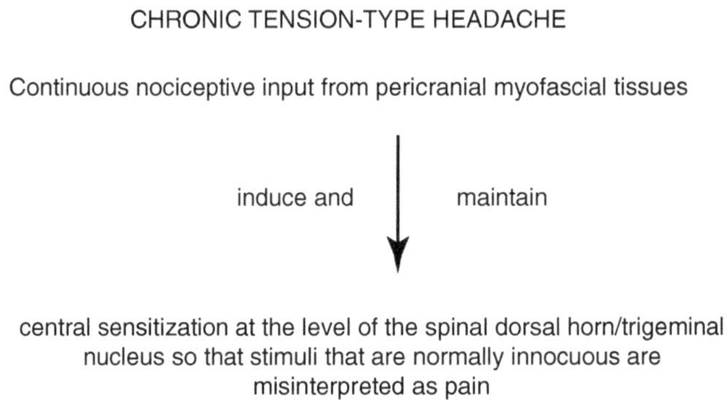

Fig. 12.1 The proposed pathophysiological model of chronic tension-type headache delineates two major aims for future research: (i) to identify the source of peripheral nociception in order to prevent the development of central sensitization in patients with episodic tension-type headache and (ii) to reduce established central sensitization in patients with chronic tension-type headache (Reprinted with permission from Bendtsen [23])

Sarchielli et al. [62] reported increased platelet NOS activity in patients with CTTH possibly reflecting central upregulation of NOS. These data support the role of central sensitization in the pathophysiology of chronic TTH. Moreover, these findings suggest that inhibition of NO and thereby central sensitization may become a novel approach in the future treatment of CTTH.

12.9 Conclusions

Pericranial myofascial mechanisms are probably of importance in ETTH, whereas sensitization of nociceptive pathways or central sensitization in the central nervous system due to prolonged nociceptive stimuli from pericranial myofascial tissues seems to be responsible for the conversion of ETTH to CTTH (Fig. 12.1). Future studies should focus on how (1) to identify the source of peripheral nociception in order to prevent the development of central sensitization and thereby the transformation of episodic into chronic TTH and (2) to reduce established central sensitization.

References

1. Rasmussen BK, Jensen R, Schroll M, Olesen J (1991) Epidemiology of headache in a general population – a prevalence study. J Clin Epidemiol 44(11):1147–1157
2. Schwartz BS, Stewart WF, Lipton RB (1997) Lost workdays and decreased work effectiveness associated with headache in the workplace. J Occup Environ Med 39(4):320–327
3. Rasmussen BK (1993) Migraine and tension-type headache in a general population: precipitating factors, female hormones, sleep pattern and relation to lifestyle. Pain 53(1):65–72

4. Spierings EL, Ranke AH, Honkoop PC (2001) Precipitating and aggravating factors of migraine versus tension-type headache. Headache 41(6):554–558
5. Cathcart S, Petkov J, Pritchard D (2008) Effects of induced stress on experimental pain sensitivity in chronic tension-type headache sufferers. Eur J Neurol 15(6):552–558
6. Cathcart S, Winefield AH, Lushington K, Rolan P (2009) Effect of mental stress on cold pain in chronic tension-type headache sufferers. J Headache Pain 10(5):367–373
7. Cathcart S, Winefield AH, Lushington K, Rolan P (2010) Noxious inhibition of temporal summation is impaired in chronic tension-type headache. Headache 50(3):403–412
8. Holroyd KA, Stensland M, Lipchik GL, Hill KR, O'Donnell FS, Cordingley G (2000) Psychosocial correlates and impact of chronic tension-type headaches. Headache 40(1):3–16
9. Janke EA, Holroyd KA, Romanek K (2004) Depression increases onset of tension-type headache following laboratory stress. Pain 111(3):230–238
10. Lipchik GL, Holroyd KA, O'Donnell FJ, Cordingley GE, Waller S, Labus J et al (2000) Exteroceptive suppression periods and pericranial muscle tenderness in chronic tension-type headache: effects of psychopathology, chronicity and disability. Cephalalgia 20(7):638–646
11. Heckman BD, Holroyd KA (2006) Tension-type headache and psychiatric comorbidity. Curr Pain Headache Rep 10(6):439–447
12. Ostergaard S, Russell MB, Bendtsen L, Olesen J (1997) Comparison of first degree relatives and spouses of people with chronic tension headache. BMJ 314(7087):1092–1093
13. Russell MB, Ostergaard S, Bendtsen L, Olesen J (1999) Familial occurrence of chronic tension-type headache. Cephalalgia 19(4):207–210
14. Ulrich V, Gervil M, Olesen J (2004) The relative influence of environment and genes in episodic tension-type headache. Neurology 62(11):2065–2069
15. Ashina S, Bendtsen L, Lyngberg AC, Lipton RB, Hajiyeva N, Jensen R (2014) Prevalence of neck pain in migraine and tension-type headache: a population study. Cephalalgia 22
16. Aaseth K, Grande RB, Lundqvist C, Russell MB (2014) Pericranial tenderness in chronic tension-type headache: the Akershus population-based study of chronic headache. J Headache Pain 15(1):58. doi:10.1186/1129-2377-15-58
17. Bendtsen L, Jensen R, Olesen J (1996) Decreased pain detection and tolerance thresholds in chronic tension-type headache. Arch Neurol 53(4):373–376
18. Jensen R, Rasmussen BK, Pedersen B, Olesen J (1993) Muscle tenderness and pressure pain thresholds in headache. A population study. Pain 52:193–199
19. Jensen R, Bendtsen L, Olesen J (1998) Muscular factors are of importance in tension-type headache. Headache 38(1):10–17
20. Langemark M, Olesen J (1987) Pericranial tenderness in tension headache. A blind, controlled study. Cephalalgia 7:249–255
21. Lipchik GL, Holroyd KA, Talbot F, Greer M (1997) Pericranial muscle tenderness and exteroceptive suppression of temporalis muscle activity: a blind study of chronic tension-type headache. Headache 37(6):368–376
22. Ashina M, Bendtsen L, Jensen R, Sakai F, Olesen J (1999) Muscle hardness in patients with chronic tension-type headache: relation to actual headache state. Pain 79(2–3):201–205
23. Bendtsen L (2000) Central sensitization in tension-type headache – possible pathophysiological mechanisms. Cephalalgia 20(5):486–508
24. Sakai F, Ebihara S, Akiyama M, Horikawa M (1995) Pericranial muscle hardness in tension-type headache. A non-invasive measurement method and its clinical application. Brain 118(Pt 2):523–531
25. Fernandez-de-Las-Penas C, Alonso-Blanco C, Cuadrado ML, Gerwin RD, Pareja JA (2006) Myofascial trigger points and their relationship to headache clinical parameters in chronic tension-type headache. Headache 46(8):1264–1272
26. Fernandez-de-Las-Penas C, Alonso-Blanco C, Cuadrado ML, Pareja JA (2006) Myofascial trigger points in the suboccipital muscles in episodic tension-type headache. Man Ther 11(3):225–230
27. Fernandez-de-Las-Penas C, Cuadrado ML, Pareja JA (2007) Myofascial trigger points, neck mobility, and forward head posture in episodic tension-type headache. Headache 47(5):662–672

28. Jensen R, Olesen J (1996) Initiating mechanisms of experimentally induced tension-type headache. Cephalalgia 16(3):175–182
29. Christensen MB, Bendtsen L, Ashina M, Jensen R (2005) Experimental induction of muscle tenderness and headache in tension-type headache patients. Cephalalgia 25(11):1061–1067
30. Jensen R (1999) Pathophysiological mechanisms of tension-type headache: a review of epidemiological and experimental studies. Cephalalgia 19(6):602–621
31. Hubbard DR, Berkoff GM (1993) Myofascial trigger points show spontaneous needle EMG activity. Spine (Phila Pa 1976) 18(13):1803–1807
32. Ashina M, Stallknecht B, Bendtsen L, Pedersen JF, Galbo H, Dalgaard P et al (2002) In vivo evidence of altered skeletal muscle blood flow in chronic tension-type headache. Brain 125(Pt 2):320–326
33. Mork H, Ashina M, Bendtsen L, Olesen J, Jensen R (2003) Induction of prolonged tenderness in patients with tension-type headache by means of a new experimental model of myofascial pain. Eur J Neurol 10(3):249–256
34. Ashina M, Stallknecht B, Bendtsen L, Pedersen JF, Schifter S, Galbo H et al (2003) Tender points are not sites of ongoing inflammation -in vivo evidence in patients with chronic tension-type headache. Cephalalgia 23(2):109–116
35. Bovim G (1992) Cervicogenic headache, migraine, and tension-type headache. Pressure-pain threshold measurements. Pain 51(2):169–173
36. Jensen R (1996) Mechanisms of spontaneous tension-type headaches: an analysis of tenderness, pain thresholds and EMG. Pain 64(2):251–256
37. Gobel H, Weigle L, Kropp P, Soyka D (1992) Pain sensitivity and pain reactivity of pericranial muscles in migraine and tension-type headache. Cephalalgia 12(3):142–151
38. Headache Classification Subcommittee of the International Headache Society (2004) The International Classification of Headache Disorders: 2nd edition. Cephalalgia 24(Suppl 1):9–160
39. Schmidt-Hansen PT, Svensson P, Bendtsen L, Graven-Nielsen T, Bach FW (2007) Increased muscle pain sensitivity in patients with tension-type headache. Pain 129(1–2):113–121
40. Ashina S, Bendtsen L, Ashina M, Magerl W, Jensen R (2006) Generalized hyperalgesia in patients with chronic tension-type headache. Cephalalgia 26(8):940–948
41. Langemark M, Jensen K, Jensen TS, Olesen J (1989) Pressure pain thresholds and thermal nociceptive thresholds in chronic tension-type headache. Pain 38(2):203–210
42. Langemark M, Bach FW, Jensen TS, Olesen J (1993) Decreased nociceptive flexion reflex threshold in chronic tension-type headache. Arch Neurol 50(10):1061–1064
43. Sandrini G, Antonaci F, Pucci E, Bono G, Nappi G (1994) Comparative study with EMG, pressure algometry and manual palpation in tension-type headache and migraine. Cephalalgia 14(6):451–457
44. Schoenen J, Bottin D, Hardy F, Gerard P (1991) Cephalic and extracephalic pressure pain thresholds in chronic tension-type headache. Pain 47(2):145–149
45. Ashina S, Babenko L, Jensen R, Ashina M, Magerl W, Bendtsen L (2005) Increased muscular and cutaneous pain sensitivity in cephalic region in patients with chronic tension-type headache. Eur J Neurol 12(7):543–549
46. Lindelof K, Jung K, Ellrich J, Jensen R, Bendtsen L (2010) Low-frequency electrical stimulation induces long-term depression in patients with chronic tension-type headache. Cephalalgia 30(7):860–867
47. Sandrini G, Rossi P, Milanov I, Serrao M, Cecchini AP, Nappi G (2006) Abnormal modulatory influence of diffuse noxious inhibitory controls in migraine and chronic tension-type headache patients. Cephalalgia 26(7):782–789
48. Mork H, Ashina M, Bendtsen L, Olesen J, Jensen R (2003) Experimental muscle pain and tenderness following infusion of endogenous substances in humans. Eur J Pain 7(2):145–153
49. Lindelof K, Ellrich J, Jensen R, Bendtsen L (2009) Central pain processing in chronic tension-type headache. Clin Neurophysiol 120(7):1364–1370
50. Buchgreitz L, Lyngberg AC, Bendtsen L, Jensen R (2006) Frequency of headache is related to sensitization: a population study. Pain 123(1–2):19–27

51. Buchgreitz L, Lyngberg AC, Bendtsen L, Jensen R (2008) Increased pain sensitivity is not a risk factor but a consequence of frequent headache: a population-based follow-up study. Pain 137(3):623–630
52. Schmidt-Wilcke T, Leinisch E, Straube A, Kampfe N, Draganski B, Diener HC et al (2005) Gray matter decrease in patients with chronic tension type headache. Neurology 65(9):1483–1486
53. Buchgreitz L, Egsgaard LL, Jensen R, Arendt-Nielsen L, Bendtsen L (2008) Abnormal pain processing in chronic tension-type headache: a high-density EEG brain mapping study. Brain 131(Pt 12):3232–3238
54. Pielsticker A, Haag G, Zaudig M, Lautenbacher S (2005) Impairment of pain inhibition in chronic tension-type headache. Pain 118(1–2):215–223
55. Lyngberg AC, Rasmussen BK, Jorgensen T, Jensen R (2005) Prognosis of migraine and tension-type headache: a population-based follow-up study. Neurology 65(4):580–585
56. Ashina S, Lyngberg A, Jensen R (2010) Headache characteristics and chronification of migraine and tension-type headache: a population-based study. Cephalalgia 30(8): 943–952
57. Bendtsen L, Jensen R, Hindberg I, Gammeltoft S, Olesen J (1997) Serotonin metabolism in chronic tension-type headache. Cephalalgia 17(8):843–848
58. Bach FW, Langemark M, Ekman R, Rehfeld JF, Schifter S, Olesen J (1994) Effect of sulpiride or paroxetine on cerebrospinal fluid neuropeptide concentrations in patients with chronic tension-type headache. Neuropeptides 27(2):129–136
59. Ashina M, Bendtsen L, Jensen R, Schifter S, Jansen-Olesen I, Olesen J (2000) Plasma levels of calcitonin gene-related peptide in chronic tension-type headache. Neurology 55(9): 1335–1340
60. Ashina M, Bendtsen L, Jensen R, Ekman R, Olesen J (1999) Plasma levels of substance P, neuropeptide Y and vasoactive intestinal polypeptide in patients with chronic tension-type headache. Pain 83(3):541–547
61. Alam Z, Coombes N, Waring RH, Williams AC, Steventon GB (1998) Plasma levels of neuroexcitatory amino acids in patients with migraine or tension headache. J Neurol Sci 156(1):102–6, S0022510X98000239 [pii]
62. Sarchielli P, Alberti A, Floridi A, Gallai V (2002) L-Arginine/nitric oxide pathway in chronic tension-type headache: relation with serotonin content and secretion and glutamate content. J Neurol Sci 198(1–2):9–15
63. Lindelof K, Bendtsen L (2009) Memantine for prophylaxis of chronic tension-type headache – a double-blind, randomized, crossover clinical trial. Cephalalgia 29(3):314–21. doi:10.1111/j.1468-2982.2008.01720.x, CHA1720 [pii]
64. Langemark M, Bach FW, Ekman R, Olesen J (1995) Increased cerebrospinal fluid Met-enkephalin immunoreactivity in patients with chronic tension-type headache. Pain 63(1):103–7, 0304-3959(95)00020-S [pii]
65. Mazzotta G, Sarchielli P, Gaggioli A, Gallai V (1997) Study of pressure pain and cellular concentration of neurotransmitters related to nociception in episodic tension-type headache patients. Headache 37(9):565–571
66. Bendtsen L, Jensen R (2000) Amitriptyline reduces myofascial tenderness in patients with chronic tension-type headache. Cephalalgia 20(6):603–610
67. Bendtsen L, Jensen R (2011) Treating tension-type headache – an expert opinion. Expert Opin Pharmacother 12(7):1099—109. doi:10.1517/14656566.2011.548806
68. Ashina S, Bendtsen L, Jensen R (2004) Analgesic effect of amitriptyline in chronic tension-type headache is not directly related to serotonin reuptake inhibition. Pain 108(1–2):108–114
69. Ashina M, Bendtsen L, Jensen R, Olesen J (2000) Nitric oxide-induced headache in patients with chronic tension-type headache. Brain 123(Pt 9):1830–1837
70. Ashina M, Lassen LH, Bendtsen L, Jensen R, Olesen J (1999) Effect of inhibition of nitric oxide synthase on chronic tension-type headache: a randomised crossover trial. Lancet 353(9149):287–289

71. Ashina M, Bendtsen L, Jensen R, Lassen LH, Sakai F, Olesen J (1999) Possible mechanisms of action of nitric oxide synthase inhibitors in chronic tension-type headache. Brain 122(Pt 9):1629–1635

Suggested Reading

Ashina M (2004) Neurobiology of chronic tension-type headache. Cephalalgia 24(3):161–172

Bendtsen L (2000) Central sensitization in tension-type headache – possible pathophysiological mechanisms. Cephalalgia 20(5):486–508

Bendtsen L, Fernández-de-la-Peñas C (2011) The role of muscles in tension-type headache. Curr Pain Headache Rep 15(6):451–8

Bezov D, Ashina S, Jensen R, Bendtsen L (2011) Pain perception studies in tension-type headache. Headache 51(2):262–271

Chapter 13
Pathophysiology of Cluster Headache: Current Status and Future Directions

Mark Obermann and Manjit Matharu

13.1 Introduction

Typical clinical features of cluster headache (CH) include trigeminal distribution of pain, circadian and circannual rhythmicity, and ipsilateral cranial autonomic features [1]. The striking circadian and circannual periodicity led to the suggestion that the hypothalamus, which is the structure involved in the human biological clock system, plays a pivotal role in the pathophysiology of this disorder [2]. Several studies using neuroimaging techniques or measuring hormone levels supported the hypothesis of a hypothalamic involvement [2–7]. Animal studies added further evidence regarding this hypothesis [8]. Based on previous data, even invasive treatment methods such as hypothalamic deep brain stimulation (DBS) were justified. More recent studies point towards a complex neural network performance deficit in CH with complex interactions and multiple influences that will have to be unravelled in the future [9, 10].

M. Obermann (✉)
Department of Neurology, University of Duisburg-Essen,
Hufelandstr. 55, Essen, NA 45122, Germany
e-mail: mark.obermann@uni-due.de

M. Matharu
Headache Group, Institute of Neurology and The National Hospital
for Neurology and Neurosurgery, Queen Square, London WC1N 3BG, UK
e-mail: m.matharu@uclmail.net

M. Ashina, P. Geppetti (eds.), *Pathophysiology of Headaches: From Molecule to Man*, Headache, DOI 10.1007/978-3-319-15621-7_13

13.2 Symptomatic Cluster Headache

Cluster headache is generally a primary headache disorder but in rare cases can be the manifestation of an underlying symptomatic condition. Much can be learned from close examination of these symptomatic cases in terms of the potentially dysfunctional regions and underlying pathophysiology. To avoid confusion with primary CH, most authors refer to symptomatic headaches with a cluster headache phenotype or cluster-like headache (CLH), but the expression symptomatic CH is used as well. Multiple different pathologies have been identified and associated with CLHs. From 1975 to 2008, 156 CLH cases were published. Many of these reports did not provide sufficient information though approximately one-half appear to be good mimics of primary CH, fulfilling the ICHD-2 criteria [11].

With regard to aetiology, most of the reported pathologies were of vascular (38 %), tumoural (26 %), and inflammatory or infectious (14 %) origin as well as headaches related to trauma (9 %).

With regard to localisation of the pathology, many cases with close proximity to the hypothalamus were published during the last decades. These include the pituitary [12–14], sella [15], third ventricle [16], cavernous sinus [17], anterior communicating artery [12], or sphenoid sinus [18]. But there are nearly as many cases which fulfil the ICHD-2 criteria for CH with pathologies in diverse locations such as herpes zoster ophthalmicus [19], upper cervical meningioma [20], facial herpes simplex [21], frontal skull fracture [22], carotid artery aneurysm [12], fronto-temporal-parietal subdural haematoma [22], epidermoid tumour of the posterior fossa [23], vertebral aneurysm [24], post-tonsillectomy, trigeminal neurinoma [25], dental extraction [26], foreign body in the maxillary sinus [27], and intraocular lens implant [28].

These symptomatic cases do not seem to share an obvious common pathophysiological pathway that would help us better understand the mechanisms associated with or leading to the development of CH. In some cases, it might just be coincidental coexistence of two conditions, while in others, the pathology may be just enough to lift a potential "subclinical" CH over a certain threshold to become clinically relevant. Interestingly, at least 12 of the reported cases fulfilling the ICHD-2 criteria ceased completely after surgery, thereby suggesting that at least some of the reported cases are secondary to the underlying pathology rather than coincidental coexistence of two disorders [29]. With regard to the diverse underlying lesions, Straube and colleagues have suggested that these lesions may trigger CH by producing an imbalance of the parasympathetic and sympathetic autonomic systems [30]. Whatever the mechanisms by which symptomatic CH are generated, it is worth appreciating that these mechanisms may not be relevant to the pathophysiology of primary CH.

13.3 Neuroendocrinology

Neuroendocrine studies have provided indirect evidence supportive of deranged hypothalamic function in CH. It was initially demonstrated that plasma testosterone concentrations were altered during the CH period in men [31]. Subsequently, it has been observed that there are abnormalities in the secretion of melatonin and cortisol, alterations in the secretion of luteinising hormone and prolactin, and altered responses of luteinising hormone, follicle stimulating hormone, prolactin, growth hormone, and thyroid-stimulating hormone to challenge tests in patients with CH [32]. There is preservation of the hypothalamic-pituitary axis, and the abnormalities are more supportive of a primary dysfunction of the hypothalamus. Interestingly, changes of the CSF orexin level were not observed during active CH episodes, which are considered to play a pivotal role in the pain processing of CH patients. Cavoli and colleagues measured orexin-A in ten patients with CH by radioimmunoassay. CSF orexin levels were in the normal range, and no association between clinical presentation and orexin-A level could be observed [33].

Several possibilities have been raised regarding the observed neuroendocrine alterations. First, these changes may be the result of the severe CH pain itself. Second, they may reflect a stress reaction (pain associated or independent) or, third, are induced by pain accompanying sleep disturbances. All of these possibilities would suggest that these alterations are rather unspecific phenomena. Interestingly, some of the observed hypothalamic changes can also be detected in remission periods which would imply that these changes can be considered to be specific for CH itself continuing independently of the pain and therefore might be a kind of trait marker for the disease itself.

13.4 Genetics

Data from twin and family studies have suggested that CH has a heritable component, with 2–7 % of patients reporting a positive family history for this disorder [34]. First-degree relatives of CH patients have a five- to 18-fold increased risk and second-degree relatives a one- to threefold increased risk to also get CH compared with the general population [35]. Genetic alterations within the orexinergic system of the hypothalamus have been suggested to be responsible for this observation. It has been shown that the G1246A polymorphism of the OX_2R gene (HCRTR2) increases the risk for CH [36]. However, these data were not replicated in larger CH patient populations [26]. This gene polymorphism was not observed in migraineurs [37].

13.5 Sleep

The clinically obvious association of CH with sleep remains undisputed. The majority of patients experience attacks during the night and are frequently awakened by these attacks. It was hypothesised that CH could be associated with REM (rapid eye movement) sleep phases, but this was not confirmed by polysomnography over four consecutive nights. The sleep phases were randomly distributed [38]. However, a small case-control study showed an association of CH with sleep apnoea syndrome during the active cluster period that resolved once the bout resolved. There was an increased rate of central apnoeas and a higher respiratory stress index (8.6 ± 16) compared with healthy controls (3.4 ± 2.1; $p = 0.002$). However, only one out of five patients treated with nasal continuous positive airway pressure showed any benefit in regard to the cluster headache attack frequency [39].

13.6 Structural Imaging

One of the pioneering studies showing hypothalamic involvement in CH was performed by May and colleagues [2]. The authors used the method of voxel-based morphometry (VBM), an automated, unbiased, whole brain technique. It allows comparison of structural brain images, particularly the volume or density of grey and white matter. May and colleagues investigated 25 CH patients compared with 29 healthy controls and detected isolated increased grey matter in the inferior posterior hypothalamus [2]. These VBM studies have been repeated in a larger patient population and with newer, more refined analysis algorithms. Up to now, three studies have been performed, but none of these have been able to confirm the initial finding. Matharu and colleagues investigated 66 patients suffering from CH and 96 age- and gender-matched healthy subjects. This study did not detect any hypothalamic changes at all [40]. Similar findings were reported by two later studies [9, 10]. However, both studies were able to demonstrate several changes within the central pain-processing network [29].

13.7 Functional Imaging

Functional imaging allows detection of alterations in the activity of the pain-processing networks in the brain in vivo during ongoing pain. Several functional imaging studies with positron emission tomography (PET) and functional magnetic resonance imaging (fMRI) have been performed in CH. Nitroglycerine triggered headache attacks in nine chronic CH patients resulted in a strong activation of the ipsilateral posterior hypothalamus detected by $H_2{}^{15}O$ PET [41]. This activation pattern was also observed in spontaneous CH attacks in one patient who had undergone deep brain stimulation (DBS) [6]. In four patients with episodic CH, fMRI confirmed the activation pattern within the ipsilateral posterior hypothalamus [42].

However, some authors suggested that the detected activation pattern in the functional imaging shows activation of an area close to the hypothalamus, most likely the midbrain tegmentum, rather than the hypothalamus itself [43].

13.8 Resting State fMRI

The analysis of low-frequency (<0.1 Hz) fluctuations seen on fMRI scans at rest allows detection of functionally connected brain regions, so-called resting state networks (RSNs). Synchronous variations of the BOLD signal can be measured as percentage signal change compared to the BOLD mean signal intensity over time [44–46]. The fluctuations observed by resting state analysis are thought to reflect the intrinsic property of the brain to handle the past and prepare for the future [47]. Resting state (RS) alterations have been observed in chronic pain [48]. Rocca and colleagues studied RS in 13 patients with episodic CH compared with healthy controls. Patients were studied in a pain-free state. Apart from other changes, the authors observed functional connectivity within the network starting from the hypothalamus [49].

13.9 Magnetic Resonance Spectroscopy

An additional existing imaging technique to study brain biochemistry in vivo is magnetic resonance spectroscopy. In episodic CH patients, hypothalamic N-acetylaspartate/creatine and choline/creatine levels are significantly reduced compared with healthy controls. Interestingly, changes were even detectable when the patients were outside bout, when they were not having any CH attacks anymore [4, 7]. This observation led to the suggestion that these alterations cannot simply reflect an epiphenomenon of the pain itself [7].

13.10 Deep Brain Stimulation

The clinical picture of CH and the results from imaging studies provided the rationale for hypothalamic deep brain stimulation (DBS) in the treatment of CH. It was thought that this technique might offer the possibility to "turn off the CH generator" as high-frequency hypothalamic stimulation would inhibit hypothalamic hyperactivity [50]. The stimulation area was mainly chosen by adoption of the results from the initial VBM study [2]. To assess to what extent DBS stimulation is able to abort acute CH attacks, Leone and colleagues investigated 136 CH attacks in 16 chronic CH patients [51]. Only 23 % of patients reported a reduction of pain intensity by more than 50 %, and only 16 % of headache attacks were completely terminated. These data indicated that DBS is not sufficient in the treatment of active CH attacks [51]. Further studies showed that only continuous stimulation over several weeks markedly reduces or terminates CH attacks [52, 53]. Fifty-eight patients with drug-resistant chronic CH and posterior hypothalamic region DBS have been reported in the literature so far. Leone and colleagues investigated 16 drug-resistant chronic CH patients who received hypothalamic implants over a mean period of 4 years. After the first 2 years, 83.3 % of patients had experienced a pain termination or at least significant pain reduction. After 4 years, 62 % of patients were still pain-free [54]. These results were confirmed by several other studies.

Interestingly, there were no changes in regard to long-term stimulation in electrolyte balance, sleep-wake cycle, or hormone levels of cortisol, prolactin, thyroid hormone, and thyroid-stimulating hormone, which are known to be altered in CH [55–61].

Although the evidence of the imaging studies seemed to be overwhelming, some authors raised the question of the precise anatomical localisation of the DBS. Sanchez del Rio and Linera questioned if the shown diencephalic/midbrain activity pattern corresponds to the midbrain tegmentum rather than the genuine hypothalamus [43, 62]. Although the anatomical boundaries of the hypothalamus are quite clear (anterior, lamina terminalis; posterior, posterior margin of the maxillary bodies; superior, hypothalamic sulcus; medial, third ventricle; lateral, subthalamus and internal capsule; inferior, optic chiasm, median eminence, tuber cinereum, mamillary bodies, and posterior pituitary), the functional boundaries are more vaguely determined. Matharu and colleagues re-examined the statistical parametric maps and coordinates of the activation pattern of PET studies in CH [62]. The observed activation in the diencephalon and the mesencephalon in CH is centred over the midbrain tegmentum and is close to the hypothalamus but more posteriorly [41]. In contrast, functional imaging studies in CH using BOLD-fMRI studies detected activation of the posterior and middle hypothalamus rather than the mesencephalon. The authors suggest that these differences are most likely based on methodological issues, mainly the problem of insufficient spatial resolution (fMRI 4–5 mm; PET 5–10 mm). They conclude that these data can only be interpreted in the context of other knowledge but might be, therefore, also influenced by the a priori hypothesis.

Though most investigators report that the DBS electrodes are implanted in the posterior hypothalamus, a careful examination reveals that the implantation site in these patients is in fact the midbrain tegmentum [62]. This raises the possibility that this therapy is not being targeted at the hypothalamus but at one of the pathways connecting to the hypothalamus. This may be one of the reasons DBS in cluster headache is only effective in approximately two-thirds of patients in the long term. Moreover, stimulation of the trigeminal pain-processing network by occipital nerve stimulation (ONS) in CH patients presented similar results in regard to pain reduction efficacy suggesting a rather unspecific role of both ONS and DBS in CH [63].

Additionally, positive DBS results were also observed in other pain disorders, questioning the pathophysiological concept of specific hypothalamic alteration in CH and raising some serious concerns regarding their validity and specificity. Interestingly, hypothalamic region DBS was also effective in the treatment of symptomatic trigeminal neuralgia (TN) in five multiple sclerosis patients [64]. These patients had to be therapy refractory prior to electrode implantation. Beneficial effects with pain reduction were observed in three of the patients even within the first 24 h after implantation. As long as controlled studies are lacking, the results of such studies should be interpreted with caution, and careless utilisation should be avoided. However, one can conclude based on the reported study results that DBS of the posterior hypothalamus region is not exclusively effective in CH but also shows beneficial effects in other pain conditions as well.

In contrast, there are also chronic pain conditions where hypothalamic DBS does not seem to be effective. Franzini and colleagues reported four patients with secondary neuropathic trigeminal pain who did not experience any relevant pain reduction after electrode implantation [64]. However, the reported patient population was heterogeneous without comparable clinical features, which makes an interpretation of the study results difficult.

13.11 Role of the Hypothalamus

On the basis of the clinical features of CH with trigeminal distribution of pain, circadian and circannual rhythmicity, and ipsilateral cranial autonomic symptoms in combination with the results from the neuroimaging studies, the pathophysiological importance of the hypothalamus seems to be robust and scientifically proven. In particular, structural and functional neuroimaging studies supported the hypothesis of hypothalamic alterations being involved in the pathophysiology of CH [2, 41, 42]. These data seemed to be so conclusive that even invasive therapy methods such as DBS were used to directly influence the "hypothalamic CH generator". However, other contrary findings should be taken into consideration before prematurely adopting this hypothalamic hypothesis [9, 10]. One major criticism about most of the interpretations from previous studies is that the focus was directed almost exclusively at results that support the importance of the hypothalamus in CH, while other data were often neglected or considered inconsequential. It might be useful to take a step back and have a look at the whole picture, as this strong hypothesis-driven research might have led us in the wrong direction.

Hypothalamic activation and structural changes can also be detected in other primary headache disorders such as migraine [65], hemicrania continua [66], chronic facial pain [67], and hypnic headache [68] and is not an exclusive feature of CH. Interestingly, hypothalamic changes can even be observed in totally different conditions such as angina pectoris [69], irritable bowel syndrome [70] or anorexia nervosa [71], autism [72], fragile X syndrome [73], narcolepsy [74], and Huntington's disease [75]. However, most of the neuroimaging studies that investigated pain disorders other than CH did not observe any hypothalamic alterations, but most of the other studies that investigated pain disorders did not predefine the hypothalamus as the target anatomic region which impedes the detection of more subtle activation or structural change below the threshold of statistical significance.

The exact anatomic location of the observed activations or structural alterations in CH has been attributed to different structures [9, 10] in view of the limitation of spatial resolution of the neuroimaging techniques used (PET 5–10 mm; MRI 4–5 mm). Based on these methodological limitations, it was suggested that the observed activations might be localised in the midbrain tegmentum rather than in the hypothalamus itself. This challenges the validity of some of the neuroimaging results in regard to the precise anatomic location of the changes reported.

Although neuroendocrine [32] and genetic studies [36] detected changes in CH and also seem to point at hypothalamic changes, the specificity of these observations must be questioned. HPA axis disturbances were also detected in fibromyalgia [76], chronic fatigue syndrome [77], irritable bowel syndrome [78], and migraine [79], genetic mutations were not reproducible [80].

13.12 Conclusion and Future Directions

Even though the results of different studies on CH are very diverse and partly contradictory on superficial examination, they point towards a complex neural network performance deficit in CH rather than a single locus of abnormality, albeit that the hypothalamus may play an important role in the pathophysiology of this disorder. Imaging has given some important insights into the pathophysiology of this very complex disorder but may not be able to resolve this puzzle alone. CH may be a good model condition to study the remarkable plasticity of the human brain due to its different disease conditions and its adaptation capacities to the different cluster-associated pain states. More sophisticated studies (especially longitudinal designs) are needed to address this aspect properly.

While posterior hypothalamic region DBS can be useful in some patients with medically refractory cluster headache, it may be a non-specific therapy and needs to be used cautiously, when all other treatment avenues have been exhausted. The importance of this issue is further outlined by the recent emergence of less invasive neurostimulation methods, such as occipital nerve stimulation, sphenopalatine ganglion stimulation, and vagal nerve stimulation, which should be considered prior to DBS.

The evidence available thus far has improved our knowledge of the pathophysiology of this disorder, pointing towards more complex pathophysiological model of the disease, but more research in this area is urgently needed to be able to find the way out of this complicated maze.

References

1. Headache Classification Committee of the International Headache Society I (2013) The international classification of headache disorders, 3rd edition (beta version). Cephalalgia Int J Headache 33:629–808
2. May A, Ashburner J, Buchel C, McGonigle DJ, Friston KJ, Frackowiak RS et al (1999) Correlation between structural and functional changes in brain in an idiopathic headache syndrome. Nat Med 5(7):836–838
3. Leone M, Patruno G, Vescovi A, Bussone G (1990) Neuroendocrine dysfunction in cluster headache. Cephalalgia Int J Headache 10(5):235–239
4. Lodi R, Pierangeli G, Tonon C, Cevoli S, Testa C, Bivona G et al (2006) Study of hypothalamic metabolism in cluster headache by proton MR spectroscopy. Neurology 66(8):1264–1266
5. May A, Bahra A, Buchel C, Frackowiak RS, Goadsby PJ (2000) PET and MRA findings in cluster headache and MRA in experimental pain. Neurology 55(9):1328–1335
6. Sprenger T, Boecker H, Tolle TR, Bussone G, May A, Leone M (2004) Specific hypothalamic activation during a spontaneous cluster headache attack. Neurology 62(3):516–517
7. Wang SJ, Lirng JF, Fuh JL, Chen JJ (2006) Reduction in hypothalamic 1H-MRS metabolite ratios in patients with cluster headache. J Neurol Neurosurg Psychiatry 77(5):622–625
8. Malick A, Strassman RM, Burstein R (2000) Trigeminohypothalamic and reticulohypothalamic tract neurons in the upper cervical spinal cord and caudal medulla of the rat. J Neurophysiol 84(4):2078–2112
9. Yang FC, Chou KH, Fuh JL, Huang CC, Lirng JF, Lin YY et al (2013) Altered gray matter volume in the frontal pain modulation network in patients with cluster headache. Pain 154(6):801–807
10. Absinta M, Rocca MA, Colombo B, Falini A, Comi G, Filippi M (2012) Selective decreased grey matter volume of the pain-matrix network in cluster headache. Cephalalgia Int J Headache 32(2):109–115
11. Headache Classification Committee of the International Headache Society I (2004) The international classification of headache disorders: 2nd edition. Cephalalgia Int J Headache 24(Suppl 1):9–160

12. Greve E, Mai J (1988) Cluster headache-like headaches: a symptomatic feature? A report of three patients with intracranial pathologic findings. Cephalalgia Int J Headache 8(2):79–82
13. Levy MJ, Matharu MS, Meeran K, Powell M, Goadsby PJ (2005) The clinical characteristics of headache in patients with pituitary tumours. Brain J Neurol 128(Pt 8):1921–1930
14. Negoro K, Kawai M, Tada Y, Ogasawara J, Misumi S, Morimatsu M (2005) A case of post-prandial cluster-like headache with prolactinoma: dramatic response to cabergoline. Headache 45(5):604–606
15. Hannerz J (1989) A case of parasellar meningioma mimicking cluster headache. Cephalalgia Int J Headache 9(4):265–269
16. Narbone MC, D'Amico D, Di Maria F, Arena MG, Longo M (1991) Cluster-like headache and a median intracranial calcified lesion: case report. Headache 31(10):684–685
17. Relja G, Nider G, Koscica N, Musco G, Negro C (1999) The role of cavernous sinus in cluster and other headaches. Ital J Neurol Sci 20(2 Suppl):S42–S45
18. Zanchin G, Rossi P, Licandro AM, Fortunato M, Maggioni F (1995) Clusterlike headache. A case of sphenoidal aspergilloma. Headache 35(8):494–497
19. Sacquegna T, D'Alessandro R, Cortelli P, de Carolis P, Baldrati A (1982) Cluster headache after herpes zoster ophthalmicus. Arch Neurol 39(6):384
20. Kuritzky A (1984) Cluster headache-like pain caused by an upper cervical meningioma. Cephalalgia Int J Headache 4(3):185–186
21. Joseph R, Rose FC (1985) Cluster headache and herpes simplex: an association? Br Med J 290(6482):1625–1626
22. Formisano R, Angelini A, De Vuono G, Calisse P, Fiacco F, Catarci T et al (1990) Cluster-like headache and head injury: case report. Ital J Neurol Sci 11(3):303–305
23. Levyman C, Dagua Filho Ados S, Volpato MM, Settanni FA, de Lima WC (1991) Epidermoid tumour of the posterior fossa causing multiple facial pain – a case report. Cephalalgia Int J Headache 11(1):33–36
24. West P, Todman D (1991) Chronic cluster headache associated with a vertebral artery aneurysm. Headache 31(4):210–212
25. Masson C, Lehericy S, Guillaume B, Masson M (1995) Cluster-like headache in a patient with a trigeminal neurinoma. Headache 35(1):48–49
26. Soros P, Frese A, Husstedt IW, Evers S (2001) Cluster headache after dental extraction: implications for the pathogenesis of cluster headache? Cephalalgia Int J Headache 21(5):619–622
27. Scorticati MC, Raina G, Federico M (2002) Cluster-like headache associated to a foreign body in the maxillary sinus. Neurology 59(4):643–644
28. Maggioni F, Dainese F, Mainardi F, Lisotto C, Zanchin G (2005) Cluster-like headache after surgical crystalline removal and intraocular lens implant: a case report. J Headache Pain 6(2):88–90
29. Naegel S, Holle D, Obermann M (2014) Structural imaging in cluster headache. Curr Pain Headache Rep 18(5):415
30. Straube A, Freilinger T, Ruther T, Padovan C (2007) Two cases of symptomatic cluster-like headache suggest the importance of sympathetic/parasympathetic balance. Cephalalgia Int J Headache 27(9):1069–1073
31. Kudrow L (1976) Plasma testosterone levels in cluster headache preliminary results. Headache 16(1):28–31
32. Leone M, Bussone G (1993) A review of hormonal findings in cluster headache. Evidence for hypothalamic involvement Cephalalgia : an international journal of headache 13(5):309–317
33. Cevoli S, Pizza F, Grimaldi D, Nicodemo M, Favoni V, Pierangeli G et al (2011) Cerebrospinal fluid hypocretin-1 levels during the active period of cluster headache. Cephalalgia Int J Headache 31(8):973–976
34. Bahra A, May A, Goadsby PJ (2002) Cluster headache: a prospective clinical study with diagnostic implications. Neurology 58(3):354–361
35. Russell MB (2004) Epidemiology and genetics of cluster headache. Lancet Neurol 3(5): 279–283

36. Rainero I, Gallone S, Valfre W, Ferrero M, Angilella G, Rivoiro C et al (2004) A polymorphism of the hypocretin receptor 2 gene is associated with cluster headache. Neurology 63(7):1286–1288
37. Pinessi L, Binello E, De Martino P, Gallone S, Gentile S, Rainero I et al (2007) The 1246G→2 polymorphism of the HCRTR2 gene is not associated with migraine. Cephalalgia Int J Headache 27(8):945–949
38. Zaremba S, Holle D, Wessendorf TE, Diener HC, Katsarava Z, Obermann M (2012) Cluster headache shows no association with rapid eye movement sleep. Cephalalgia Int J Headache 32(4):289–296
39. Evers S, Barth B, Frese A, Husstedt IW, Happe S (2014) Sleep apnea in patients with cluster headache: a case-control study. Cephalalgia Int J Headache 34(10):828–832
40. Matharu MS. Functional and structural neuroimaging in primary headache disorders. Phd thesis. 2006. Headache Group, Institute of Neurology and The National Hospital for Neurology and Neurosurgery, Queen Square, London
41. May A, Bahra A, Buchel C, Frackowiak RS, Goadsby PJ (1998) Hypothalamic activation in cluster headache attacks. Lancet 352(9124):275–278
42. Morelli N, Pesaresi I, Cafforio G, Maluccio MR, Gori S, Di Salle F et al (2009) Functional magnetic resonance imaging in episodic cluster headache. J Headache Pain 10(1):11–14
43. Sanchez del Rio M, Alvarez Linera J (2004) Functional neuroimaging of headaches. Lancet Neurol 3(11):645–651
44. Fox MD, Raichle ME (2007) Spontaneous fluctuations in brain activity observed with functional magnetic resonance imaging. Nat Rev Neurosci 8(9):700–711
45. Biswal B, Yetkin FZ, Haughton VM, Hyde JS (1995) Functional connectivity in the motor cortex of resting human brain using echo-planar MRI. Magn Reson Med Off J Soc Magn Reson Med Soc Magn Reson Med 34(4):537–541
46. Damoiseaux JS, Rombouts SA, Barkhof F, Scheltens P, Stam CJ, Smith SM et al (2006) Consistent resting-state networks across healthy subjects. Proc Natl Acad Sci U S A 103(37):13848–13853
47. Raichle ME, MacLeod AM, Snyder AZ, Powers WJ, Gusnard DA, Shulman GL (2001) A default mode of brain function. Proc Natl Acad Sci U S A 98(2):676–682
48. Baliki MN, Geha PY, Apkarian AV, Chialvo DR (2008) Beyond feeling: chronic pain hurts the brain, disrupting the default-mode network dynamics. J Neurosci Off J Soc Neurosci 28(6):1398–1403
49. Rocca MA, Valsasina P, Absinta M, Colombo B, Barcella V, Falini A et al (2010) Central nervous system dysregulation extends beyond the pain-matrix network in cluster headache. Cephalalgia Int J Headache 30(11):1383–1391
50. Leone M, Franzini A, Bussone G (2001) Stereotactic stimulation of posterior hypothalamic gray matter in a patient with intractable cluster headache. N Engl J Med 345(19):1428–1429
51. Leone M, Franzini A, Broggi G, Mea E, Cecchini AP, Bussone G (2006) Acute hypothalamic stimulation and ongoing cluster headache attacks. Neurology 67(10):1844–1845
52. Leone M, Franzini A, Cecchini AP, Broggi G, Bussone G (2010) Hypothalamic deep brain stimulation in the treatment of chronic cluster headache. Ther Adv Neurol Disord 3(3):187–195
53. Leone M, Proietti Cecchini A, Franzini A, Broggi G, Cortelli P, Montagna P et al (2008) Lessons from 8 years' experience of hypothalamic stimulation in cluster headache. Cephalalgia Int J Headache 28(7):787–797; discussion 798
54. Leone M, Franzini A, Broggi G, Bussone G (2006) Hypothalamic stimulation for intractable cluster headache: long-term experience. Neurology 67(1):150–152
55. Fontaine D, Lazorthes Y, Mertens P, Blond S, Geraud G, Fabre N et al (2010) Safety and efficacy of deep brain stimulation in refractory cluster headache: a randomized placebo-controlled double-blind trial followed by a 1-year open extension. J Headache Pain 11(1):23–31
56. Bartsch T, Pinsker MO, Rasche D, Kinfe T, Hertel F, Diener HC et al (2008) Hypothalamic deep brain stimulation for cluster headache: experience from a new multicase series. Cephalalgia Int J Headache 28(3):285–295

57. Owen SL, Green AL, Davies P, Stein JF, Aziz TZ, Behrens T et al (2007) Connectivity of an effective hypothalamic surgical target for cluster headache. J Clin Neurosci Off J Neurosurg Soc Aust 14(10):955–960
58. Starr PA, Barbaro NM, Raskin NH, Ostrem JL (2007) Chronic stimulation of the posterior hypothalamic region for cluster headache: technique and 1-year results in four patients. J Neurosurg 106(6):999–1005
59. Leone M, Franzini A, Broggi G, May A, Bussone G (2004) Long-term follow-up of bilateral hypothalamic stimulation for intractable cluster headache. Brain J Neurol 127(Pt 10):2259–2264
60. Franzini A, Ferroli P, Leone M, Broggi G (2003) Stimulation of the posterior hypothalamus for treatment of chronic intractable cluster headaches: first reported series. Neurosurgery 52(5):1095–1099; discussion 9–101
61. Franzini A, Ferroli P, Leone M, Bussone G, Broggi G (2004) Hypothalamic deep brain stimulation for the treatment of chronic cluster headaches: a series report. Neuromodulation J Int Neuromodulation Soc 7(1):1–8
62. Matharu MS, Zrinzo L (2010) Deep brain stimulation in cluster headache: hypothalamus or midbrain tegmentum? Curr Pain Headache Rep 14(2):151–159
63. Brewer AC, Trentman TL, Ivancic MG, Vargas BB, Rebecca AM, Zimmerman RS et al (2013) Long-term outcome in occipital nerve stimulation patients with medically intractable primary headache disorders. Neuromodulation J Int Neuromodulation Soc 16(6):557–562; discussion 63–4
64. Franzini A, Messina G, Cordella R, Marras C, Broggi G (2010) Deep brain stimulation of the posteromedial hypothalamus: indications, long-term results, and neurophysiological considerations. Neurosurg Focus 29(2):E13
65. Denuelle M, Fabre N, Payoux P, Chollet F, Geraud G (2007) Hypothalamic activation in spontaneous migraine attacks. Headache 47(10):1418–1426
66. Matharu MS, Cohen AS, McGonigle DJ, Ward N, Frackowiak RS, Goadsby PJ (2004) Posterior hypothalamic and brainstem activation in hemicrania continua. Headache 44(8):747–761
67. Kupers RC, Gybels JM, Gjedde A (2000) Positron emission tomography study of a chronic pain patient successfully treated with somatosensory thalamic stimulation. Pain 87(3):295–302
68. Holle D, Naegel S, Krebs S, Gaul C, Gizewski E, Diener HC et al (2011) Hypothalamic gray matter volume loss in hypnic headache. Ann Neurol 69(3):533–539
69. Rosen SD, Paulesu E, Frith CD, Frackowiak RS, Davies GJ, Jones T et al (1994) Central nervous pathways mediating angina pectoris. Lancet 344(8916):147–150
70. Blankstein U, Chen J, Diamant NE, Davis KD (2010) Altered brain structure in irritable bowel syndrome: potential contributions of pre-existing and disease-driven factors. Gastroenterology 138(5):1783–1789
71. Boghi A, Sterpone S, Sales S, D'Agata F, Bradac GB, Zullo G et al (2011) In vivo evidence of global and focal brain alterations in anorexia nervosa. Psychiatry Res 192(3):154–159
72. Kurth F, Narr KL, Woods RP, O'Neill J, Alger JR, Caplan R et al (2011) Diminished gray matter within the hypothalamus in autism disorder: a potential link to hormonal effects? Biol Psychiatry 70(3):278–282
73. Hoeft F, Lightbody AA, Hazlett HC, Patnaik S, Piven J, Reiss AL (2008) Morphometric spatial patterns differentiating boys with fragile X syndrome, typically developing boys, and developmentally delayed boys aged 1 to 3 years. Arch Gen Psychiatry 65(9):1087–1097
74. Kim SJ, Lyoo IK, Lee YS, Lee JY, Yoon SJ, Kim JE et al (2009) Gray matter deficits in young adults with narcolepsy. Acta Neurol Scand 119(1):61–67
75. Douaud G, Gaura V, Ribeiro MJ, Lethimonnier F, Maroy R, Verny C et al (2006) Distribution of grey matter atrophy in Huntington's disease patients: a combined ROI-based and voxel-based morphometric study. Neuroimage 32(4):1562–1575
76. Sommer C, Hauser W, Gerhold K, Joraschky P, Petzke F, Tolle T et al (2008) Etiology and pathophysiology of fibromyalgia syndrome and chronic widespread pain. Schmerz 22(3):267–282

77. Van Den Eede F, Moorkens G (2008) HPA-axis dysfunction in chronic fatigue syndrome: clinical implications. Psychosomatics 49(5):450
78. Chang L, Sundaresh S, Elliott J, Anton PA, Baldi P, Licudine A et al (2009) Dysregulation of the hypothalamic-pituitary-adrenal (HPA) axis in irritable bowel syndrome. Neurogastroenterol Motil Off J Eur Gastrointest Motil Soc 21(2):149–159
79. Peres MF, Sanchez del Rio M, Seabra ML, Tufik S, Abucham J, Cipolla-Neto J et al (2001) Hypothalamic involvement in chronic migraine. J Neurol Neurosurg Psychiatry 71(6): 747–751
80. Baumber L, Sjostrand C, Leone M, Harty H, Bussone G, Hillert J et al (2006) A genome-wide scan and HCRTR2 candidate gene analysis in a European cluster headache cohort. Neurology 66(12):1888–1893

Chapter 14
Pathophysiology of Medication Overuse Headache: Current Status and Future Directions

Signe Bruun Munksgaard and Frank Porreca

14.1 Introduction

Migraine is thought to be the world's third most common neurological disorder and is ranked as seventh among the leading causes of disability [1, 2]. Its economic impact is measured in billions of dollars [3, 4], and it accounts for 2.9 % of work time lost due to disability [2]. Notably, these high rankings for the burden of migraine occur in spite of the fact that treatments are available, clearly suggesting that current migraine therapeutics are inadequate [2, 5].

The introduction of triptans represented a significant advance in the therapeutic management of migraine, and triptans as a class are considered the drugs of choice for management of migraine [6–8]. However, the frequent use of triptans can lead to the conversion of episodic migraine into a chronic condition, frequently referred to as medication overuse headache (MOH) [9, 10]. MOH was recognized before the introduction of triptans from patients overusing ergots, opioids, and other analgesics [11–13]. The overuse of analgesics in the treatment of cluster headache or tension-type headache (TTH) [14] can also lead to MOH [15, 16]. Importantly, MOH is more disabling than episodic headache and much more difficult to treat [15–17]. MOH is a chronic headache, affecting patients mainly between 20 and 50 years of age and thus in their most productive years [18–21].

S.B. Munksgaard, MD, PhD (✉)
Department of Neurology, Danish Headache Center, Glostrup Hospital, University of Copenhagen, Glostrup, Denmark
e-mail: signebm@gmail.com

F. Porreca, PhD (✉)
Department of Pharmacology and Anesthesiology, Arizona Health Sciences Center, University of Arizona, Tucson, AZ, USA
e-mail: frankp@u.arizona.edu

M. Ashina, P. Geppetti (eds.), *Pathophysiology of Headaches: From Molecule to Man*, Headache, DOI 10.1007/978-3-319-15621-7_14

The treatments employed in MOH are primarily a combination of information on the disease and further detoxification from the overused drug combined with medical prophylactic treatment and sometimes behavioral therapy. Simple information can be sufficient for MOH patients who do not suffer from comorbidities and who overuse medications that do not cause severe withdrawal symptoms. They are often able to reduce their medication intake and thus experience a reduction in headache frequency to episodic headache [22, 23]. In many patients, especially those with comorbidities or previous relapse to MOH, these treatments are not effective in reducing headache frequency. Current prophylactic migraine treatments arise not from rational, evidence-based approaches, but from serendipity and presumptions of efficacy based on the success of related compounds within a pharmacologic class [18–21]. The limited therapeutic efficacy of currently available prophylactic treatments against chronic migraine and the high relapse rate in MOH underscore a strong medical need for the understanding of basic mechanisms underlying MOH.

14.2 Clinical Description

The possibility that aggressive analgesic therapy can lead to enhanced pain has long been recognized in the clinical management of headache and gave rise to terms such as rebound headache, medication misuse headache, or transformed migraine, later defined as MOH [24]. MOH is characterized by 15 or more headache days/month that result from excessive use of antimigraine drugs or analgesics. The current consensus diagnostic criteria for MOH are summarized in Table 14.1. The general worldwide prevalence of MOH is estimated to be at least 1 % in adults and 0.5 % in adolescents [25–27], and approximately 33 % of individuals reporting chronic daily headache meet the criteria for the overuse of medication [25, 28, 29]. Clinical surveys indicate that only patients predisposed to headache will develop MOH when overusing analgesics [30] and that patients with MOH most commonly have migraine, followed by TTH [14]. The risk of developing MOH increases with lower socioeconomic status and female sex [31–33]; the male/female ratio is 1:3.5 [34]. The prevalence of psychiatric disorders such as obsessive-compulsive disorders, depression, and anxiety is greatly increased in persons with MOH [31, 35–38]. These psychiatric disorders are described as significant risk factors for developing

Table 14.1 The ICHD-3 criteria for medication overuse headache

A. Headache present on 15 or more days/month in a patient with a preexisting headache disorder
B. Regular overuse for more than 3 months of one or more drugs that can be taken for acute and/or symptomatic treatment of headache
1. Ergotamine, triptans, opioids, or combination analgesics on 10 or more days/month
2. Simple analgesics on 15 or more days/month
3. Any combination of acute/symptomatic drugs on 10 or more days/month without overuse of any single class alone
C. Not better accounted for by any other ICHD-3 diagnosis

MOH but also complicate treatment and increase the relapse rate to MOH after withdrawal [31, 35–38]. Other risk factors include stressful life events, sleep disturbances, obesity, increased caffeine consumption, and elevated baseline headache frequency (10 or more per month) [39–42].

Both underlying headache type and overused drug appear to contribute to the pathogenesis of MOH. Most patients with MOH have migraine, and most (>95 %) of migraine sufferers regularly use acute medications that include analgesics, migraine-specific medications (triptans, ergot medications), opioids, or a combination thereof [43]. The choice and frequency of use of acute medications have a major influence on migraine prognosis [44, 45]. Acute medications, particularly opioids, barbiturate-containing combination analgesics, as well as triptans and ergots, potentiate the risk of progression [30, 46–50]. The intensity and frequency of recurrent headaches was increased with opiate use and diminished when opiate administration was terminated [11, 12]. Triptans tend to produce MOH over a shorter dosing regimen than either ergots or analgesics [51]. In a prospective study, it was found that the mean interval between first dose and MOH for triptans was 1.7 years, whereas the mean interval for ergots and analgesics was 2.7 years and 4.8 years, respectively [51]. The duration of withdrawal headache was less severe and shorter in patients overusing triptans (4.1 days) than in patients overusing ergots (6.7 days) or analgesics (9.5 days) [13]. Additionally, the MOH headache pattern reflected both the underlying headache type and the medication overused. Triptans were more likely to produce a daily migraine-like headache or an increase in migraine frequency, whereas simple analgesics and ergots were more likely to produce a TTH-like headache [13, 51]. Studies on subgroups of MOH patients overusing different types of drugs [13, 52–54] show different responses in neuroimaging and cortical potentials.

MOH across different medications appears to share some common neurobiological pathways, including those that modulate motivation, reward, and behavioral control [55]. A large proportion of patients with chronic daily headache with concomitant medication overuse fulfilled the diagnostic criteria for substance dependence [56]. Dependence-like behavior is more frequently observed in patients with chronic headaches without MOH who smoke, who are obese, and who use sleep medications and tranquilizers [31]. Approximately two-thirds of MOH patients were considered to be dependent on acute treatments of headaches, and many of these individuals had migraine as preexisting primary headache, and most of them overused opioid analgesics. The severity of dependence predicted a poorer outcome of treatment for MOH [22, 57–59]. A correlation has been observed between high severity of dependence scale (SDS) score before treatment and a poor outcome of treatment for MOH [58, 57, 60].

14.3 Neurophysiological Mechanisms of MOH

The mechanisms behind the development of MOH are largely unknown, but both human and animal studies indicate drug-induced modifications of peripheral and central pain transmission and modulatory pathways. Cutaneous allodynia, especially when present at extracephalic sites in premonitory phases of migraine, during a migraine attack, and in the post-drome period, is a clinical sign that suggests the occurrence of central sensitization in humans [61–64]. Individuals with migraine present a greater prevalence of cutaneous allodynia than do those with nonmigraine headaches, and patients with MOH are more likely to develop allodynia than individuals with episodic migraine [15, 63, 65–67]. The mechanisms that trigger a migraine attack are largely unknown but are thought to reflect a disorder of the brain [68, 69]. Evaluating mechanisms that lead to pain in migraine headache is difficult as the pain is intermittent and there is no tissue injury to serve as an obvious trigger [69]. Increasing evidence points ultimately to engagement and sensitization of the trigeminovascular system in the genesis of pain resulting from a migraine episode. Migraine sufferers were found to have increased excitability of the trigeminal nociceptive pathway, both during a migraine episode and during the interictal period [63, 70–72].

Perivascular stimulation of the dura results in pain referred to the head [73, 74]. Activation of trigeminal afferent fibers terminating in the dura can release excitatory mediators accompanied by neurogenic vasodilation of dural blood vessels, further release of pronociceptive mediators, degranulation of mast cells, and extravasation of plasma proteins, thus sensitizing the peripheral terminals of trigeminal nociceptors [74–76]. This cascade of events can also result in enhanced nociceptive transmission into the trigeminal nucleus caudalis and promoting central sensitization of second-order neurons in this brainstem nucleus [71, 72]. Consequently, nociceptive inputs are transmitted to higher brain centers including the thalamus, hypothalamus, and cortical sites, manifesting as migraine pain [62, 71, 72, 77]. Because migraineurs are most vulnerable to develop MOH, and MOH commonly resembles migraine in quality, it is likely that migraine and MOH might share some neural mechanisms. Extrapolation of potential mechanisms of MOH from animal models of migraine-related pain thus seems reasonable.

The application of inflammatory mediators to the dura mater of rodents and the resulting trigeminal sensitization have been used as an animal model of migraine pain [78–80]. Dural inflammation has resulted in electrophysiologic, neurochemical, and behavioral evidence of central sensitization [61, 81, 82]. The development and progression of central sensitization has been shown by the progressive spread of cutaneous allodynia, i.e., enhanced neuronal responses from stimuli applied at cephalic and extracephalic sites [71, 81, 83, 84]. Additionally, dural inflammation was accompanied by enhanced descending facilitation demonstrated by increased discharge of "ON" cells in the rostral ventromedial medulla (RVM). Inactivation of this area with microinjection of local anesthetics abolishes cutaneous allodynia [81]. The generalized spread, and delayed appearance, of cutaneous allodynia implicated a role for central modulation in this preclinical migraine model that is reminiscent of cutaneous allodynia observed in many humans during a migraine attack [61].

Exposure of rats to either opioids or triptans over a period of days produced persistent biochemical changes that appear relevant to promoting a sensitized state of nociceptive transmission [81, 85–88]. During the 7-day period of triptan or opioid infusion, the stimulus required to elicit the orbital and paw withdrawal reflex was shown to decrease gradually, demonstrating cutaneous allodynia that showed a time-dependent reversal toward predrug baseline levels after the infusion was discontinued. These observations suggested that the pain system can be modulated by acute medication and that it can normalize after detoxification [88]. Following recovery to baseline sensory thresholds, animals with prior triptan treatment have increased sensitivity to provocative triggers including environmental stimuli [88–91]. The persistent hypersensitivity to provocative triggers observed following pre-treatment of animals with triptans or opioids was termed "latent sensitization" [88, 92]. Challenge of rats with either opioid or triptan-induced latent sensitization with either bright light stress or a nitric oxide donor was shown to produce a delayed and generalized cutaneous allodynia that is detected in the periorbital region as well as in the hind paw [88, 92]. The delayed and generalized cutaneous allodynia was blocked by inactivation of the RVM with local anesthetics supporting a role of descending pain modulatory systems in promoting central sensitization in this model of MOH, similar to that observed with acute activation of dural nociceptors with inflammatory mediators [80]. Evidence supports a persistent sensitization as these animals previously treated with either triptans or opioids maintain increased sensitivity to human migraine triggers such as stress and nitric oxide donors long after discontinuation of drug administration [88, 92–94].

The mechanisms underlying enhanced sensitivity to innocuous and provocative stimuli are not fully known but may be related to medication-induced adaptations in both primary afferents and central pain transmission pathways. Persistent increased labeling for both CGRP and neuronal NOS (nNOS) in identified dural afferents of the trigeminal ganglia has been observed following pretreatment with either morphine or triptans [90, 92]. Notably, however, the apparent expression of CGRP and nNOS in trigeminal ganglion cells persists long after discontinuation of either opiate or triptan exposure (for at least 2 weeks) and resolution of cutaneous allodynia [88]. These persistent changes in CGRP and nNOS could underlie the sensitization to provocative triggers [90]. Treatment with selective inhibitors of nNOS was demonstrated to be capable of blocking stress-induced cutaneous allodynia [88]. Rats with latent sensitization resulting from either morphine or triptans also showed increased release of CGRP following challenge with provocative triggers [88] consistent with observations during migraine attack in humans [95, 96]; but see [97]. These observations are consistent with provocative studies in migraineurs where nitroglycerin elicits attacks that are indistinguishable from spontaneous migraine that are accompanied by increased blood levels of CGRP [95, 96, 98]. These observations are also consistent with the suggestion that MOH may be associated with a state of central sensitization in afflicted individuals.

Evidence also exists supporting medication-induced changes in cortical excitability. While prophylactic treatments for migraine, such as topiramate and valproate, reduce the frequency of cortical spreading depression (CSD) events [19], sustained exposure to paracetamol has recently been shown to increase the frequency of CSD events induced by cortical potassium chloride as well as c-fos expression in trigeminal nucleus caudalis [99, 100]. Likewise, pretreatment of rats with sumatriptan was shown to significantly decrease the threshold for electrically induced CSD [93]. Topiramate normalized the decreased CSD threshold as well as stress-induced behavioral withdrawal thresholds in sumatriptan-treated rats compared to saline-treated animals. Additionally, both CSD and environmental stress increased c-fos expression in trigeminal nucleus caudalis (TNC) of sumatriptan- but not saline-treated rats, and these effects were blocked by topiramate. Sumatriptan exposure thus produces long-lasting increased susceptibility to stimuli that could be associated with migraine attack that includes both lowered CSD threshold and enhanced consequences of CSD events (increased activation of TNC) and may represent an underlying biological mechanism of medication overuse headache related to triptans.

Other studies have also implicated medication-induced changes in brain neurotransmitter systems. Exposure to paracetamol and triptans for 15–30 days alters the serotonin system in the rat brain [89, 101–103], and this may also be the case in patients with MOH. In MOH patients, a lower serotonin level than controls and a more rapid uptake in platelets have been reported [104, 105]. Serotonin is important for cortical pain processing and plays a pivotal role in affective disorders that are often found in patients with MOH [31, 35]. A lower endocannabinoid level [23] and a faster degradation of endocannabinoids [106, 107] have been demonstrated in MOH patients with migraine as primary headache. A role of endocannabinoids may be to inhibit transmission from nociceptive afferents [108] suggesting a possible impaired pain inhibition in MOH patients.

14.4 Pain Perception in MOH

Different methods have been used to evaluate pain sensitivity in patients with MOH, generally showing an increase in sensitivity with higher headache frequency [109, 110]. Alterations in pain perception between MOH patients and healthy controls support the presence of a sensitized state in patients with MOH [52, 107, 109, 111, 112]. One study found that patients with MOH are more sensitive to pressure pain than patients with chronic TTH and chronic migraine without medication overuse [109]. Perrotta et al. have demonstrated decreased thresholds for eliciting nociceptive withdrawal reflexes in MOH patients before detoxification compared with healthy volunteers and found that the stimulation needed to elicit these reflexes increased toward baseline after detoxification both after 10 days and after 2 months indicating decreased sensitization after withdrawal [107, 112]. An additional study on pain perception in MOH patients showed that the intensity of pressure pain above the pain threshold continued to decrease during the first year after detoxification [113].

Preclinical evidence suggests that enhanced pain perception of migraineurs could be linked to alterations in descending pain modulatory mechanisms. In rats with opiate- or triptan-induced latent sensitization, inactivation of the RVM with local anesthetics blocks cutaneous allodynia associated with provocative stimuli [88, 90, 92]. An impairment of the diffuse noxious inhibitory controls (i.e., "DNIC") known in humans as conditioned pain modulation (CPM) has also been noted [114]. The loss of DNIC is consistent with many clinical observations made with patients with functional pain conditions, including migraine [115, 116], TTH [115, 117], and MOH [112]. Patients with MOH show heightened sensitivity to electrically evoked reflexes, accompanied by increased pain rating and diminished CPM [112]. Studies with rodents showed that persistent morphine exposure reduced the activation threshold of TNC neurons while expanding their receptive fields, indicating the presence of central sensitization [91]. In control animals, activation of these neurons by application of stimuli to their receptive fields in the ophthalmic region was inhibited by placing the tail of control rats in hot water, demonstrating the DNIC response [91]. In contrast, the DNIC response was lost in animals treated with morphine [91]. Importantly, the apparent loss of DNIC in animals treated with morphine could be reinstated by inactivation of the RVM [91]. This suggests that enhanced descending facilitation could present as a loss of inhibition. Distinguishing loss of inhibition from enhanced facilitation has been difficult to dissect in humans. Collectively, however, such studies suggest that an abnormality of pain modulatory circuits results in a net loss of inhibition, possibly due to decreased inhibition or increased facilitation or both, and this loss is likely to be important in the development of MOH.

The possibility of dysfunction of descending pain modulatory pathways as contributing factors to MOH is also supported by imaging studies. In patients with MOH and migraine as the primary headache, changes in both cortical and midbrain pain-related areas have been observed [118–121]. One study using functional magnetic resonance imaging (fMRI) found reduced activity in the right supramarginal gyrus and in the superior and inferior parietal cortex, which normalized 6 months after detoxification [118], implying a reversible change in the pain system caused by the medication overuse. Another study showed increased gray matter volume in the periaqueductal gray (PAG), a structure highly important in the descending pain response, and reduced gray matter volume in several cortical pain-related structures in migraine patients with MOH [121]. In patients with a significant reduction in headache frequency after treatment, the PAG returned to normal [121].

14.5 Future Perspectives

The mechanisms that underlie MOH are just beginning to be uncovered. Future work may begin to explore mechanisms behind MOH with TTH as the underlying headache; TTH is the most common primary headache type and is an important target for future studies. The development of central sensitization in MOH and the differences in MOH patients according to the effect of detoxification could be a target for future studies assessing pain perception. Work from animal models showing increased effectiveness of headache triggers following induction of latent sensitization [94] has yet to determine numerous variables of relevance to patients. It remains unknown whether sensitization is completely reversible, whether the threshold for developing MOH after withdrawal of medication is decreased, and if the critical frequency of medication intake could be reduced. This is supported by the high relapse rate in MOH patients after treatment. Future clinical strategies of MOH prevention should take these into account. Treatment with onabotulinumtoxinA has proven superior to placebo in reducing headache frequency in a post hoc subgroup analysis on patients with MOH [122]. The mechanisms behind the effect on chronic migraine with and without MOH are still unknown and should be target for further investigation as the use of onabotulinumtoxinA for chronic migraine with MOH is increasing. The development in brain imaging will undoubtedly be valuable for further assessing alterations in pain-processing structures in MOH patients and the possible changes from before to after withdrawal. Research in the pathophysiology behind chronic migraine and TTH without medication overuse will aid the investigation into the relationship and differences between chronic headache with and without medication overuse.

Acknowledgements The authors gratefully acknowledge the assistance and helpful suggestions of Drs. Michael Ossipov, Jennifer Xie (University of Arizona), and Ian Meng (University of New England) in the preparation of the manuscript as well as the support of NIH (NS069572).

References

1. Vos T, Flaxman AD, Naghavi M, Lozano R, Michaud C, Ezzati M et al (2012) Years lived with disability (YLDs) for 1160 sequelae of 289 diseases and injuries 1990–2010: a systematic analysis for the Global Burden of Disease Study 2010. Lancet 380(9859):2163–2196
2. Steiner TJ, Stovner LJ, Birbeck GL (2013) Migraine: the seventh disabler. Headache J Head Face Pain 53(2):227–229
3. Andlin-Sobocki P, Jonsson B, Wittchen HU, Olesen J (2005) Cost of disorders of the brain in Europe. Eur J Neurol 12(Suppl 1):1–27
4. Goadsby PJ (2009) Pathophysiology of migraine. Neurol Clin 27(2):335–360
5. Burden WHOaLT (2011) Atlas of headache disorders and resources in the world 2011. WHO, Geneva
6. Hoffmann J, Goadsby PJ (2014) Emerging targets in migraine. CNS Drugs 28(1):11–17
7. Dodick DW, Silberstein S, Dahlof CG (2002) Is there a preferred triptan? Headache 42(1):1–7
8. Goadsby PJ, Sprenger T (2010) Current practice and future directions in the prevention and acute management of migraine. Lancet Neurol 9(3):285–298
9. Ghiotto N, Sances G, Galli F, Tassorelli C, Guaschino E, Sandrini G et al (2009) Medication overuse headache and applicability of the ICHD-II diagnostic criteria: 1-year follow-up study (CARE I protocol). Cephalalgia 29(2):233–243
10. Olesen J, Bousser MG, Diener HC, Dodick D, First M, Goadsby PJ et al (2006) New appendix criteria open for a broader concept of chronic migraine. Cephalalgia 26(6):742–746
11. Isler H (1982) Migraine treatment as a cause of chronic migraine. In: Rose FC (ed) Advances in migraine research and therapy. Raven, New York, pp 159–164
12. Schofferman J (1993) Long-term use of opioid analgesics for the treatment of chronic pain of nonmalignant origin. J Pain Symptom Manage 8(5):279–288
13. Katsarava Z, Fritsche G, Muessig M, Diener HC, Limmroth V (2001) Clinical features of withdrawal headache following overuse of triptans and other headache drugs. Neurology 57(9):1694–1698
14. Rapoport A, Stang P, Gutterman DL, Cady R, Markley H, Weeks R et al (1996) Analgesic rebound headache in clinical practice: data from a physician survey. Headache 36(1):14–19

15. Dodick D, Silberstein S (2006) Central sensitization theory of migraine: clinical implications. Headache 46(Suppl 4):S182–S191
16. Dodick DW (2006) Clinical practice. Chronic daily headache. N Engl J Med 354(2):158–165
17. Katsarava Z, Manack A, Yoon M-S, Obermann M, Becker H, Dommes P et al (2011) Chronic migraine: classification and comparisons. Cephalalgia 31(5):520–529
18. Galletti F, Cupini LM, Corbelli I, Calabresi P, Sarchielli P (2009) Pathophysiological basis of migraine prophylaxis. Prog Neurobiol 89(2):176–192
19. Ayata C, Jin H, Kudo C, Dalkara T, Moskowitz MA (2006) Suppression of cortical spreading depression in migraine prophylaxis. Ann Neurol 59(4):652–661
20. Ayata C (2009) Spreading depression: from serendipity to targeted therapy in migraine prophylaxis. Cephalalgia 29(10):1095–1114
21. Dodick DW (2009) Tonabersat for migraine prevention: new life or last gasp? Lancet Neurol 8(8):693–695
22. Grande RB, Aaseth K, Saltyte Benth J, Gulbrandsen P, Russell MB, Lundqvist C (2009) The Severity of Dependence Scale detects people with medication overuse: the Akershus study of chronic headache. J Neurol Neurosurg Psychiatry 80(7):784–789
23. Rossi C, Pini LA, Cupini ML, Calabresi P, Sarchielli P (2008) Endocannabinoids in platelets of chronic migraine patients and medication-overuse headache patients: relation with serotonin levels. Eur J Clin Pharmacol 64(1):1–8
24. Silberstein SD, Liu D (2002) Drug overuse and rebound headache. Curr Pain Headache Rep 6(3):240–247
25. Colas R, Munoz P, Temprano R, Gomez C, Pascual J (2004) Chronic daily headache with analgesic overuse: epidemiology and impact on quality of life. Neurology 62(8):1338–1342
26. Dyb G, Holmen TL, Zwart JA (2006) Analgesic overuse among adolescents with headache: the Head-HUNT-Youth Study. Neurology 66(2):198–201
27. Zwart JA, Dyb G, Hagen K, Svebak S, Stovner LJ, Holmen J (2004) Analgesic overuse among subjects with headache, neck, and low-back pain. Neurology 62(9):1540–1544
28. Castillo J, Munoz P, Guitera V, Pascual J (1999) Kaplan Award 1998. Epidemiology of chronic daily headache in the general population. Headache 39(3):190–196
29. Chakravarty A (2003) Chronic daily headaches: clinical profile in Indian patients. Cephalalgia 23(5):348–353
30. Williams L, O'Connell K, Tubridy N (2008) Headaches in a rheumatology clinic: when one pain leads to another. Eur J Neurol 15(3):274–277
31. Hagen K, Linde M, Steiner TJ, Stovner LJ, Zwart JA (2012) Risk factors for medication-overuse headache: an 11-year follow-up study. The Nord-Trondelag Health Studies. Pain 153(1):56–61
32. Jonsson P, Linde M, Hensing G, Hedenrud T (2012) Sociodemographic differences in medication use, health-care contacts and sickness absence among individuals with medication-overuse headache. J Headache Pain 13(4):281–290
33. Evers S, Marziniak M (2010) Clinical features, pathophysiology, and treatment of medication-overuse headache. Lancet Neurol 9(4):391–401
34. Diener HC, Silberstein SD (2006) Medication overuse headaches. Lippincott Williams & Wilkins, Philadelphia
35. Atasoy HT, Atasoy N, Unal AE, Emre U, Sumer M (2005) Psychiatric comorbidity in medication overuse headache patients with pre-existing headache type of episodic tension-type headache. Eur J Pain 9(3):285–291
36. da Silva A, Costa EC, Gomes JB, Leite FM, Gomez RS, Vasconcelos LP et al (2010) Chronic headache and comorbidities: a two-phase, population-based. Cross-Sectional Study. Headache 50(8):1306–1312
37. Zebenholzer K, Thamer M, Wober C (2012) Quality of life, depression, and anxiety 6 months after inpatient withdrawal in patients with medication overuse headache an observational study. Clin J Pain 28(4):284–290

38. Lucas C, Lanteri-Minet M, Massiou H, Radat F, Pradalier A, Nachit-Ouinekh F, El HA (2007) Medical and psychological characteristics of patients with chronic daily migrainous features: compared with migraineurs in the general French population. GRIM 3 survey. In: Jensen R, Diener HC, Olesen J (eds) Headache clinics – organization, patients and treatment. Oxford University Press, London
39. Diener HC, Dodick DW, Goadsby PJ, Lipton RB, Olesen J, Silberstein SD (2012) Chronic migraine – classification, characteristics and treatment. Nat Rev Neurol 8(3):162–171
40. Ashina S, Lyngberg A, Jensen R (2010) Headache characteristics and chronification of migraine and tension-type headache: a population-based study. Cephalalgia 30(8):943–954
41. Bigal ME, Lipton RB (2009) The epidemiology, burden, and comorbidities of migraine. Neurol Clin 27(2):321–334
42. Bigal ME, Lipton RB, Holland PR, Goadsby PJ (2007) Obesity, migraine, and chronic migraine: possible mechanisms of interaction. Neurology 68(21):1851–1861
43. Zwart JA, Dyb G, Hagen K, Svebak S, Holmen J (2003) Analgesic use: a predictor of chronic pain and medication overuse headache: the Head-HUNT Study. Neurology 61(2):160–164
44. Bigal ME, Lipton RB (2008) Excessive acute migraine medication use and migraine progression. Neurology 71(22):1821–1828
45. Diamond S, Bigal ME, Silberstein S, Loder E, Reed M, Lipton RB (2007) Patterns of diagnosis and acute and preventive treatment for migraine in the United States: results from the American Migraine Prevalence and Prevention study. Headache 47(3):355–363
46. Bahra A, Walsh M, Menon S, Goadsby PJ (2003) Does chronic daily headache arise de novo in association with regular use of analgesics? Headache 43(3):179–190
47. Bigal ME, Lipton RB (2008) Clinical course in migraine: conceptualizing migraine transformation. Neurology 71(11):848–855
48. Paemeleire K, Bahra A, Evers S, Matharu MS, Goadsby PJ (2006) Medication-overuse headache in patients with cluster headache. Neurology 67(1):109–113
49. Scher AI, Stewart WF, Liberman J, Lipton RB (1998) Prevalence of frequent headache in a population sample. Headache 38(7):497–506
50. Wilkinson SM, Becker WJ, Heine JA (2001) Opiate use to control bowel motility may induce chronic daily headache in patients with migraine. Headache 41(3):303–309
51. Limmroth V, Katsarava Z, Fritsche G, Przywara S, Diener HC (2002) Features of medication overuse headache following overuse of different acute headache drugs. Neurology 59(7):1011–1014
52. Coppola G, Curra A, Di Lorenzo C, Parisi V, Gorini M, Sava SL et al (2010) Abnormal cortical responses to somatosensory stimulation in medication-overuse headache. BMC Neurol 10:126
53. Fumal A, Laureys S, Di Clemente L, Boly M, Bohotin V, Vandenheede M et al (2006) Orbitofrontal cortex involvement in chronic analgesic-overuse headache evolving from episodic migraine. Brain J Neurol 129(Pt 2):543–550
54. Curra A, Coppola G, Gorini M, Porretta E, Bracaglia M, Di Lorenzo C et al (2011) Drug-induced changes in cortical inhibition in medication overuse headache. Cephalalgia Int J Headache 31(12):1282–1290
55. Calabresi P, Cupini LM (2005) Medication-overuse headache: similarities with drug addiction. Trends Pharmacol Sci 26(2):62–68
56. Fuh JL, Wang SJ, Lu SR, Juang KD (2005) Does medication overuse headache represent a behavior of dependence? Pain 119(1–3):49–55
57. Radat F, Creac'h C, Guegan-Massardier E, Mick G, Guy N, Fabre N et al (2008) Behavioral dependence in patients with medication overuse headache: a cross-sectional study in consulting patients using the DSM-IV criteria. Headache 48(7):1026–1036
58. Lundqvist C, Grande RB, Aaseth K, Russell MB (2012) Dependence scores predict prognosis of medication overuse headache: a prospective cohort from the Akershus study of chronic headache. Pain 153(3):682–686
59. Corbelli I, Caproni S, Eusebi P, Sarchielli P (2012) Drug-dependence behaviour and outcome of medication-overuse headache after treatment. J Headache Pain 13(8):653–660

60. Radat F, Lanteri-Minet M, Nachit-Ouinekh F, Massiou H, Lucas C, Pradalier A et al (2009) The GRIM2005 study of migraine consultation in France. III: psychological features of subjects with migraine. Cephalalgia 29(3):338–350
61. Burstein R (2001) Deconstructing migraine headache into peripheral and central sensitization. Pain 89(2–3):107–110
62. Burstein R, Collins B, Jakubowski M (2004) Defeating migraine pain with triptans: a race against the development of cutaneous allodynia. Ann Neurol 55(1):19–26
63. Burstein R, Cutrer MF, Yarnitsky D (2000) The development of cutaneous allodynia during a migraine attack clinical evidence for the sequential recruitment of spinal and supraspinal nociceptive neurons in migraine. Brain 123(Pt 8):1703–1709
64. Charles A (2013) The evolution of a migraine attack – a review of recent evidence. Headache 53(2):413–419
65. Bigal ME, Ashina S, Burstein R, Reed ML, Buse D, Serrano D et al (2008) Prevalence and characteristics of allodynia in headache sufferers: a population study. Neurology 70(17):1525–1533
66. Lipton RB, Bigal ME (2008) Toward an epidemiology of refractory migraine: current knowledge and issues for future research. Headache 48(6):791–798
67. Bigal ME, Ferrari M, Silberstein SD, Lipton RB, Goadsby PJ (2009) Migraine in the triptan era: lessons from epidemiology, pathophysiology, and clinical science. Headache 49(Suppl 1):S21–S33
68. Akerman S, Holland PR, Goadsby PJ (2011) Diencephalic and brainstem mechanisms in migraine. Nat Rev Neurosci 12(10):570–584
69. Dussor G, Yan J, Xie JY, Ossipov MH, Dodick DW, Porreca F (2014) Targeting TRP channels for novel migraine therapeutics. ACS Chem Neurosci 5(11):1085–1096
70. Moulton EA, Burstein R, Tully S, Hargreaves R, Becerra L, Borsook D (2008) Interictal dysfunction of a brainstem descending modulatory center in migraine patients. PLoS One 3(11):e3799
71. Noseda R, Burstein R (2013) Migraine pathophysiology: anatomy of the trigeminovascular pathway and associated neurological symptoms, cortical spreading depression, sensitization, and modulation of pain. Pain 154(Suppl 1):S44–S53
72. Bernstein C, Burstein R (2012) Sensitization of the trigeminovascular pathway: perspective and implications to migraine pathophysiology. J Clin Neurol 8(2):89–99
73. Wolff HG (1948) Headache and other head pain. Oxford University Press, London
74. Goadsby PJ, Charbit AR, Andreou AP, Akerman S, Holland PR (2009) Neurobiology of migraine. Neuroscience 161(2):327–341
75. Levy D, Burstein R, Strassman AM (2006) Mast cell involvement in the pathophysiology of migraine headache: a hypothesis. Headache 46(Suppl 1):S13–S18
76. Goadsby PJ, Edvinsson L (1993) The trigeminovascular system and migraine: studies characterizing cerebrovascular and neuropeptide changes seen in humans and cats. Ann Neurol 33(1):48–56
77. Noseda R, Jakubowski M, Kainz V, Borsook D, Burstein R (2011) Cortical projections of functionally identified thalamic trigeminovascular neurons: implications for migraine headache and its associated symptoms. J Neurosci 31(40):14204–14217
78. Burstein R, Jakubowski M (2004) Analgesic triptan action in an animal model of intracranial pain: a race against the development of central sensitization. Ann Neurol 55(1):27–36
79. De Felice M, Eyde N, Dodick D, Dussor GO, Ossipov MH, Fields HL et al (2013) Capturing the aversive state of cephalic pain preclinically. Ann Neurol 74(2):257–265
80. Edelmayer RM, Ossipov MH, Porreca F (2012) An experimental model of headache-related pain. In: Luo ZD (ed) Methods in molecular biology. Pain research methods and protocols. Springer, New York, pp 109–120. 851. 2012/02/22 ed
81. Edelmayer RM, Vanderah TW, Majuta L, Zhang ET, Fioravanti B, De Felice M et al (2009) Medullary pain facilitating neurons mediate allodynia in headache-related pain. Ann Neurol 65(2):184–193
82. Oshinsky ML, Gomonchareonsiri S (2007) Episodic dural stimulation in awake rats: a model for recurrent headache. Headache 47(7):1026–1036

83. Burstein R, Yamamura H, Malick A, Strassman AM (1998) Chemical stimulation of the intracranial dura induces enhanced responses to facial stimulation in brain stem trigeminal neurons. J Neurophysiol 79(2):964–982
84. Yamamura H, Malick A, Chamberlin NL, Burstein R (1999) Cardiovascular and neuronal responses to head stimulation reflect central sensitization and cutaneous allodynia in a rat model of migraine. J Neurophysiol 81(2):479–493
85. Woolf CJ (1981) Intrathecal high dose morphine produces hyperalgesia in the rat. Brain Res 209(2):491–495
86. Vanderah TW, Suenaga NM, Ossipov MH, Malan TP Jr, Lai J, Porreca F (2001) Tonic descending facilitation from the rostral ventromedial medulla mediates opioid-induced abnormal pain and antinociceptive tolerance. J Neurosci 21(1):279–286
87. Xie JY, Herman DS, Stiller CO, Gardell LR, Ossipov MH, Lai J et al (2005) Cholecystokinin in the rostral ventromedial medulla mediates opioid-induced hyperalgesia and antinociceptive tolerance. J Neurosci 25(2):409–416
88. De Felice M, Ossipov MH, Wang R, Lai J, Chichorro J, Meng I et al (2010) Triptan-induced latent sensitization: a possible basis for medication overuse headache. Ann Neurol 67(3):325–337
89. Srikiatkhachorn A, Tarasub N, Govitrapong P (2000) Effect of chronic analgesic exposure on the central serotonin system: a possible mechanism of analgesic abuse headache. Headache 40(5):343–350
90. De Felice M, Porreca F (2009) Opiate-induced persistent pronociceptive trigeminal neural adaptations: potential relevance to opiate-induced medication overuse headache. Cephalalgia 29(12):1277–1284
91. Okada-Ogawa A, Porreca F, Meng ID (2009) Sustained morphine-induced sensitization and loss of diffuse noxious inhibitory controls in dura-sensitive medullary dorsal horn neurons. J Neurosci 29(50):15828–15835
92. De Felice M, Ossipov MH, Wang R, Dussor G, Lai J, Meng ID et al (2010) Triptan-induced enhancement of neuronal nitric oxide synthase in trigeminal ganglion dural afferents underlies increased responsiveness to potential migraine triggers. Brain J Neurol 133(Pt 8): 2475–2488
93. Green AL, Gu P, De Felice M, Dodick D, Ossipov MH, Porreca F (2013) Increased susceptibility to cortical spreading depression in an animal model of medication-overuse headache. Cephalalgia Int J Headache 34(8):594–604
94. De Felice M, Ossipov MH, Porreca F (2011) Persistent medication-induced neural adaptations, descending facilitation, and medication overuse headache. Curr Opin Neurol 24(3):193–196
95. Goadsby PJ, Edvinsson L, Ekman R (1990) Vasoactive peptide release in the extracerebral circulation of humans during migraine headache. Ann Neurol 28(2):183–187
96. Juhasz G, Zsombok T, Modos EA, Olajos S, Jakab B, Nemeth J et al (2003) NO-induced migraine attack: strong increase in plasma calcitonin gene-related peptide (CGRP) concentration and negative correlation with platelet serotonin release. Pain 106(3):461–470
97. Tvedskov JF, Lipka K, Ashina M, Iversen HK, Schifter S, Olesen J (2005) No increase of calcitonin gene-related peptide in jugular blood during migraine. Ann Neurol 58(4):561–568
98. Sarchielli P, Alberti A, Codini M, Floridi A, Gallai V (2000) Nitric oxide metabolites, prostaglandins and trigeminal vasoactive peptides in internal jugular vein blood during spontaneous migraine attacks. Cephalalgia 20(10):907–918
99. Ayata C, Moskowitz MA (2006) Cortical spreading depression confounds concentration-dependent pial arteriolar dilation during N-methyl-D-aspartate superfusion. Am J Physiol Heart Circ Physiol 290(5):H1837–H1841
100. Supornsilpchai W, le Grand SM, Srikiatkhachorn A (2010) Cortical hyperexcitability and mechanism of medication-overuse headache. Cephalalgia Int J Headache 30(9):1101–1109

101. Supornsilpchai W, le Grand SM, Srikiatkhachorn A (2010) Involvement of pro-nociceptive 5-HT2A receptor in the pathogenesis of medication-overuse headache. Headache 50(2):185–197
102. Dobson CF, Tohyama Y, Diksic M, Hamel E (2004) Effects of acute or chronic administration of anti-migraine drugs sumatriptan and zolmitriptan on serotonin synthesis in the rat brain. Cephalalgia Int J Headache 24(1):2–11
103. Reuter U, Salomone S, Ickenstein GW, Waeber C (2004) Effects of chronic sumatriptan and zolmitriptan treatment on 5-HT receptor expression and function in rats. Cephalalgia Int J Headache 24(5):398–407
104. Ayzenberg I, Obermann M, Leineweber K, Franke L, Yoon MS, Diener HC et al (2008) Increased activity of serotonin uptake in platelets in medication overuse headache following regular intake of analgesics and triptans. J Headache Pain 9(2):109–112
105. Srikiatkhachorn A, Anthony M (1996) Platelet serotonin in patients with analgesic-induced headache. Cephalalgia Int J Headache 16(6):423–426
106. Cupini LM, Costa C, Sarchielli P, Bari M, Battista N, Eusebi P et al (2008) Degradation of endocannabinoids in chronic migraine and medication overuse headache. Neurobiol Dis 30(2):186–189
107. Perrotta A, Arce-Leal N, Tassorelli C, Gasperi V, Sances G, Blandini F et al (2012) Acute reduction of anandamide-hydrolase (FAAH) activity is coupled with a reduction of nociceptive pathways facilitation in medication-overuse headache subjects after withdrawal treatment. Headache 52(9):1350–1361
108. Akerman S, Holland PR, Goadsby PJ (2007) Cannabinoid (CB1) receptor activation inhibits trigeminovascular neurons. J Pharmacol Exp Ther 320(1):64–71
109. Zappaterra M, Guerzoni S, Cainazzo MM, Ferrari A, Pini LA (2011) Basal cutaneous pain threshold in headache patients. J Headache Pain 12(3):303–310
110. Buchgreitz L, Lyngberg AC, Bendtsen L, Jensen R (2006) Frequency of headache is related to sensitization: a population study. Pain 123(1–2):19–27
111. Ayzenberg I, Obermann M, Nyhuis P, Gastpar M, Limmroth V, Diener HC et al (2006) Central sensitization of the trigeminal and somatic nociceptive systems in medication overuse headache mainly involves cerebral supraspinal structures. Cephalalgia 26(9):1106–1114
112. Perrotta A, Serrao M, Sandrini G, Burstein R, Sances G, Rossi P et al (2010) Sensitisation of spinal cord pain processing in medication overuse headache involves supraspinal pain control. Cephalalgia Int J Headache 30(3):272–284
113. Munksgaard SB, Bendtsen L, Jensen RH (2013) Modulation of central sensitisation by detoxification in MOH: results of a 12-month detoxification study. Cephalalgia Int J Headache 33(7):444–453
114. Yarnitsky D, Granot M, Granovsky Y (2014) Pain modulation profile and pain therapy: between pro- and antinociception. Pain 155(4):663–665
115. Cathcart S, Petkov J, Winefield AH, Lushington K, Rolan P (2010) Central mechanisms of stress-induced headache. Cephalalgia 30(3):285–295
116. de Tommaso M, Difruscolo O, Sardaro M, Libro G, Pecoraro C, Serpino C et al (2007) Effects of remote cutaneous pain on trigeminal laser-evoked potentials in migraine patients. J Headache Pain 8(3):167–174
117. Sandrini G, Tassorelli C, Ghiotto N, Nappi G (2006) Uncommon primary headaches. Curr Opin Neurol 19(3):299–304
118. Grazzi L, Chiapparini L, Ferraro S, Usai S, Andrasik F, Mandelli ML et al (2010) Chronic migraine with medication overuse pre-post withdrawal of symptomatic medication: clinical results and FMRI correlations. Headache 50(6):998–1004
119. Ferraro S, Grazzi L, Mandelli ML, Aquino D, Di Fiore D, Usai S et al (2012) Pain processing in medication overuse headache: a functional magnetic resonance imaging (fMRI) study. Pain Med 13(2):255–262

120. Chiapparini L, Grazzi L, Ferraro S, Mandelli ML, Usai S, Andrasik F et al (2009) Functional-MRI evaluation of pain processing in chronic migraine with medication overuse. Neurol Sci 30(Suppl 1):S71–S74
121. Riederer F, Marti M, Luechinger R, Lanzenberger R, von Meyenburg J, Gantenbein AR et al (2012) Grey matter changes associated with medication-overuse headache: correlations with disease related disability and anxiety. World J Biol Psychiatry 13(7):517–525
122. Silberstein SD, Blumenfeld AM, Cady RK, Turner IM, Lipton RB, Diener HC, Aurora SK, Sirimanne M, DeGryse RE, Turkel CC, Dodick DW (2013) OnabotulinumtoxinA for treatment of chronic migraine: PREEMPT 24-week pooled subgroup analysis of patients who had acute headache medication overuse at baseline. J Neurol Sci 31(1–2):48–56

MIX
Papier aus verantwortungsvollen Quellen
Paper from responsible sources
FSC® C105338

If you have any concerns about our products,
you can contact us on
ProductSafety@springernature.com

In case Publisher is established outside the EU,
the EU authorized representative is:
Springer Nature Customer Service Center GmbH
Europaplatz 3, 69115 Heidelberg, Germany

Printed by Libri Plureos GmbH
in Hamburg, Germany